UNITEXT - La Matematica per il 3+2

Volume 91

More information about this series at http://www.springer.com/series/5418

Marco Manetti

Topology

 Springer

Marco Manetti
Department of Mathematics
Sapienza - Università di Roma
Rome
Italy

Translated by Simon G. Chiossi, UFBA—Universidade Federal da Bahia, Salvador (Brazil).

Translation from the Italian language edition: *Topologia*, Marco Manetti, © Springer-Verlag Italia, Milano 2014. All rights reserved.

ISSN 2038-5722   ISSN 2038-5757   (electronic)
UNITEXT - La Matematica per il 3+2
ISBN 978-3-319-16957-6   ISBN 978-3-319-16958-3   (eBook)
DOI 10.1007/978-3-319-16958-3

Library of Congress Control Number: 2014945348

Springer Cham Heidelberg New York Dordrecht London
© Springer International Publishing Switzerland 2015

Cover design: Simona Colombo, Giochi di Grafica, Milano, Italy

Printed on acid-free paper

Springer International Publishing AG Switzerland is part of Springer Science+Business Media (www.springer.com)

Preface

To the Student

This textbook offers a primer in general topology (point-set topology), together with an introduction to algebraic topology. It is meant primarily for students with a mathematical background that is usually taught in the first year of undergraduate degrees in Mathematics and Physics.

Point-set topology is the language in which a considerable part of mathematics is written. It is not an accident that the original name 'analytic topology' was replaced by 'general topology', a more apt term for that part of topology that is used by the vast majority of mathematicians and is fundamental in many areas of mathematics. Over time its unabated employment has had a constant polishing effect on its theorems and definitions, thus rendering it an extraordinarily elegant subject. There is no doubt that point-set topology has a significant formative value, in that it forces the brain—and trains it at the same time—to handle extremely abstract objects, defined solely by axioms. In studying on this book, you will experience hands-on that the point-set topology resembles a language more than a theory. There are endless terms and definitions to be learnt, a myriad of theorems whose proof is often rather easy, only occasionally exceeding 20 lines. There are, obviously, also deep and far-from-trivial results, such as the theorems of Baire, Alexander and Tychonov.

The part on algebraic topology, details of which we will give in Chap. 9 together with the mandatory motivations, is devoted to the study of homotopy, fundamental groups and covering spaces.

I included around 500 exercises in the text: trying to solve them with dedication is the best way to attain a firm hold on the matter, adapt it to your own way of thinking and also learn to develop original ideas. Some exercises are solved directly in the text, either in full or almost. They are called '*Examples*', and their importance should not be underestimated: understanding them is the correct way to make abstract notions concrete. Exercises marked with $\heartsuit$, instead, are solved in Chap. 16.

It is a matter of fact that the best way to learn a new subject is by attending lectures, or studying on books, and trying to understand definitions, theorems and the interrelationships properly. At the same time you should solve the exercises, without the fear of making mistakes, and then compare the solutions with the ones in the text, or those provided by teacher, classmates or the internet.

This book also proposes a number of exercises marked with 🖐, which I personally believe to be harder than the typical exam question. These exercises should therefore be taken as endeavours to intelligence, and incentives to be creative: they require that we abandon ourselves to new synergies of ideas and accept to be guided by subtler analogies, rather than trail patiently along a path paved by routine ideas and standard suggestions.

To the Lecturer

In the academic years 2004–2005 and 2005–2006, I taught a lecture course called 'Topology' for the Bachelor's degree in Mathematics at University of Rome 'La Sapienza'. The aim was to fit the newly introduced programme specifications for mathematical teaching in that part of the syllabus traditionally covered in 'Geometry 2' course of the earlier 4-year degrees. The themes were carefully chosen so to keep into account on one side the formative and cultural features of the single topics, on the other their usefulness in the study of mathematics and research alike. Some choices certainly break with a long-standing and established tradition of topology teaching in Italy, and with hindsight I suspect they might have been elicited by my own research work in algebra and algebraic geometry. I decided it would be best to get straight to the point and state key results and definitions as early as possible, thus fending off the terato(po)logical aspects.

From the initial project to the final layout of my notes, I tried to tackle the conceptual obstacles gradually, and make both theory and exercises as interesting and entertaining as possible for students. Whether I achieved these goals the reader will tell.

The background necessary to benefit from the book is standard, as taught in first-year Maths and Physics undergraduate courses. Solid knowledge of the language of sets, of linear algebra, basic group theory, the properties of real functions, series and sequences from 'Calculus' are needed. The second chapter is dedicated to the arithmetic of cardinal numbers and Zorn's lemma, two pivotal prerequisites that are not always addressed during the first year: it will be up to the lecturer to decide— after assessing the students' proficiency—whether to discuss these topics or not.

The material present here is more than sufficient for 90 hours of lectures and exercise classes, even if, nowadays, mathematics syllabi tend to allocate far less time to topology. In order to help teachers decide what to skip I indicated with the symbol ∿ ancillary topics, which may be left out at first reading. It has to be said, though, that Chaps. 3–6 (with the exception of the sections displaying ∿), form the backbone of point-set topology and, as such, should not be excluded.

The bibliography is clearly incomplete and lists manuals that I found most useful, plus a selection of research articles and books where the willing student can find further information about the topics treated, or mentioned in passing, in this volume.

Acknowledgments

I am grateful to Ciro Ciliberto and Domenico Fiorenza for reading earlier versions and for the tips they gave me. I would like to thank Francesca Bonadei of Springer-Verlag Italia for helping with the final layout, and all the students ('victims') of my lectures on topology for the almost-always-useful and relevant observations that allowed me to amend and improve the book.

This volume is based on the second Italian edition. Simon G. Chiossi has done an excellent work of translation; he also pointed out a few inaccuracies and proposed minor improvements. To him and the staff at Springer, I express here my heartfelt gratitude.

Future updates can be found at
http://www.mat.uniroma1.it/people/manetti/librotopology.html.

Rome, February 2015   Marco Manetti

Contents

Chapter 1
Geometrical Introduction to Topology

Let us start off with an excerpt from the introduction to Chapter V in *What is mathematics?* by R. Courant and H. Robbins.

'In the middle of the nineteenth century there began a completely new development in geometry that was soon to become one of the great forces in modern mathematics. The new subject, called analysis situs or topology, has as its object the study of the properties of geometrical figures that persist even when the figures are subjected to deformations so drastic that all their metric and projective properties are lost.' (...) 'When Bernhard Riemann (1826–1866) came to Göttingen as a student, he found the mathematical atmosphere of that university town filled with keen interest in these strange new geometrical ideas. Soon he realized that here was the key to the understanding of the deepest properties of analytic functions of a complex variable.'

The expression 'deformations so drastic' is rather vague, and as a matter of fact modern topology studies several classes of transformations of geometrical figures, the most important among which are **homeomorphisms** and **homotopy equivalences**.

While precise definitions for these will be given later, in this chapter we shall encounter a preliminary, and only partial notion of homeomorphism, and discuss a few examples. We will stay away from excessively rigorous definitions and proofs at this early stage, and try instead to rely on the geometrical intuition of the reader, with the hope that in this way we might help newcomers acquire the basic ideas.

1.1 A Bicycle Ride Through the Streets of Rome

Problem 1.1 On a bright Sunday morning Mr. B. decides to go on a bike ride *that crosses every bridge in Rome once, and only once*. Knowing that he can decide where to start from and where to end, will Mr. B. be able to accomplish his wish?

We remind those who aren't familiar with the topography of Rome that the city is divided by the river Tiber and its affluent Aniene in four regions, one being the Tiberine island, lying in the middle of the river and joined to either riverbank by bridges.

© Springer International Publishing Switzerland 2015
M. Manetti, *Topology*, UNITEXT - La Matematica per il 3+2 91,
DOI 10.1007/978-3-319-16958-3_1

Fig. 1.1 The topology of the
bridges of Rome

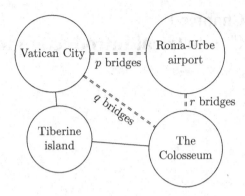

In order to answer the question we needn't go around the streets of Rome on two
wheels testing possibilities, nor mess up a street map with a marker.

The problem can be visualised by drawing four circles on a sheet of paper, one for
each region, and indicating how many bridges connect any two regions (Fig. 1.1).

A configuration of this kind contains all the information we need to solve the
problem. If you agree with this statement, then you are **thinking topologically**. In
other words you have understood that whether the proposed bicycle ride is possible doesn't depend on metric or projective properties like the bridges' length, their
architectural structure, the extension of mainland areas and so on. But if you insist
on solving the problem using a street map, you should imagine the map is drawn on
a thin rubber sheet. Think of stretching, twisting and rumpling it, as much as you
like, without tearing it nor making different points touch one another by folding: you
will agree that the answer stays the same.

Problem 1.2 In the old town of Königsberg, on the river Pregel, there are two islets
and seven bridges as in Fig. 1.2. The story goes that Mr. C. wanted to *cross on foot
every bridge in Königsberg only once*, and could start and finish his walk at any point.
Would have Mr. C. been able to do so?

Fig. 1.2 The bridges of old
Königsberg

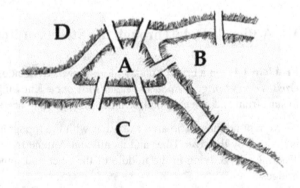

Here's another problem of topological flavour:

Problem 1.3 Can one drive from Rome to Venice by never leaving the motorway?

As the motorway goes through Rome and Venice, the problem is truly a question about the **connectedness** of the Italian road system: a network of roads is connected if one can drive from any one point to any other. Here, too, the answer is clearly independent of how long the single stretches of road are, or how many turns or slopes there are etc.

A useful mathematical concept for solving the previous problems is that of a **graph**. In Euclidean space a graph is a non-empty set of V points, called **nodes** or **vertices**, certain pairs of which are joined by S segments called **edges**. Edges are not necessarily straight, but can be parts of circles, parabolas, ellipses, or more generally 'regular' arcs that do not pass through the same point in space more than once. Another assumption is that distinct edges meet only at nodes. A **walk of length** p in a graph is a sequence $v_0, v_1, \ldots, v_p$ of nodes and a sequence $l_1, l_2, \ldots, l_p$ of edges, with l_i joining v_{i-1} to v_i for any i. The nodes v_0 and v_p are the endpoints of the walk. A graph is said to be **connected** if there is a walk beginning and ending at any two given nodes u, w.

If u is a node in a graph Γ, we call **degree of u in Γ** the number of edges containing u, counting twice edges with both endpoints in u. It's straightforward that the sum of the degrees of all nodes equals twice the number of edges; in particular, any graph must have an even number of odd-degree nodes.

A graph is called **Eulerian** if there's a walk that visits each edge exactly once: in this case the length of the walk equals the number of edges in the graph.

Going back to the bridges' problem, no matter whether in Rome or Königsberg, let's construct a graph Γ having mainland areas as nodes and bridges as edges. The crossing problem has a positive answer if and only if Γ is an Eulerian graph (Fig. 1.3).

Theorem 1.1 *In any Eulerian graph there are at most two nodes of odd degree. More specifically, the answer to the problem of the Königsberg's bridges is 'no'.*

Proof Choose a sequence of nodes $v_0, \ldots, v_S$ and a sequence of edges $l_1, \ldots, l_S$ that give a walk visiting every edge precisely once. Every node u other than v_0 and v_S has degree equal twice the number of indices i such that $u = v_i$.   □

Fig. 1.3 The Königsberg bridges' graph

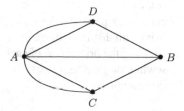

Fig. 1.4 Tralfamadore's
zoo, with shed location D

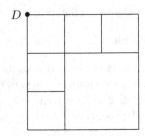

Exercises

1.4 ($\heartsuit$) Suppose Γ is a connected graph where each node has even degree. Prove that every graph obtained by removing one edge from Γ is still connected.

1.5 Let $v_1, \ldots, v_n$ be the nodes of a graph Γ and A the adjacency matrix: this is the $n \times n$-matrix whose entry a_{ij} equals the number of edges joining v_i to v_j. Explain the geometrical meaning of the entries of the powers of A.

1.6 Tralfamadore's zoo has six thematic areas surrounded by walking paths, as in Fig. 1.4. Altogether there are 16 pathways that are long 1 km and two that are long 2 km. Once appointed director of the zoo, Mister V. realised that the zoo's janitor was going around 24 km every day to keep sidewalks tidy, and blamed him for wasting time. The caretaker said that to sweep every path, starting from his shed and then making a way back, he really needed to go around as much. Who's right according to you?

1.2 Topological Sewing

To sharpen our topological intuition let's pay a visit to the topological tailor: this is a pro whose skills allow him to work a very thin fabric with incredible elastic features, that can be folded, modelled, stretched and deformed at pleasure without tearing it. The simplest handicrafts involve these operations:

1. cutting out a piece X of material in the shape of a convex polygon with $n \geq 2$ sides;
2. choosing k pairs of sides of X, $2k \leq n$, and marking sides belonging to different pairs with distinct letters;
3. deciding the orientation of each of the $2k$ sides chosen in 2;
4. sewing, or glueing, or pasting together the sides in each couple, respecting the orientation.

For instance, if we glue two opposite sides of a square that are oriented in the same way we obtain a (hollow) cylinder,

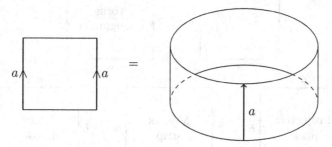

while if we glue two adjacent sides, oriented starting from the common vertex, we end up with a disc.

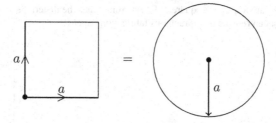

The famous Möbius strip arises by glueing together two non-consecutive, oppositely oriented sides.

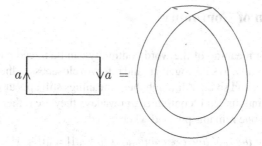

We leave it to the reader to try and visualise the other possible outcomes of pasting a square's edges in this sort of way (Fig. 1.5 and Exercise 1.7).

Exercises

1.7 (♡) Say whether Fig. 1.6—after glueing sides as indicated—will produce a disc, a cylinder, a Möbius strip or something completely different. Explain your answer.

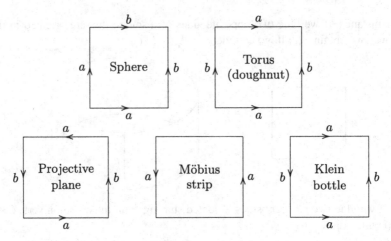

Fig. 1.5 Handiworks arising from a square of fabric. Sometimes the material's elastic features are not relevant, and one can replace the miraculous fabric with paper

Fig. 1.6 What's this?

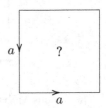

1.3 The Notion of Continuity

In mathematics the meaning of the word continuity can be traced back to the Latin verb *continēre* (literally, to keep together) and refers to closeness, adherence, forming an unbroken whole, and so on. Although these meanings still appear in some dictionaries under 'continuous' and 'continuity', nowadays they are rather old-fashioned, to the point where one remains puzzled when asked:

Which points on the real line are continuous to $]0, 1[= \{0 < x < 1\}$?

The same question, phrased in a less obsolete language, becomes the more intelligible:

Which points on the real line are adherent (attached/stick) to $]0, 1[$?

In the space $\mathbb{R}^n$ of n-tuples $(x_1, \ldots, x_n)$ of real numbers the notion of continuity (=adherence) is easy to render.

Definition 1.2 A point $x \in \mathbb{R}^n$ is called **adherent** to a subset $A \subset \mathbb{R}^n$ if it is possible to find points of A that lie arbitrarily close to x.

Any real number, for example, is adherent to the set of rationals, while any number in $[0, 1] = \{0 \le x \le 1\}$ adheres to the interval $]0, 1[\subset \mathbb{R}$.

The distance between two points $x = (x_1, \ldots, x_n)$ and $y = (y_1, \ldots, y_n)$ in $\mathbb{R}^n$ is given by Pythagoras' formula

$$d(x, y) = \sqrt{(x_1 - y_1)^2 + \cdots + (x_n - y_n)^2}.$$

Equivalently, $d(x, y) = \|x - y\|$, where $\|x\| = \sqrt{(x \cdot x)}$ is the norm associated to the standard dot product $(x \cdot y) = \sum x_i y_i$.

Lemma 1.3 *The distance d satisfies the triangle inequality*

$$d(x, y) \le d(x, z) + d(z, y)$$

for any $x, y, z \in \mathbb{R}^n$.

Proof Setting $x - z = u$, $z - y = v$, the triangle inequality amounts to $\|u + v\| \le \|u\| + \|v\|$. Since both sides are non-negative, the latter inequality is the same as

$$0 \le (\|u\| + \|v\|)^2 - \|u + v\|^2 = 2(\|u\|\|v\| - (u \cdot v)).$$

As $\|u\|\|v\| \ge 0$, it suffices to prove the so-called **Cauchy-Schwarz inequality**:

$$\|u\|^2\|v\|^2 - (u \cdot v)^2 \ge 0.$$

When $v = 0$ the Cauchy-Schwarz inequality is trivially true. So we can assume $v \ne 0$ and consider the vector

$$w = u\|v\| - (u \cdot v)\frac{v}{\|v\|}.$$

An easy computation shows that

$$0 \le \|w\|^2 = \|u\|^2\|v\|^2 - (u \cdot v)^2. \qquad \square$$

The distance enables us to better explain the notion of a set of points that is 'arbitrarily close' to a given point, and thus understand adherence. More precisely, *a point p is adherent to a set A if and only if for any real number $\delta > 0$ there exists a point $x \in A$ such that $d(p, x) < \delta$.*

The intuitive idea of a tear in the fabric should hint at a failure of adherence: if a point $x \in \mathbb{R}^n$ is adherent to a subset A, we say that a map $f: \mathbb{R}^n \to \mathbb{R}^m$ tears x away from A when the point $f(x)$ is not adherent to $f(A)$.

Continuous maps are those that do not cause tears, slits and the like, hence those that preserve the continuity relationship (=adherence) between points and subsets.

Definition 1.4 Let $X \subset \mathbb{R}^n$ and $Y \subset \mathbb{R}^m$ be subsets. A map $f: X \to Y$ is called **continuous** if, for any subset $A \subset X$ and any point $x \in X$ adherent to A, the point $f(x)$ adheres to $f(A)$.

Don't worry if this definition of a continuous map doesn't at first sight coincide with what you have learnt in high school or during analysis classes. We'll explain in the sequel that our definition is equivalent to the ones you are familiar with.

An immediate consequence of the definition above is that constant maps, translations in $\mathbb{R}^n$ and more generally, isometries (distance-preserving functions) are continuous.

Quite often one doesn't rely on Definition 1.4 to check whether a function is continuous, but prefers instead to use general properties of continuity and reduce to a basic list of remarkable continuous maps. In the rest of the section we'll denote with capital letters $X, Y, Z \ldots$ subsets of Euclidean spaces.

(**C1**) *Let* $f: X \to Y$ *be continuous and* $W \subset X,\ Z \subset Y$ *subsets such that* $f(W) \subset Z$. *Then the restriction* $f: W \to Z$ *is continuous.*

(**C2**) *If* $f: X \to Y$ *and* $g: Y \to Z$ *are continuous, then also* $gf: X \to Z$ *is continuous.*

(**C3**) *If* $X \subset Y$, *the inclusion map* $i: X \to Y$ *is continuous.*

Consider two sets $X \subset Y$ and a function $f: Z \to X$. Then $f: Z \to X$ is continuous if and only if $f: Z \to Y$ is continuous. In fact, if $f: Z \to Y$ is continuous, also $f: Z \to X$ is continuous by C1. Conversely, $f: Z \to Y$ is the composite of $f: Z \to X$ with the continuous (by C3) inclusion $X \to Y$. All this to say that to establish whether a function is continuous we can reduce to consider functions $f: X \to \mathbb{R}^n$.

(**C4**) *Let* $f_i: X \to \mathbb{R}$, $i = 1, \ldots, n$, *denote the components of a function* $f: X \to \mathbb{R}^n$, *i.e.* $f(x) = (f_1(x), \ldots, f_n(x))$, $x \in X$. *Then* f *is continuous if and only if the* $f_1, \ldots, f_n$ *are all continuous.*

The next list is borrowed from analysis and calculus courses:

(**C5**) The following maps are continuous:

1. *every linear map* $\mathbb{R}^n \to \mathbb{R}$;
2. *the multiplication* $\mathbb{R}^2 \to \mathbb{R}$, $(x, y) \mapsto xy$;
3. *the inverse* $\mathbb{R} - \{0\} \to \mathbb{R} - \{0\}$, $x \mapsto x^{-1}$;
4. *the absolute value* $\mathbb{R} \to \mathbb{R}$, $x \mapsto |x|$;
5. *the exponential function, the logarithm, the sine, the cosine and all other trigonometric functions, on the respective domains;*
6. *the functions* $\mathbb{R}^2 \to \mathbb{R}$, $(x, y) \mapsto \max(x, y)$ *and* $(x, y) \mapsto \min(x, y)$.

Conditions C1,..., C5 imply the continuity of many other functions. For instance, if $f, g: X \to \mathbb{R}$ are continuous then also $f+g$, fg, and f/g (provided g is not zero on X) are continuous. The reason is that the map $(f, g): X \to \mathbb{R}^2, x \mapsto (f(x), g(x))$, is continuous and $f + g$ is the composite of (f, g) with the linear map $(x, y) \mapsto x + y$;

fg is the composite of (f, g) with the product map $(x, y) \mapsto xy$. Analogously for f/g. Along this line of thought it is easy to show by induction that any polynomial expression involving continuous maps is continuous. There are countless examples like these and we shall not overindulge.

Intimately related to adherence and continuity is the concept of **closed sets**.

Definition 1.5 A subset $C \subset X$ in $\mathbb{R}^n$ is called **closed** in X if it coincides with the set of points of X that adhere to C. Equivalently, C is closed in X if, for any $x \in X - C = \{x \in X \mid x \notin C\}$, a number $\delta > 0$ exists such that $d(x, y) \geq \delta$ for every $y \in C$.

Example 1.6 Take $X \subset \mathbb{R}^n$, a number $r > 0$ and a point $x \in \mathbb{R}^n$. The set

$$C = \{x' \in X \mid d(x, x') \leq r\}$$

is closed in X. If $y \in X - C$, in fact, $\delta = d(x, y) - r > 0$. By the triangle inequality

$$d(y, x') \geq d(x, y) - d(x', x) \geq \delta$$

for every $x' \in C$, and so y is not adherent to C.

Example 1.7 Consider a subset $X \subset \mathbb{R}^n$ and a continuous map $f : X \to \mathbb{R}$. The set $C = \{x \in X \mid f(x) = 0\}$ is closed in X. To see this take $x \in X$ adherent to C; then $f(x)$ is adherent to $\{0\}$, so $f(x) = 0$. This means $x \in C$ and C is closed. The same argument also proves that the pre-image of a closed set $Z \subset \mathbb{R}$, i.e. $\{x \in X \mid f(x) \in Z\}$, is closed in X.

Theorem 1.8 (Glueing lemma) *Let $f : X \to Y$ be a function, A, B two closed subsets in X such that $X = A \cup B$ and the restrictions $f : A \to Y$, $f : B \to Y$ are continuous. Then f is continuous.*

Proof Take a subset $C \subset X$ and a point $x \in X$ adherent to C. We have to prove that $f(x)$ is adherent to $f(C)$. Observe that x is adherent to at least one between $C \cap A$ and $C \cap B$. If not, in fact, there would exist positive numbers δ_A and δ_B such that

$$d(x, y) \geq \delta_A \quad \text{for every } y \in C \cap A,$$

$$d(x, y) \geq \delta_B \quad \text{for every } y \in C \cap B.$$

If δ is the minimum between δ_A and δ_B, we would have $d(x, y) \geq \delta$ for every $y \in C$, so x would not be adherent to C. Just to fix ideas, assume x is adherent to $C \cap A$. Then x must also adhere to A, and since A is closed, we get $x \in A$. Now, the restriction $f : A \to Y$ is continuous, so $f(x)$ is adherent to $f(A \cap C)$. Since $f(A \cap C) \subset f(C)$, *a fortiori* $f(x)$ is adherent to $f(C)$. $\square$

The set of complex numbers $\mathbb{C}$ is in one-to-one correspondence with the plane $\mathbb{R}^2$. The distance of two complex numbers z, u coincides with the modulus of their difference $|z - u|$. The identification extends to n dimensions, hence maps $\mathbb{C}^n$ bijectively to $\mathbb{R}^{2n}$, and everything we said so far is also valid for subsets in $\mathbb{C}^n$ and maps between them.

Exercises

1.8 Consider the 8 types of **interval**, where a, b are real numbers with $a < b$:

1. $]a, b[= \{x \in \mathbb{R} \mid a < x < b\}$;
2. $[a, b[= \{x \in \mathbb{R} \mid a \leq x < b\}$;
3. $]a, b] = \{x \in \mathbb{R} \mid a < x \leq b\}$;
4. $[a, b] = \{x \in \mathbb{R} \mid a \leq x \leq b\}$;
5. $] - \infty, a[= \{x \in \mathbb{R} \mid x < a\}$;
6. $] - \infty, a] = \{x \in \mathbb{R} \mid x \leq a\}$;
7. $]a, +\infty[= \{x \in \mathbb{R} \mid a < x\}$;
8. $[a, +\infty[= \{x \in \mathbb{R} \mid a \leq x\}$.

Say which ones are closed subsets in $\mathbb{R}$ according to Definition 1.5.

1.9 Which of the following sets are closed in $\mathbb{R}^2$?

$$\{(x, y) \mid x^2 + 2y^2 = 1\}, \qquad \{(x, y) \mid 0 \leq x \leq 1, \ 0 \leq y < 1\},$$

$$\{(x, y) \mid 0 \leq x, \ 0 \leq y\}, \qquad \{(x, y) \mid 0 \leq x \leq 1, \ x + y \leq 1\},$$

$$\{(x, y) \mid 0 < x^2 + y^2 \leq 1\}, \qquad \{(x, y) \mid 0 \leq x, \ 0 \leq y \leq \sin(x)\}.$$

1.10 Establish which of the following sets are closed in $\mathbb{C}$ ($i = \sqrt{-1}$):

$$\{z \in \mathbb{C} \mid z^2 \in \mathbb{R}\}, \quad \{z \in \mathbb{C} \mid |z^2 - z| \leq 1\},$$

$$\{2^n + i2^n \mid n \in \mathbb{Z}\}, \quad \{2^{-n} + i2^n \mid n \in \mathbb{Z}\}.$$

1.11 Let $X \subset \mathbb{R}^n$ be a given subset, $A, Y \subset X$ and A closed in X. Prove that $A \cap Y$ is closed in Y.

1.12 Prove that, for a function $f : X \to Y$, the following assertions are equivalent:

1. f is continuous;
2. for any closed set C in Y the set $\{x \in X \mid f(x) \in C\}$ is closed in X.

1.13 Prove that in $\mathbb{R}^n$:

1. the union of two closed sets is closed;
2. the intersection of any number of closed sets is a closed set.

Fig. 1.7 The Mandelbrot set, well known in the world of 'fractals'

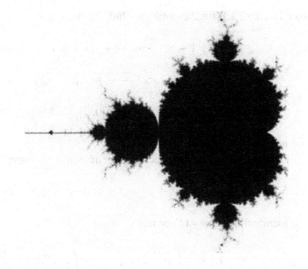

1.14 (Mandelbrot set) Given a complex number $z \in \mathbb{C}$ define recursively a sequence $\{p_n(z)\} \subset \mathbb{C}$ by $p_0(z) = z$, $p_{n+1}(z) = p_n(z)^2 + z$. Prove that:

1. if the sequence $p_n(z)$ is bounded, then $|p_n(z)| \leq 2$ for every $n \geq 0$; (Hint: if not, call n the smallest integer such that $|p_n(z)| = 2(1+a)$, $a > 0$, and show that $|p_{n+s}(z)| \geq 2(1+a)^{s+1} \geq |p_0(z)|$ for any $s > 0$.)
2. for any $n \geq 0$ the map $z \mapsto p_n(z)$ is continuous;
3. let $M \subset \mathbb{C}$ be the set of complex numbers z such that $\{p_n(z)\}$ is a bounded sequence. Prove that M is closed.

The set M (depicted in Fig. 1.7) is called **Mandelbrot set**.

1.4 Homeomorphisms

Definition 1.9 A **homeomorphism** is a continuous and bijective map with continuous inverse. Two subsets in $\mathbb{R}^n$ are called **homeomorphic** if there is a homeomorphism mapping one to the other.

In the eyes of the topologist homeomorphic spaces are indistinguishable. For instance, he won't see any difference between the four intervals

$$]0, 1[, \quad]0, 2[, \quad]0, +\infty[, \quad]-\infty, +\infty[$$

(see Exercise 1.8 for the notation). Indeed, the maps

$$f: \;]-\infty, +\infty[\; \longrightarrow \;]0, +\infty[\qquad f(x) = e^x,$$

$$g: \;]0, +\infty[\; \longrightarrow \;]0, 1[\qquad g(x) = e^{-x},$$

$$h: \;]0, 1[\; \longrightarrow \;]0, 2[\qquad h(x) = 2x$$

are homeomorphisms.

Through a topologist's spectacles a circle and a square:

$$S^1 = \{(x, y) \in \mathbb{R}^2 \mid x^2 + y^2 = 1\} \quad \text{and} \quad P = \{(x, y) \in \mathbb{R}^2 \mid |x| + |y| = 1\}$$

are identical. It is easy to see that

$$f: S^1 \to P, \qquad f(x, y) = \left(\frac{x}{|x| + |y|}, \frac{y}{|x| + |y|} \right)$$

and

$$g: P \to S^1, \qquad g(x, y) = \left(\frac{x}{\sqrt{x^2 + y^2}}, \frac{y}{\sqrt{x^2 + y^2}} \right)$$

are both continuous and inverse to one another, see Fig. 1.8.

Let us introduce some commonly used symbols. For any integer $n \geq 0$ we set

$$D^n = \{x \in \mathbb{R}^n \mid \|x\| \leq 1\}, \text{ called closed unit } \mathbf{ball} \text{ of dimension } n,$$

$$S^n = \{x \in \mathbb{R}^{n+1} \mid \|x\| = 1\}, \text{ called unit } \mathbf{n\text{-sphere}}.$$

Moreover, for any point $x \in \mathbb{R}^n$ and any real $r > 0$ one calls

$$B(x, r) = \{y \in \mathbb{R}^n \mid d(x, y) < r\} \text{ the } \mathbf{open \; ball} \text{ with centre } x \text{ and radius } r.$$

Fig. 1.8 Topological
equivalence of circle and
square

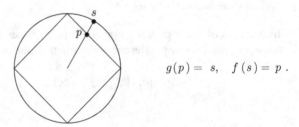

$$g(p) = s, \quad f(s) = p \, .$$

Example 1.10 Affine transformations in $\mathbb{R}^2$ are homeomorphisms. This implies, in particular, that all triangles are homeomorphic to one another, and so are parallelograms. Consider the square Q and the triangle T defined, in coordinates, by

$$Q = \{(x, y) \mid |x| + |y| \leq 1\}, \qquad T = \{(x, y) \in Q \mid y \leq 0\}.$$

Then

$$f: Q \to T, \qquad f(x, y) = \left(x, \frac{1}{2}(y + |x| - 1) \right),$$

is a homeomorphism that fixes the common boundary sides.

Example 1.11 The following spaces are homeomorphic:

- the punctured plane $X = \mathbb{R}^2 - \{(0, 0)\}$;
- the round cylinder $Y = \{(x, y, z) \in \mathbb{R}^3 \mid x^2 + y^2 = 1\}$;
- the 1-sheeted hyperboloid $Z = \{(x, y, z) \in \mathbb{R}^3 \mid x^2 + y^2 = 1 + z^2\}$.

To prove this fact we may consider mappings

$$f: Y \to X \qquad f(x, y, z) = (xe^z, ye^z) \quad \text{and}$$

$$g: Z \to Y \qquad g(x, y, z) = \left(\frac{x}{\sqrt{1 + z^2}}, \frac{y}{\sqrt{1 + z^2}}, z \right).$$

The reader can check that f and g are invertible with continuous inverses.

Example 1.12 Open balls in $\mathbb{R}^n$ are homeomorphic to the whole $\mathbb{R}^n$. Namely, for any $p \in \mathbb{R}^n$ and $r > 0$ the map $x \mapsto rx + p$ induces a homeomorphism between $B(0, 1)$ and $B(p, r)$, while

$$g: \mathbb{R}^n \to B(0, 1), \qquad g(x) = \frac{x}{\sqrt{1 + \|x\|^2}}$$

is a homeomorphism with inverse $g^{-1}(y) = \dfrac{y}{\sqrt{1 - \|y\|^2}}$.

Example 1.13 (*Stereographic projection*) Let $N = (1, 0, \ldots, 0)$ be the 'North pole' of the sphere $S^n = \{(x_0, \ldots, x_n) \in \mathbb{R}^{n+1} \mid \sum x_i^2 = 1\}$. The stereographic projection

$$f: S^n - \{N\} \to \mathbb{R}^n$$

is defined by identifying $\mathbb{R}^n$ with the hyperplane $H \subset \mathbb{R}^{n+1}$ of equation $x_0 = 0$ and setting $f(x)$ to be the intersection of H with the line through x and N. In coordinates, f and its inverse read

$$f(x_0, \ldots, x_n) = \frac{1}{1 - x_0}(x_1, \ldots, x_n),$$

$$f^{-1}(y_1, \ldots, y_n) = \left(\frac{\sum_i y_i^2 - 1}{1 + \sum_i y_i^2}, \frac{2y_1}{1 + \sum_i y_i^2}, \ldots, \frac{2y_n}{1 + \sum_i y_i^2} \right).$$

The functions f and f^{-1} are continuous, so the stereographic projection is a homeomorphism. An easy exercise in proportions shows that if we identify $\mathbb{R}^n$ not with H but with the hyperplane $x_0 = c$, $c \neq 1$, f gets multiplied by $1 - c$.

Example 1.14 Let $h: \mathbb{R}^n \to \mathbb{R}$ be a continuous map, and consider

$$X = \{(x_0, \ldots, x_n) \in \mathbb{R}^{n+1} \mid x_0 \leq 0\},$$

$$Y = \{(x_0, \ldots, x_n) \in \mathbb{R}^{n+1} \mid x_0 \leq h(x_1, \ldots, x_n)\}.$$

Then

$$f: X \to Y, \qquad f(x_0, \ldots, x_n) = (x_0 + h(x_1, \ldots, x_n), x_1, \ldots, x_n)$$

is a homeomorphism (Fig. 1.9).

Example 1.15 The complement of a circle in $\mathbb{R}^3$ is homeomorphic to the complement of a line plus a point. Call $e_1, e_2, e_3 \in \mathbb{R}^3$ the canonical basis and consider the circle

$$K = \{x \in \mathbb{R}^3 \mid (x \cdot e_3) = 0, \ \|x - e_1\|^2 = 1\}$$

through the origin. Since the inversion map

$$r: \mathbb{R}^3 - \{0\} \to \mathbb{R}^3 - \{0\}, \qquad r(x) = \frac{x}{\|x\|^2}$$

is a homeomorphism that coincides with r^{-1}, the space $\mathbb{R}^3 - K$ is homeomorphic to $\mathbb{R}^3 - (\{0\} \cup r(K - \{0\}))$. Now it suffices to show that $r(K - \{0\})$ is a line not passing through the origin. This can be done using either elementary geometry

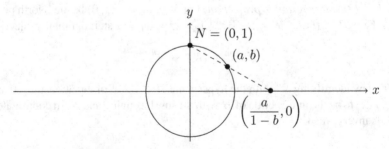

Fig. 1.9 The stereographic projection $f: S^1 - \{N\} \to \mathbb{R}^1$

(Euclid's theorems) or analytical techniques. By setting $y = r(x)$, in fact, we have $x = y/\|y\|^2$, so the plane $(x \cdot e_3) = 0$ is mapped to $(y \cdot e_3) = 0$ while the sphere $\|x - e_1\|^2 = 1$ becomes the locus

$$0 = \|x - e_1\|^2 - 1 = \|x\|^2 - 2(x \cdot e_1) = \frac{1}{\|y\|^2} - 2\left(\frac{y}{\|y\|^2} \cdot e_1\right) = \frac{1 - 2(y \cdot e_1)}{\|y\|^2}.$$

All in all, $r(K - \{0\})$ equals the intersection of the planes $(y \cdot e_3) = 0$ and $2(y \cdot e_1) = 1$.

Example 1.16 A complex 2×2-matrix

$$\begin{pmatrix} a & b \\ c & d \end{pmatrix}$$

is, by definition, an element of the special unitary group $SU(2, \mathbb{C})$ if

$$\begin{pmatrix} a & b \\ c & d \end{pmatrix} \begin{pmatrix} \bar{a} & \bar{c} \\ \bar{b} & \bar{d} \end{pmatrix} = I \quad \text{and} \quad ad - bc = 1,$$

that is to say, if

$$ad - bc = |a|^2 + |b|^2 = |c|^2 + |d|^2 = 1, \qquad \bar{a}c + d\bar{b} = \bar{c}a + b\bar{d} = 0.$$

Multiplying $\bar{a}c + d\bar{b} = 0$ by a and substituting ad with $1 + bc$ gives the constraint $(|a|^2 + |b|^2)c + \bar{b} = 0$, from which $c = -\bar{b}$. A similar computation gives $d = \bar{a}$, so special unitary matrices of order 2 are those of the form

$$\begin{pmatrix} a & b \\ -\bar{b} & \bar{a} \end{pmatrix}, \quad \text{with } (a, b) \in S^3 = \{(a, b) \in \mathbb{C}^2 \mid |a|^2 + |b|^2 = 1\}.$$

Thus we proved that $SU(2, \mathbb{C})$ is homeomorphic to the sphere S^3.

Example 1.17 The set $X \subset \mathbb{R}^3 \times \mathbb{R}^3$ of pairs of orthonormal vectors is homeomorphic to the set $Y \subset \mathbb{C}^3$ of vectors (z_1, z_2, z_3) such that

$$z_1^2 + z_2^2 + z_3^2 = 0, \qquad |z_1|^2 + |z_2|^2 + |z_3|^2 = 2 \ .$$

If x_i and y_i denote the real and imaginary parts of z_i, in fact, the previous relations read

$$\sum_{i=1}^3 x_i^2 + \sum_{i=1}^3 y_i^2 = 2, \qquad \sum_{i=1}^3 x_i^2 - \sum_{i=1}^3 y_i^2 = 0, \qquad \sum_{i=1}^3 x_i y_i = 0 \ ,$$

a system expressing the orthonormality of (x_1, x_2, x_3) and (y_1, y_2, y_3).

Example 1.18 Consider complex numbers $p, q \in \mathbb{C}$ with the same modulus $|p| = |q| = r \geq 0$. For any $\delta > 0$ there exists a homeomorphism $f: \mathbb{C} \to \mathbb{C}$ such that $f(p) = q$ and $f(x) = x$ whenever $||x| - r| > \delta$. Given $\alpha \in \mathbb{R}$ we set $R(\alpha, x) = (\cos \alpha + i \sin \alpha)x$. Choose an angle $\alpha \in \mathbb{R}$ so that $R(\alpha, p) = q$, and pick a continuous map $g: \mathbb{R} \to \mathbb{R}$ satisfying $g(r) = \alpha$ and $g(t) = 0$ for $t \notin [r - \delta, r + \delta]$ (e.g., $g(t) = \alpha \max(0, 1 - |t - r|/\delta)$). The continuous map

$$f: \mathbb{C} \to \mathbb{C}, \qquad f(x) = R(g(|x|), x),$$

is a homeomorphism, with inverse $f^{-1}(y) = R(-g(|y|), y)$, that fulfils the requirements.

Example 1.19 Let $A, B \subset \mathbb{C}$ be finite subsets containing the same number of points. Then $\mathbb{C} - A$ and $\mathbb{C} - B$ are homeomorphic. Since composites of homeomorphisms are homeomorphisms, an induction argument on the number of elements of $A - B$ reduces the claim to the following fact.

Let $x, y, v_1, \ldots, v_s$ be distinct points in $\mathbb{C}$. There exists a homeomorphism $f: \mathbb{C} \to \mathbb{C}$ such that $f(x) = y$ and $f(v_i) = v_i$, for any i.

Up to translations we can assume (exercise) that $|x| = |y| \neq |v_i|$ for all i. Write $r = |x| = |y|$ and take $\delta > 0$ so that $|v_i| \notin [r - \delta, r + \delta]$ for every i. We conclude by applying Example 1.18.

Topological spaces are most certainly not all homeomorphic to each other. For instance $\mathbb{R}$ is not homeomorphic to $[0, 1]$: a hypothetical homeomorphism $[0, 1] \to \mathbb{R}$ would have to be onto, and thus violate Weierstraß's theorem on the existence of maxima for continuous maps defined on closed, bounded intervals.

Deciding whether two Euclidean subsets are homeomorphic or not is not always easy: at times the problem becomes so difficult that not even the most brilliant mathematicians can solve it. This textbook offers the simplest topological methods that enable one to handle and solve that problem in a number of interesting situations. It'll be shown, later on, that the spaces in Example 1.19 are homeomorphic only in case A and B have the same cardinality.

Exercises

1.15 Prove that if X is homeomorphic to Y and Z is homeomorphic to W, then $X \times Z$ is homeomorphic to $Y \times W$.

1.16 Verify that orthogonal projection in $\mathbb{R}^3$ induces a homeomorphism between the paraboloid $z = x^2 + y^2$ and the plane $z = 0$.

1.17 ($\heartsuit$) Find a homeomorphism between $S^n \times \mathbb{R}$ and $\mathbb{R}^{n+1} - \{0\}$.

1.18 ($\heartsuit$) Find a homeomorphism between the hypercube $I^n = [0, 1]^n \subset \mathbb{R}^n$ and the unit ball D^n.

1.19 Find a subset of $\mathbb{R}^3$ homeomorphic to $S^1 \times S^1$.

1.20 Describe a homeomorphism from $S^2 \times S^2$ to a subset of $\mathbb{R}^5$. (Hint: identify $S^2 \times S^2$ with a subset of S^5.)

1.21 Show that the mapping

$$f: \{x \in \mathbb{R}^n \mid \|x\| < 1\} \to \{(x, y) \in \mathbb{R}^n \times \mathbb{R} \mid \|x\|^2 + y^2 = 1, \ y < 1\},$$

$$f(x) = (2x\sqrt{1 - \|x\|^2}, 2\|x\|^2 - 1),$$

is a homeomorphism. What does f have to do with the homeomorphisms of Examples 1.12 and 1.13?

1.22 Prove that the quadric in $\mathbb{R}^n$

$$x_1^2 + \cdots + x_p^2 - x_{p+1}^2 - \cdots - x_{p+q}^2 = 1 \qquad \text{(where } p + q \leq n\text{)}$$

is homeomorphic to $S^{p-1} \times \mathbb{R}^{n-p}$. (Hint: Example 1.11.)

1.23 We have seen that $]0, 1[$ is homeomorphic to $\mathbb{R}$, and hence not homeomorphic to $[0, 1]$. Consider the map $f:]0, 1[\times [0, 1[\to [0, 1] \times [0, 1[$, shown in Fig. 1.10:

$$f(x, y) = \begin{cases} \left(\dfrac{y}{3}, 1 - 3x\right) & \text{if } 3x + y \leq 1, \\[2mm] \left(x + (1 - 2x)\dfrac{2y - 1}{2y + 1}, y\right) & \text{if } 1 - y \leq 3x \leq 2 + y, \\[2mm] \left(1 - \dfrac{y}{3}, 3x - 2\right) & \text{if } 3x - y \geq 2. \end{cases}$$

Check, even using the glueing lemma (Theorem 1.8), that f is a homeomorphism.

1.24 Consider a continuous map $h: \mathbb{R}^n \to]0, +\infty[$ and the two spaces

$$X = \{(x_0, \dots, x_n) \in \mathbb{R}^{n+1} \mid 0 \leq x_0 \leq 1\}$$

and

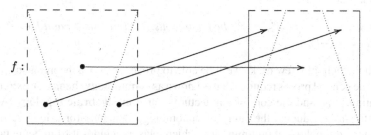

Fig. 1.10 In topology the cancellation rule does not hold: if $X \times Z$ is homeomorphic to $Y \times Z$, we cannot say X is homeomorphic to Y

$$Y = \{(x_0, \ldots, x_n) \in \mathbb{R}^{n+1} \mid 0 \leq x_0 \leq h(x_1, \ldots, x_n)\}.$$

Show that

$$f \colon X \to Y, \qquad f(x_0, \ldots, x_n) = (x_0 h(x_1, \ldots, x_n), x_1, \ldots, x_n),$$

is a homeomorphism.

1.25 Prove that the lower half-plane $X = \{(x, y) \in \mathbb{R}^2 \mid y \leq 0\}$ is homeomorphic to $Y = X \cup ([-1, 1] \times [0, 1])$. (Hint: show, with Example 1.10 and the glueing lemma, that Y is homeomorphic to $X \cup \{(x, y) \mid y \leq 1 - |x|\}$, then use Exercise 1.24.)

1.26 ($\heartsuit$) Check that

$$\{z \in \mathbb{C} \mid z = a + ib, \, b > 0\} \to \{z \in \mathbb{C} \mid |z| < 1\}, \qquad z \mapsto \frac{z - i}{z + i}$$

is a homeomorphism and describe the inverse.

1.5 Facts Without Proof

Here is a short compilation of interesting and/or paramount results, whose proof goes beyond our scope:

Theorem 1.20 *No sphere S^n is homeomorphic to a ball D^m, $n, m \geq 1$.*

Theorem 1.21 *Two products of spheres $S^{n_1} \times \cdots \times S^{n_k}$ and $S^{m_1} \times \cdots \times S^{m_h}$ are homeomorphic only if $h = k$ and $n_i = m_i \, \forall i$ (up to permuting indices).*

Theorem 1.22 $\mathbb{R}^n$ *is homeomorphic to* $\mathbb{R}^m$ *if and only if $n = m$.*

Theorem 1.23 *Every continuous map $f \colon D^n \to D^n$ has at least one fixpoint.*

Theorem 1.24 *Let $f \colon S^{2n} \to S^{2n}$ be continuous. Then there exists a point $x \in S^{2n}$ such that $f(x) = \pm x$.*

Theorem 1.25 *Let $p \colon S^n \to \mathbb{R}^n$ be continuous. There exists a point $l \in S^n$ such that $p(l) = p(-l)$.*

With the help of a few tricks we'll be able to prove the above theorems in special cases. The general proofs require what is known as **cohomology theory**, an extremely fascinating topic and the core of any lecture course in **algebraic topology**. Albeit not particularly arduous, the study of cohomology demands for a long series of algebraic and topological preliminaries, which, alas, precludes it from being taught at an undergraduate level.

Interested readers will find the proofs of Theorems 1.20–1.24 (or equivalent versions thereof) in any introductory text on singular homology and cohomology, such as [Do80, Ma91, GH81, Vi94] for instance. Theorem 1.25 is attributed to Borsuk and requires more refined techniques than the others: [Fu95] contains one such proof. Alternative arguments for Borsuk's theorem, that don't rely on cohomology, can be found in [Du66, GP74].

Exercises

1.27 Use Theorem 1.25 to prove Theorems 1.20 and 1.22.

1.28 Show that Theorem 1.24 doesn't hold for continuous maps $f: S^1 \to S^1$, nor more generally for continuous maps $f: S^{2n+1} \to S^{2n+1}$.

1.29 Even though we'll prove Theorem 1.25 when $n = 1$ later on, it could be fun to try and find a naïve argument (still $n = 1$) based on this observation: if two mice live in an infinite pipe that has no side exits, and at 8 pm each mouse is where the other one was at 8 am, at some moment they must have met.

References

[Du66] Dugundji, J.: Topology. Allyn and Bacon, Inc., Boston (1966)

[Do80] Dold, A.: Lectures on Algebraic Topology, 2nd edn. Springer, Berlin (1980)

[Fu95] Fulton, W.: Algebraic Topology. Springer, Berlin (1995)

[GP74] Guillemin, V., Pollack, A.: Differential Topology. Prentice-Hall Inc., Englewood Cliffs (1974)

[GH81] Greenberg, M., Harper, J.: Algebraic Topology: A First Course. Addison-Wesley (1981)

[Ma91] Massey, W.: Basic Course in Algebraic Topology. Springer, Berlin (1991)

[Vi94] Vick, J.W.: Homology Theory. Springer, Berlin (1994)

Chapter 2
Sets

In this book we'll always work within the so-called naïve set theory and completely avoid the axiomatic backgrounds. This choice has the advantage of leaving the reader free to pick his own preferred notions of 'set' and 'element', for instance ones suggested by conventional wisdom, or those learnt during undergraduate lectures on algebra, analysis and geometry. The only exception will concern the axiom of choice, whose content is less evident from the point of view of elementary logic, or to the layman.

We'll deliberately adopt the approach of the ostrich, that buries its head in the sand so not to see the paradoxes[1] to which this MO may lead. A healthy and low-cost rule of thumb—that also helps to by-pass the classical paradoxes—is to eschew statements of the sort *the set of all sets such that* ..., and prefer rather *the family of sets* .... This, by the way, also rids the language of tedious and lacklustre repetitions. The same principle holds when talking about families of objects, so we will say: *the class of all families* ..., *the collection of all classes* ... and so on.

2.1 Notations and Basic Concepts

If X is a set we'll write $x \in X$ if x belongs to X, that is, if x is an element of X. We'll indicate with $\emptyset$ the empty set, while the symbols $\{*\}$ and $\{\infty\}$ will both denote the **singleton**, the set with only one element. A set is called **finite** if it contains a finite number of elements, and we'll write $|X| = n$ if X has exactly n elements. A set that isn't finite is called **infinite**.

If A and B are sets we write $A \subset B$ if A is contained in B, i.e. when every element of A is an element of B. We use $A \subset B$, $A \neq B$, or $A \subsetneq B$, in case A is strictly contained in B. The set A is said to meet or intersect B (then A and B meet, or intersect) if their intersection $A \cap B$ isn't empty.

---

[1]Russell's paradox being the most widely known, see Sect. 8.1.

© Springer International Publishing Switzerland 2015
M. Manetti, *Topology*, UNITEXT - La Matematica per il 3+2 91,
DOI 10.1007/978-3-319-16958-3_2

Example 2.1 A consequence of the definition of $\subset$ is that $\emptyset \subset A$, for any set A. Even if this doesn't convince you completely you must accept it anyhow, at least as a convention. In this way, given any property $\mathfrak{p}$ defined on the elements of A, it makes sense to write

$$\{a \in A \mid \mathfrak{p}(a)\} \subset A,$$

where $\{a \in A \mid \mathfrak{p}(a)\}$ denotes the set of elements in A for which $\mathfrak{p}$ is true. Similarly, for any set A there is a unique map $\emptyset \to A$. Sometimes it may be better to think of the singleton as the set of maps from the empty set to itself.

The set of elements in A that do not belong to B is denoted $A - B = \{x \in A \mid x \notin B\}$. If $x_1, \ldots, x_n$ belong to a set X, $\{x_1, \ldots, x_n\}$ will indicate the subset of X whose elements are precisely $x_1, \ldots, x_n$.

We write $f : X \to Y$ or $X \xrightarrow{f} Y$ to mean that f is a map from X to Y, and $x \mapsto y$ to mean $y = f(x)$. Given a map $f : X \to Y$ and a subset $A \subset Y$ one defines

$$f^{-1}(A) = \{x \in X \mid f(x) \in A\}.$$

Subsets in X of the form $f^{-1}(A)$ are called **saturated** with respect to f. If $y \in Y$ we call $f^{-1}(\{y\})$ the **fibre** of f over y. Often, slightly abusing notations, $f^{-1}(y)$ is used to mean $f^{-1}(\{y\})$.

If $f : X \to Y$ is a map between sets and A, B are subsets in X, it is easy to check that $f(A \cup B) = f(A) \cup f(B)$, while it is usually not true that $f(A \cap B) = f(A) \cap f(B)$. What can be said in this respect is the following straightforward fact.

Proposition 2.2 (Projection formula) *Let $f : X \to Y$ be a map and $A \subset X$, $B \subset Y$ subsets. Then*

$$f(A \cap f^{-1}(B)) = f(A) \cap B.$$

To say that a map $f : X \to Y$ is one-to-one (a.k.a. injective, or an injection) one uses sometimes an arrow with hook $f : X \hookrightarrow Y$, or $X \overset{f}{\hookrightarrow} Y$. The symbol used for onto, or surjective, maps (i.e. surjections) is occasionally the double-headed arrow $f : X \twoheadrightarrow Y$ or $X \overset{f}{\twoheadrightarrow} Y$.

If $\mathcal{A}$ is a family of sets, the symbols

$$\cup\{A \mid A \in \mathcal{A}\} \quad \text{and} \quad \bigcup_{A \in \mathcal{A}} A$$

both denote their union. Similar notations hold for the intersection, using $\cap$ instead of $\cup$. We use the coproduct symbol $\coprod$ to denote disjoint unions: for any family of sets $\mathcal{A}$ there is an obvious surjective map

$$\coprod\{A \mid A \in \mathcal{A}\} \to \bigcup\{A \mid A \in \mathcal{A}\}$$

that is one-to-one if and only if $A \cap B = \emptyset$ for every $A, B \in \mathcal{A}$.

A family $\mathcal{A}$ is called **indexed** if there exists an onto map $A : I \to \mathcal{A}$, where I is the set of **indices**. One usually writes A_i instead of $A(i)$ and $\cup\{A_i \mid i \in I\}$ instead of $\cup\{A \mid A \in \mathcal{A}\}$. **Parameters** and **parametrised family** are synonyms of indices and indexed family, respectively.

De Morgan's laws prescribe that the set-theoretical operation of complementation swaps the roles of union and intersection: in other words, if $A, B \subset X$,

$$X - (A \cup B) = (X - A) \cap (X - B), \qquad X - (A \cap B) = (X - A) \cup (X - B).$$

This is true more generally for every family $\{A_i \mid i \in I\}$ of subsets of X: in formulas

$$X - \left(\bigcup_{i \in I} A_i \right) = \bigcap_{i \in I} (X - A_i), \qquad X - \left(\bigcap_{i \in I} A_i \right) = \bigcup_{i \in I} (X - A_i).$$

The **Cartesian product** of a finite number of sets $X_1, \ldots, X_n$ is written

$$X_1 \times \cdots \times X_n, \quad \text{or} \quad \prod_{i=1}^{n} X_i,$$

and is, by definition, the set of n-tuples $(x_1, \ldots, x_n)$ where $x_i \in X_i$ for every i. Given a set X and any natural number $n \in \mathbb{N}$ the Cartesian product of X with itself n times is indicated with X^n.

Let A be a subset of $X \times Y$ and B a non-empty subset of Y. It makes sense to consider the **quotient** $(A : B) \subset X$, which we define as

$$(A : B) = \{x \in X \mid \{x\} \times B \subset A\}.$$

In other terms, $(A : B)$ is the largest subset of X such that $(A : B) \times B \subset A$; in particular $(X \times Y : Y) = X$.

Exercises

2.1 Prove Proposition 2.2.

2.2 Prove that for any three sets A, B, C one has

$$A \cap (B \cup C) = (A \cap B) \cup (A \cap C) \quad \text{and} \quad A \cup (B \cap C) = (A \cup B) \cap (A \cup C).$$

2.3 Show that if $\{A_i \mid i \in I\}$ and $\{B_j \mid j \in J\}$ are arbitrary families of sets, the **distributive laws** of union and intersection hold:

$$\left(\bigcup_{i \in I} A_i\right) \cap \left(\bigcup_{j \in J} B_j\right) = \bigcup_{\substack{i \in I \\ j \in J}} (A_i \cap B_j),$$

$$\left(\bigcap_{i \in I} A_i\right) \cup \left(\bigcap_{j \in J} B_j\right) = \bigcap_{\substack{i \in I \\ j \in J}} (A_i \cup B_j).$$

2.4 ($\heartsuit$) Let $f : X \to Y$ be a map and A, B subsets of X. Tell which of the following statements are always true, and which are not:

1. $f(A \cap B) \supset f(A) \cap f(B)$;
2. $f(A \cap B) \subset f(A) \cap f(B)$;
3. $f(X - A) \subset f(X) - f(A)$;
4. $f(X - A) \supset f(X) - f(A)$;
5. $f^{-1}(f(A)) \subset A$;
6. $f^{-1}(f(A)) \supset A$.

2.2 Induction and Completeness

We denote by:

- $\mathbb{N} = \{1, 2, 3, \ldots\}$ the set of natural numbers (positive integers);
- $\mathbb{N}_0 = \{0, 1, 2, 3, \ldots\}$ the set of non-negative integers;
- $\mathbb{Z} = \{0, \pm 1, \pm 2, \pm 3, \ldots\}$ the ring of integers;
- $\mathbb{Z}/n$ the group of integers modulo n;
- $\mathbb{Q}, \mathbb{R}, \mathbb{C}$ the fields of rational, real and complex numbers.

We'll suppose that readers are familiar with the induction principle and other properties of natural numbers.

Induction principle. *Let*

$$P : \mathbb{N} \to \{\text{true, false}\}$$

be a map such that $P(1) = $ true, *and* $P(n) = $ true *each time* $P(n - 1) = $ true. *Then* $P(n) = $ true *for every* n.

Besides induction, we'll often make use of two statements equivalent to it.

Well-ordering principle.[2] *Every non-empty subset of* $\mathbb{N}$ *contains a smallest element.*

Principle of recursive definition. *Take a non-empty set* X *and maps*

$$r_n : X^n \to X$$

---

[2]Less frequently called 'least-integer principle'.

for every $n \in \mathbb{N}$. *Then for each* $x \in X$ *there exists a unique map* $f : \mathbb{N} \to X$ *such that*

$$f(1) = x, \qquad f(n+1) = r_n(f(1), f(2), \ldots, f(n)) \quad \forall n \geq 1.$$

There are plenty typical and well-known applications of the latter (the factorial of a natural number, Pascal's triangle,[3] Fibonacci numbers etc.). Let us now discuss another consequence, to be used later.

Lemma 2.3 *Let* $X \subset \mathbb{N}$ *be an infinite subset. There exists a strictly increasing bijection* $f : \mathbb{N} \to X$.

Proof Given a finite subset $Y \subset X$, its complement $X - Y$ isn't empty, so it has a smallest element. It is enough to define f recursively by

$$f(1) = \min(X), \qquad f(n+1) = \min(X - \{f(1), f(2), \ldots, f(n)\}).$$

$\square$

Concerning real numbers, a particularly important property is the following.

Completeness of the real numbers. *Every non-empty set of real numbers that is bounded from above has a least upper bound.*

To refresh the memory we remind that $X \subset \mathbb{R}$ is said bounded from above (or upper bounded) if there exists $M \in \mathbb{R}$ such that $M \geq x$ for all $x \in X$. A real number $s \in \mathbb{R}$ is called a **least upper bound** or **supremum** of X, written $s = \sup(X)$, if:

1. $s \geq x$ for every $x \in X$;
2. for any $\varepsilon > 0$ there exists $x \in X$ such that $x \geq s - \varepsilon$.

Swapping $\geq$ with $\leq$ gives the notions of a set bounded from below and of greatest lower bound/infimum $\inf(X)$. For intervals we'll adopt the notations of Exercise 1.8.

Exercises

2.5 Using the well-ordering principle prove that for every surjective map $f : \mathbb{N} \to X$ there is a map $g : X \to \mathbb{N}$ such that $fg(x) = x$ for every $x \in X$.

2.6 ($\heartsuit$) Define $3^{\sqrt{2}}$ using solely the following notions: powers with integer exponent, well-ordering principle and completeness of the reals.

2.7 ($\heartsuit$) Use the principle of recursive definition to show that for any real number x there is a bijection $g : \mathbb{N} \to \mathbb{N}$ such that

$$\lim_{n \to \infty} \sum_{i=1}^{n} \frac{(-1)^{g(i)}}{g(i)} = x.$$

---

[3] Known in Italy as Tartaglia's triangle.

2.3 Cardinality

Definition 2.4 Two sets X, Y are said to have the same **cardinality**, written $|X| = |Y|$, if a bijective map $X \to Y$ exists.

Clearly, if $|X| = |Y|$ and $|Y| = |Z|$ then $|X| = |Z|$. Sometimes the symbol $|X|$ might be ambiguous (for instance when one has to do with absolute values); in that case the alternatives are $\text{Card}(X)$ and $\#X$.

Definition 2.5 A set is called **countably infinite** if it has the cardinality of $\mathbb{N} = \{1, 2, 3, \ldots\}$, and **countable** if it is either countably infinite or finite.[4]

By Lemma 2.3 a set is countable if and only if it has the same cardinality of a subset of $\mathbb{N}$. Let us see a few examples.

Example 2.6 The sets $\mathbb{Z}$ and $\mathbb{N}_0$ are countable. The maps $f : \mathbb{Z} \to \mathbb{N}$,

$$f(n) = \begin{cases} 2n & \text{if } n > 0, \\ 1 - 2n & \text{if } n \le 0, \end{cases}$$

and $g : \mathbb{N}_0 \to \mathbb{N}$, $g(n) = n + 1$, in fact, are bijective.

Example 2.7 The set $\mathbb{N} \times \mathbb{N}$ of pairs of natural numbers is countable. To prove it observe that $C = \{(x, y) \in \mathbb{Z}^2 \mid 0 \le x \le y\}$ is countable, since for any integer $n \ge 0$ there is a unique element $(x, y) \in C$ such that $n = x + \sum_{i=0}^{y} i$. Moreover, the map

$$\mathbb{N}_0 \times \mathbb{N}_0 \to C, \qquad (a, b) \mapsto (a, a + b)$$

is bijective. Therefore an explicit bijection between $\mathbb{N} \times \mathbb{N}$ and $\mathbb{N}$ is

$$\mathbb{N} \times \mathbb{N} \to \mathbb{N}, \qquad (a, b) \mapsto a + \sum_{i=0}^{a+b-2} i = \frac{1}{2}(a + b - 2)(a + b - 1) + a.$$

Given a set X we write $\mathcal{P}(X)$ for the family of its subsets, and $\mathcal{P}_0(X) \subset \mathcal{P}(X)$ for the family of finite subsets of X.

Example 2.8 The family $\mathcal{P}_0(X)$ of finite subsets of a countable set X is countable. In fact we can identify X with $\mathbb{N}_0$, then notice that any finite subset of $\mathbb{N}_0$ can be viewed as the set of digits equal to 1 of a suitable binary number: this fact amounts to saying that

$$\mathcal{P}_0(\mathbb{N}_0) \to \mathbb{N}_0, \qquad \{\emptyset\} \mapsto 0, \quad \{n_1, \ldots, n_k\} \mapsto \sum_{i=1}^{k} 2^{n_i},$$

is a bijection.

---

[4] The reader should be aware that this terminology is far from universal: some people call 'countable' what we defined as countably infinite, and use 'at most countable' to mean countable.

We'll prove later in the book (Corollary 2.31) that $|\mathcal{P}_0(X)| = |X|$ for any infinite set X.

Theorem 2.9 (Cantor) *If X isn't empty, there are no surjective maps $f: X \to \mathcal{P}(X)$. In particular X and $\mathcal{P}(X)$ do not have the same cardinality.*

Proof Take a map $f: X \to \mathcal{P}(X)$ and let's prove that the subset

$$S = \{x \in X \mid x \notin f(x)\} \in \mathcal{P}(X)$$

isn't contained in the range of f. Assume, by contradiction, there existed an element $s \in X$ such that $f(s) = S$. If $s \in S$, by definition of S we would have $s \notin f(s) = S$. If, instead, $s \notin S = f(s)$, the definition of S would imply $s \in f(s) = S$. Either possibility leads to a contradiction so there cannot be such an element s. $\qquad\square$

Example 2.10 Any open interval $]a, b[\subset \mathbb{R}$, with $a < b$, has the cardinality of $\mathbb{R}$. Given a real number $c \in]a, b[$, in fact, the map $f:]a, b[\to \mathbb{R}$,

$$f(x) = \begin{cases} \dfrac{x-c}{b-x} & \text{if } c \le x < b \\[2ex] \dfrac{x-c}{x-a} & \text{if } a < x \le c \end{cases}$$

is bijective.

Lemma 2.11 *Consider sets X, Y and mappings $f: X \to Y$, $g: Y \to X$. There exists a subset $A \subset X$ such that*

$$A \cap g(Y - f(A)) = \emptyset, \qquad A \cup g(Y - f(A)) = X.$$

Proof Consider the family $\mathcal{A}$ of subsets $B \subset X$ such that $B \cap g(Y - f(B)) = \emptyset$. Clearly $\mathcal{A}$ isn't empty because it contains the empty set. Let's show that $A = \cup\{B \mid B \in \mathcal{A}\}$ has the required properties. If $x \in A$ there is $B \in \mathcal{A}$ such that $x \in B$, so $x \notin g(Y - f(B))$ and therefore $x \notin g(Y - f(A))$. If there existed $x \notin A \cup g(Y - f(A))$, then setting $C = A \cup \{x\}$ we would have $g(Y - f(C)) \subset g(Y - f(A))$ and then $C \cap g(Y - f(C)) = \emptyset$, against the definition of A. $\qquad\square$

From Lemma 2.11 follows a useful criterion to decide whether two sets have the same cardinality.

Theorem 2.12 (Cantor-Schröder-Bernstein) *Let X, Y be sets. If there exist two injective maps $f: X \to Y$ and $g: Y \to X$, then X and Y have the same cardinality.*

Proof According to Lemma 2.11 there exists a subset $A \subset X$ such that, putting $B = Y - f(A)$,

$$A \cap g(B) = \emptyset \quad \text{and} \quad A \cup g(B) = X.$$

The definition of B also tells that $B \cap f(A) = \emptyset$ and $B \cup f(A) = Y$, while the injectivity of f and g implies that the maps $f: A \to f(A)$ and $g: B \to g(B)$ are bijections. To finish it suffices to remark that

$$h: X \to Y, \qquad h(x) = \begin{cases} f(x) & \text{if } x \in A \\ g^{-1}(x) & \text{if } x \in g(B) \end{cases}$$

is a bijection.   □

If X and Y are sets we write $|X| \le |Y|$ if an injection $X \to Y$ exists. We have just proved the following properties:

Reflexivity: $|X| \le |X|$ *for any set X.*
Anti-symmetry: $|X| \le |Y|$ and $|Y| \le |X|$ *imply* $|X| = |Y|$.
Transitivity: $|X| \le |Y|$ and $|Y| \le |Z|$ *imply* $|X| \le |Z|$.

We'll write $|X| \ge |Y|$ to mean $|Y| \le |X|$.

Remark 2.13 It is possible to prove, and we shall do so later (Proposition 2.25), that the cardinalities of any two sets can always be compared, i.e. that for any X, Y either $|X| \le |Y|$ or $|Y| \le |X|$.

Given sets X, Y we denote by X^Y the set of all maps $f: Y \to X$. For example $\mathbb{R}^{\mathbb{N}}$ is the set of real sequences $a_1, a_2, \ldots$. There exists a natural bijection

$$\alpha: (X^Y)^Z \to X^{Y \times Z}, \qquad \alpha(g)(y, z) = g(z)(y).$$

Proposition 2.14 *The sets $\mathbb{R}$, $\mathbb{N}^{\mathbb{N}}$, $\mathbb{R}^{\mathbb{N}}$, $2^{\mathbb{N}}$ and $\mathcal{P}(\mathbb{N})$ have the same cardinality. In particular, the set of real numbers $\mathbb{R}$ isn't countable.*

Proof Before we get started with the proof, let's remind that for any X there is a natural one-to-one correspondence between $\mathcal{P}(X)$ and the set of maps from X to $\{0, 1\}$: it associates to $f: X \to \{0, 1\}$ the subset $A = \{x \in X \mid f(x) = 1\}$. Hence $\mathcal{P}(X)$ and 2^X have the same cardinality.

Let now C be the set of strictly increasing sequences $\{f_n\}$ of natural numbers. The bijection

$$\alpha: \mathbb{N}^{\mathbb{N}} \to C, \qquad \alpha(f)_n = \sum_{i=1}^{n} f(i),$$

shows that C has the same cardinality of $\mathbb{N}^{\mathbb{N}}$. The mapping

$$\beta:]1, +\infty[\to C, \qquad \beta(x)_n = \text{ integer part of } 10^n x,$$

is one-to-one, so $|\mathbb{R}| \leq |C|$. But also $\gamma \colon C \to \mathbb{R}$ defined by

$$\gamma(f) = \sum_{n=1}^{+\infty} \frac{1}{10^{f_n}} = \sup\left\{\sum_{n=1}^{N} \frac{1}{10^{f_n}} \mid N \in \mathbb{N}\right\}$$

is injective, whence $|\mathbb{R}| \geq |C|$. By the Cantor-Schröder-Bernstein theorem $|\mathbb{R}| = |C| = |\mathbb{N}^{\mathbb{N}}|$.

Due to the bijection between $\mathbb{N}$ and $\mathbb{N} \times \mathbb{N}$ the sets $\mathbb{N}^{\mathbb{N}}$ and $\mathbb{N}^{\mathbb{N} \times \mathbb{N}}$ have the same cardinality, so $|\mathbb{N}^{\mathbb{N} \times \mathbb{N}}| = |(\mathbb{N}^{\mathbb{N}})^{\mathbb{N}}| = |\mathbb{R}^{\mathbb{N}}|$. Similarly

$$|\mathcal{P}(\mathbb{N})| = |2^{\mathbb{N}}| = |2^{\mathbb{N} \times \mathbb{N}}| = |(2^{\mathbb{N}})^{\mathbb{N}}| = |\mathcal{P}(\mathbb{N})^{\mathbb{N}}|,$$

and the natural injections $2^{\mathbb{N}} \to \mathbb{N}^{\mathbb{N}}$ and $\mathbb{N} \to \mathcal{P}(\mathbb{N})$ entail

$$|\mathbb{N}^{\mathbb{N}}| \leq |\mathcal{P}(\mathbb{N})^{\mathbb{N}}| = |2^{\mathbb{N}}| \leq |\mathbb{N}^{\mathbb{N}}|. \qquad \square$$

Exercises

2.8 Describe a bijective map between the intervals $[0, 1[$ and $[0, +\infty[$.

2.9 Show that if $A \subset \mathbb{R}$ contains a non-empty open interval, then $|A| = |\mathbb{R}|$.

2.10 Prove that rational numbers are countable.

2.11 ($\heartsuit$) Let X be the set of finite sequences of natural numbers. Prove X is countable.

2.12 ($\spadesuit$, $\heartsuit$) Prove the existence of an infinite, non-countable family of subsets $C_i \subset \mathbb{N}$ such that $C_i \cap C_j$ has finite cardinality for every $i \neq j$.

2.4 The Axiom of Choice

Suppose we have a surjection $g \colon Y \to X$ between two sets and we want to construct a map $f \colon X \to Y$ such that $g(f(x)) = x$ for every x. We can take an element $x_1 \in X$ and *choose* an element $f(x_1)$ in the non-empty set $g^{-1}(\{x_1\})$, then take $x_2 \in X - \{x_1\}$ and *choose* $f(x_2)$ in the non-empty set $g^{-1}(\{x_2\})$ and so on. If X is finite the process will at some point end, and furnish the required map f. But if X is infinite, without further information there's no elementary reason to guarantee we can make the *infinitely many choices* necessary to define f. This is why we have to add to our arsenal—if we want to make some mathematical progress—the celebrated axiom of choice, which we'll state in two equivalent versions.

Axiom of choice (V1) *If* $g\colon Y \to X$ *is a surjective map between sets, there exists a mapping* $f\colon X \to Y$ *such that* $g(f(x)) = x$ *for every* $x \in X$.

The axiom of choice is, in some sense, neither true nor false, and each one of us may or not believe in it (and accept the consequences). For more information and explanations on this matter we refer to [To03]. However, the axiom of choice is accepted by the overwhelming majority of mathematicians, and we'll follow suit without further ado.

Proposition 2.15 *Given non-empty sets* X *and* Y, $|X| \leq |Y|$ *if and only if there exists an onto map* $Y \twoheadrightarrow X$.

Proof If there is a 1-1 map $f\colon X \to Y$ between non-empty sets, then clearly there is also an onto map $g\colon Y \to X$: it is enough to fix an element $x_0 \in X$ and define $g(y) = f^{-1}(y)$ if $y \in f(X)$ and $g(y) = x_0$ otherwise.

Vice versa, if $g\colon Y \to X$ is onto, by the axiom of choice we also have $f\colon X \to Y$ with $g(f(x)) = x$ for every $x \in X$. Such an f is one-to-one, for if $f(x_1) = f(x_2)$ then $x_1 = g(f(x_1)) = g(f(x_2)) = x_2$. $\qquad\square$

A **relation** on a set X is a subset $\mathfrak{R} \subset X \times X$. It is customary to write $x\mathfrak{R}y$ whenever $(x, y) \in \mathfrak{R}$.

An **equivalence relation** on X is a relation $\sim$ satisfying these properties:

Reflexivity: $x \sim x$ *for every* $x \in X$.
Symmetry: $x \sim y$ *implies* $y \sim x$.
Transitivity: $x \sim y$ *and* $y \sim z$ *imply* $x \sim z$.

Definition 2.16 Let $\sim$ be an equivalence relation on a set X. The **equivalence class** (or **coset**) of an element $x \in X$ is the subset

$$[x] \subset X, \qquad [x] = \{y \in X \mid y \sim x\}.$$

Equivalence classes determine the equivalence relation uniquely, and the above properties read:

Reflexivity: $x \in [x]$ *for every* $x \in X$.
Symmetry: $x \in [y]$ *implies* $y \in [x]$.
Transitivity: $x \in [y]$ *and* $y \in [z]$ *imply* $x \in [z]$.

It's not hard to see that $[x] \cap [y] \neq \emptyset$ forces $x \sim y$ and $[x] = [y]$. The set of equivalence classes $X/\!\sim\, = \{[x] \mid x \in X\}$ is called the **quotient set** of X under the relation $\sim$. The well-defined map

$$\pi\colon X \to X/\!\sim, \qquad \pi(x) = [x],$$

is called **quotient map**. According to the axiom of choice there exists a map $f\colon X/\!\sim \to X$ such that $f([x]) \in [x]$ for every equivalence class. The range of f is a subset $S \subset X$ that meets each equivalence class in exactly one point.

Example 2.17 Consider the set $\mathbb{R}^{n+1} - \{0\}$ of non-zero vectors with $n+1$ components and define on it $x \sim y$ iff $x = ty$ for some $t \in \,]0, +\infty[$. It is straightforward to see that $\sim$ is an equivalence relation, and the sphere S^n meets each equivalence class in one point.

Lemma 2.18 *Let $\sim$ be an equivalence relation on X, $\pi: X \to X/\sim$ the quotient map and $f: X \to Y$ another mapping. The following facts are equivalent:*

1. the map f is constant on equivalence classes, i.e. $f(x) = f(y)$ whenever $x \sim y$;
2. there is a unique map $g: X/\sim \, \to Y$ such that $f = g\pi$.

Proof Exercise.   □

In certain cases the following version of the axiom of choice is more suitable.

Axiom of choice (V2) *Let $X = \cup \{X_i \mid i \in I\}$ be the union of a family of non-empty sets X_i indexed by a set I. Then there exists a map $f: I \to X$ such that $f(i) \in X_i$ for every i.*

Let's show that the two versions V1, V2 of the axiom of choice are equivalent. If $X = \cup\{X_i \mid i \in I\}$, we can consider the set $Y = \{(x, i) \in X \times I \mid x \in X_i\}$ and the projections $p: Y \to X$, $q: Y \to I$. By assumption p and q are onto, and V1 ensures the existence of $h: I \to Y$ such that the composite qh is the identity map on I. Then it is enough to take $f = ph: I \to X$.

Conversely, if $g: X \to I$ is onto $X_i = g^{-1}(i) = \{x \in X \mid g(x) = i\}$ is non-empty for every $i \in I$. As $X = \cup_i X_i$, by V2 there exists $f: I \to X$ such that $f(i) \in X_i$, i.e. $gf(i) = i$ for every $i \in I$.

Example 2.19 We exploit the axiom of choice to prove that the countable union of countable sets is countable. Let $\{X_n \mid n \in \mathbb{N}\}$ be a countable family of countable sets; for any n the set of injections $X_n \hookrightarrow \mathcal{N}$ isn't empty and the axiom of choice allows to pick, for every $n \in \mathbb{N}$, a map $f_n: X_n \to \mathbb{N}$. Now consider

$$\mu: \cup_n X_n \to \mathbb{N}, \qquad \mu(x) = \min\{n \mid x \in X_n\},$$

$$\phi: \cup_n X_n \to \mathbb{N} \times \mathbb{N}, \qquad \phi(x) = (\mu(x), f_{\mu(x)}(x)).$$

It is immediate to see that ϕ is one-to-one, and therefore $\cup_n X_n$ is countable.

Remark 2.20 The axiom of choice permits to prove a number of counter-intuitive results, and this is one of the reasons why it raises eyebrows. The most striking examples are probably the *Hausdorff paradox* (see Exercise 14.9), and the *Banach-Tarski paradox* [Wa93]. Two subsets $A, B \subset \mathbb{R}^3$ are called scissors-congruent if there are sets $A_1, \ldots, A_n$ and direct isometries (roto-translations) $\theta_1, \ldots, \theta_n$ of $\mathbb{R}^3$ such that A and B are the disjoint unions of $A_1, \ldots, A_n$ and $\theta_1(A_1), \ldots, \theta_n(A_n)$ respectively. The Banach-Tarski theorem states that the closed unit ball in $\mathbb{R}^3$ is scissors-congruent to the disjoint union of two closed unit balls.

Lacking the axiom of choice, on the other hand, a great deal of mathematics inevitably falls apart. For example the countable union of countable sets might no longer be countable without the axiom of choice, see [To03, p. 228].

Exercises

2.13 Given a set X call $\mathcal{P}(X)'$ the family of non-empty subsets in X. A *choice map* on X is by definition a map $s \colon \mathcal{P}(X)' \to X$ such that $s(A) \in A$ for any non-empty subset $A \subset X$. Prove that the axiom of choice is equivalent to the existence of a choice map on any given set.

2.14 ($\heartsuit$) Let $X = \coprod \{X_i \mid i \in I\}$ and $Y = \coprod \{Y_i \mid i \in I\}$ be disjoint unions of families indexed by the same set I. If X_i has the cardinality of Y_i for every $i \in I$, prove that X has the same cardinality of Y.

2.15 Prove that the axiom of choice is equivalent to **Zermelo's postulate**: *let $\mathcal{A}$ be a family of non-empty disjoint sets. There exists a set C such that $C \cap A$ consists of one element, for each $A \in \mathcal{A}$.*

2.16 ($\heartsuit$) Given a set X and a countable family $\mathcal{B}$ of subsets in X, denote by $\mathcal{T}$ the collection of subsets of X that can be written as unions of elements of $\mathcal{B}$. Prove that for any family $\mathcal{A} \subset \mathcal{T}$ there is a countable subfamily $\mathcal{A}' \subset \mathcal{A}$ such that

$$\bigcup_{U \in \mathcal{A}'} U = \bigcup_{V \in \mathcal{A}} V.$$

2.17 Use the axiom of choice to prove the following statement rigorously.

Let $f \colon X \to Y$ be an onto map. If for every $y \in Y$ the set $f^{-1}(y)$ is countably infinite, there exists a bijection $X \to Y \times \mathbb{N}$ whose composition with the projection $Y \times \mathbb{N} \to Y$ equals f.

2.18 Find the cardinality of the set of bijections $\mathbb{N} \to \mathbb{N}$ (Hint: for every subset $A \subset \mathbb{N}$ with at least 2 elements, there exists a bijection $f \colon \mathbb{N} \to \mathbb{N}$ such that $f(n) = n$ iff $n \notin A$).

2.5 Zorn's Lemma

Before it can be employed to tackle many mathematical problems, the axiom of choice needs to be formulated in other, less intuitive ways. Among its various incarnations (see [Ke55, p. 33]), the most renowned is—without doubt—**Zorn's lemma**. Whereas the axiom of choice has to do with equivalence relations, Zorn's lemma deals with order relations.

An **ordering** on a set X is a relation $\leq$ satisfying three properties:

Reflexivity: $x \leq x$ *for every* $x \in X$.
Anti-symmetry: *if* $x \leq y$ *and* $y \leq x$ *then* $x = y$.
Transitivity: *if* $x \leq y$ *and* $y \leq z$ *then* $x \leq z$.

Other names of an ordering are **order relation**, or simply **order**. If $\leq$ is an order relation one defines $x < y$ to mean $x \leq y$ and $x \neq y$.

Example 2.21 Let $A, B \subset Y$ be sets. Define

$$A \leq B \quad \text{if} \quad A \subset B.$$

The relation $\leq$ is an ordering on $\mathcal{P}(Y)$, called *inclusion*.

An order relation on X is called a **total ordering** if either $x \leq y$ or $y \leq x$ for every $x, y \in X$. An **ordered set** is a set equipped with an ordering, and a **totally ordered set** is a set with a total ordering.[5]

Any subset in an ordered set becomes ordered under the induced inclusion relation.

Definition 2.22 Let $(X, \leq)$ be an ordered set.

1. A subset $C \subset X$ is called a **chain** if $x \leq y$ or $x \geq y$ for any $x, y \in C$. Put differently, $C \subset X$ is a chain if and only if C is totally ordered by the induced order relation.
2. Let $C \subset X$ be a subset. A point $x \in X$ is called an **upper bound** for C if $x \geq y$ for every $y \in C$.
3. A point $m \in X$ is a **maximal element** in X if it is the unique upper bound of itself, i.e. $\{x \in X \mid m \leq x\} = \{m\}$.

Theorem 2.23 (Zorn's lemma) *Let* $(X, \leq)$ *be a non-empty ordered set. If every chain in X is upper bounded, then X contains maximal elements.*

We warn the young and inexperienced reader that the consequences of Zorn's lemma are far more instructive than its proof. For this reason one might want to skip the proof at first. A proper argument, which requires the axiom of choice, will be given in Sect. 8.2, just for the sake of completeness.

---

[5] Also these names are not universally accepted: some people call 'ordering' our total ordering, and 'partial ordering' our ordering. **Poset** (standing for 'partially ordered set') is another term for an ordered set.

Let's show that the axiom of choice can be deduced from Zorn's lemma. The proof we'll provide is the most straightforward and useful means of applying Zorn's lemma, whence we shall be very thorough. Consider the first version of the axiom of choice (V1), and let X, Y be non-empty, $g: Y \to X$ onto. We'll show that Zorn's lemma yields the existence of a map $f: X \to Y$ such that $g(f(x)) = x$ for every $x \in X$. For this let's introduce the set S made by pairs (E, f) where:

1. $E \subset X$ is a subset;
2. $f: E \to Y$ is such that $gf(x) = x$ for every $x \in E$.

The set S isn't empty, for it contains the pair $(\emptyset, \emptyset \hookrightarrow Y)$. Elements in S can be ordered by *extension*: $(E, h) \leq (F, k)$ iff k extends h, that is to say $E \subset F$ and $h(x) = k(x)$ for every $x \in E$. We claim that any chain $C \subset S$ is bounded from above. Consider the set

$$A = \bigcup_{(E,h) \in C} E$$

and define $a: A \to Y$ like this: if $x \in A$ there is $(E, h) \in C$ such that $x \in E$, and we set $a(x) = h(x)$. The map a is well defined: if $(F, k) \in C$ and $x \in F$ then $(E, h) \leq (F, k)$ or $(E, h) \geq (F, k)$ because C is a chain. In either case $x \in E \cap F$ and $h(x) = k(x)$. It is clear that $(A, a) \in S$ is an upper bound for C.

By Zorn's lemma there is a maximal element $(U, f) \in S$, so it suffices to show $U = X$. If that were not the case, there would exist $y \in Y$ such that $g(y) \notin U$ and the pair $(U \cup \{g(y)\}, f')$, extending (U, f) and such that $f'(g(y)) = y$, would belong to S, contradicting the maximality of (U, f).

Corollary 2.24 *Let $(X, \leq)$ be an ordered set whose every chain has at least one upper bound. Then for any $a \in X$ there exists a maximal element m in X such that $m \geq a$.*

Proof Apply Zorn's lemma to $\{x \in X \mid x \geq a\}$.   □

Proposition 2.25 *Cardinalities are always comparable: if X and Y are sets, then either $|X| \leq |Y|$ or $|Y| \leq |X|$.*

Proof Consider the family A of subsets $A \subset X \times Y$ for which the projections $p: A \to X$ and $q: A \to Y$ are one-to-one. As $\emptyset$ belongs to A, this isn't empty and is ordered by the inclusion. Every chain $C \subset A$ is bounded: considering the obvious element $C = \cup\{A \mid A \in C\}$, the maps $p: C \to X$ and $q: C \to Y$ are injective and thus C bounds C.

By Zorn's lemma A has a maximal element; let's show that one of the projections $p: A \to X$ and $q: A \to Y$ is onto. If this were not true there would exist $x \in X - p(A)$ and $y \in Y - q(A)$; therefore $A \cup \{(x, y)\} \in A$, contradicting maximality.

Now, if $p: A \to X$ is onto then $|X| = |A| \leq |Y|$. Similarly if $q: A \to Y$ is onto, $|Y| = |A| \leq |X|$.   □

Exercises

2.19 Let X be an ordered set with the property that every non-empty subset has a maximum and a minimum. Prove that X is finite and totally ordered.

2.20 Consider, on a set X, a relation $\prec$ such that:

1. for every $x, y \in X$, at least one between $x \prec y$, $y \prec x$ is false;
2. if $x \prec y$ and $y \prec z$, then $x \prec z$.

Prove that

$$x \leq y \quad \Longleftrightarrow \quad x \prec y \text{ or } x = y$$

defines an ordering.

2.21 Let $f : X \to X$ be a map on a set X and denote by

$$f^0 = Id_X, \quad f^1 = f, \quad f^2 = f \circ f, \quad \ldots, \quad f^n = f \circ f^{n-1}, \quad \ldots$$

the iterations of f. Consider the relation '$x \leq y$ *if there exists $n \geq 0$ such that* $x = f^n(y)$'. Show that $\leq$ is an ordering on X if and only if, whenever $f^n(x) = x$ for some $x \in X$ and an integer $n > 0$, we have $f(x) = x$.

2.22 Given a vector space V (not necessarily finite-dimensional) consider the set $G(V)$ of its vector subspaces, ordered by inclusion (Example 2.21). Prove that for every subset $X \subset V$ containing 0 the family

$$\{F \in G(V) \mid F \subset X\}$$

contains maximal elements.

2.23 (Tukey's lemma, $\heartsuit$) Let X be a set and $\mathcal{B}$ a non-empty family of subsets of X such that $A \subset X$ belongs in $\mathcal{B}$ iff every finite subset $B \subset A$ belongs in $\mathcal{B}$. Show that $\mathcal{B}$ has maximal elements for the inclusion.

2.24 Use Zorn's lemma to show that any set admits a total ordering.

2.25 Let V be a vector space over a field $\mathbb{K}$. For any subset $A \subset V$ call $L(A) \subset V$ the linear closure of A, i.e. the intersection of all vector subspaces containing A; note that $L(\emptyset) = \{0\}$. A subset $A \subset V$ is called a **set of generators** of V if $L(A) = V$, and **linearly independent** if $v \notin L(A - \{v\})$ for every $v \in A$.

(1) Prove that $L(A)$ coincides with the set of finite linear combinations

$$a_1 v_1 + \cdots + a_n v_n, \quad a_i \in \mathbb{K}, \quad v_i \in A,$$

of elements of A.
(2) Show that a set of vectors is linearly independent if and only if every finite subset of it is linearly independent.

(3) Let $A \subset V$ be linearly independent and $v \in V$ a vector. Prove that $A \cup \{v\}$ is linearly independent if and only if $v \notin L(A)$.

Call $\mathcal{B}$ the family of all linearly independent subsets of V, ordered by inclusion: $A \leq B$ if $A \subset B$. An (unordered) **basis** of V is a maximal element of $\mathcal{B}$.

(4) Prove that bases do exist, and that for every basis B one has $L(B) = V$, i.e. any basis is a set of generators. Then show that any linearly independent set is contained in at least one basis, and further, any set of generators contains at least one basis.

Let $B \subset V$ be a set of generators and $A \subset V$ a linearly independent set. To show $|A| \leq |B|$ we consider the set $\mathcal{C}$ of pairs (S, f), where $S \subset A$ and $f : S \to B$ is an injection such that $(A - S) \cup f(S)$ is linearly independent. We also impose $A \cap B \subset S$ and $f(v) = v$ for any $v \in A \cap B$. Now let's order $\mathcal{C}$ by extension: $(S, f) \leq (T, g)$ if $S \subset T$ and g extends f.

(5) Prove that $\mathcal{C}$ isn't empty, has maximal elements and that if (S, f) is maximal, then $S = A$. (Hint: if there exists $v \in A - S$, then B isn't contained in $L((A - (S \cup \{v\})) \cup f(S))$.)
(6) Show that any two bases of a vector space have the same cardinality.

2.6 The Cardinality of the Product

In this section we'll apply the axiom of choice and Zorn's lemma to show that any infinite set has the same cardinality of its Cartesian powers.

Lemma 2.26 *Every infinite set contains a countably infinite subset.*

Proof We shall prove that any infinite set X admits a 1-1 map $f : \mathbb{N} \to X$. The range $f(\mathbb{N})$ will be the required countable subset. Call $\mathcal{P}_0(X)$ the family of finite subsets of X; for every $A \in \mathcal{P}_0(X)$ the complement $X - A$ isn't empty and we can write

$$X = \cup\{X - A \mid A \in \mathcal{P}_0(X)\}.$$

The axiom of choice ensures the existence of $g : \mathcal{P}_0(X) \to X$ such that $g(A) \notin A$ for every finite subset $A \subset X$. Hence we can define a 1-1 map $f : \mathbb{N} \to X$ recursively by setting

$$f(1) = g(\emptyset) \quad \text{and} \quad f(n) = g(\{f(1), f(2), \ldots, f(n-1)\}) \quad \text{for } n > 1.$$

$\square$

Lemma 2.27 *If A is infinite and B is countable, $|A \cup B| = |A|$.*

Proof It's not restrictive to assume $A \cap B = \emptyset$. By Lemma 2.26 there exists a countably infinite subset $C \subset A$. Note that $C \cup B$ is countable, i.e. it has the same cardinality of C, and so $A = C \cup (A - C)$ has the same cardinality of $A \cup B = (C \cup B) \cup (A - C)$. □

Lemma 2.28 $|X \times \mathbb{N}| = |X|$ *for any infinite set* X.

Proof Consider the family $\mathcal{A}$ consisting of pairs (E, f) with $E \subset X$ and $f : E \times \mathbb{N} \to E$ injective. We know X contains countably infinite subsets, so $\mathcal{A}$ isn't empty. We can order $\mathcal{A}$ using the extension relation, and declare $(E, f) \leq (H, g)$ iff $E \subset H$ and g extends f. It isn't hard to show that any chain is bounded, and Zorn's lemma then tells there is a maximal element (A, f).

Now we prove $|A| = |X|$. By contradiction, suppose $|A| < |X|$, so that $X - A$ is infinite and must contain a countably infinite subset B. Choose a 1-1 map $g : B \times \mathbb{N} \to B$ and then define $h : (A \cup B) \times \mathbb{N} \to A \cup B$ by $h(x, n) = f(x, n)$ if $x \in A$ and $h(x, n) = g(x, n)$ if $x \in B$. Then h is injective and extends f, against the maximality of (A, f). □

Remark 2.29 It descends from Lemma 2.28 that if Y is infinite and $X = \cup_{i=1}^{+\infty} X_i$ is the union of a countable family where $|X_i| \leq |Y|$ for every i, then $|X| \leq |Y|$. Let in fact $A \subset \mathbb{N}$ be the set of indices i such that $X_i \neq \emptyset$; if $A = \emptyset$ there's nothing to prove, while if $A \neq \emptyset$

$$|Y| \leq |Y \times A| \leq |Y \times \mathbb{N}| = |Y|$$

and so $|Y| = |Y \times A|$. If we choose for every $i \in A$ an onto map $f_i : Y \to X_i$, it follows that $f : Y \times A \to X$, $f(y, i) = f_i(y)$, is onto.

Theorem 2.30 *Any infinite set* X *satisfies* $|X| = |X^2|$.

Proof The argument is very similar to that of Lemma 2.28. Consider the family $\mathcal{A}$ of pairs (E, f) where E is a non-empty subset in X and $f : E \times E \to E$ is one-to-one. As X contains countably infinite subsets, $\mathcal{A} \neq \emptyset$. Let's order $\mathcal{A}$ by extension, i.e. $(E, f) \leq (H, g)$ iff $E \subset H$ and g extends f. Easily, every chain is bounded, and so there is a maximal element (A, f) by Zorn's lemma. The fact that f is 1-1 tells that $|A^2| = |A|$, so it is enough to prove that $|A| = |X|$. Again by contradiction, suppose $|A| < |X|$. Then $X - A$ contains some B of the cardinality of A, and in particular

$$|B| = |B \times A| = |A \times B| = |B \times B|.$$

Since $(A \cup B) \times (A \cup B) = A \times A \cup A \times B \cup B \times A \cup B \times B$, and by Remark 2.29 there is a bijection from B to $A \times B \cup B \times A \cup B \times B$, we can extend f to $A \cup B$. But this contradicts the maximality of (A, f). □

Given a set X we define $S(X)$ to be the disjoint union of all Cartesian powers $X^n, n \in \mathbb{N}$. When X is finite and non-empty $S(X)$ is countable. We also remind that $\mathcal{P}_0(X)$ denotes the family of finite subsets in X.

Corollary 2.31 $|S(X)| = |\mathcal{P}_0(X)| = |X|$ *for any infinite set* X.

Proof For any integer $n > 0$ we have $|X^n| = |X|$. In fact, if $n > 1$ by induction we have $|X^n| = |X^{n-1} \times X| = |X \times X| = |X|$, so $S(X)$ is a countable union of sets with the cardinality of X. The natural onto map $S(X) \to \mathcal{P}_0(X) - \{\emptyset\}$ then shows that $|X| = |\mathcal{P}_0(X)|$.   □

Exercises

2.26 Prove that any infinite set can be written as disjoint union of countably infinite subsets.

2.27 Let $\mathbb{K}$ be an infinite field. Prove that the ring $\mathbb{K}[x]$ of polynomials with coefficients in $\mathbb{K}$ has the same cardinality of $\mathbb{K}$.

2.28 Let X, Y be infinite sets and $A \subset X \times Y$ a subset such that:

1. for every $x \in X$, $\{y \in Y \mid (x, y) \in A\}$ is countable;
2. for every $y \in Y$, $\{x \in X \mid (x, y) \in A\}$ isn't empty.

Prove that $|X| \geq |Y|$. (Hint: show there exists a 1-1 map $A \to X \times \mathbb{N}$.)

2.29 Let $\overline{\mathbb{Q}} \subset \mathbb{C}$ denote the set of complex numbers that are roots of polynomials with rational coefficients. Show that $\overline{\mathbb{Q}}$ is countable and deduce the existence of transcendental numbers. (Hint: Exercise 2.28. A number is transcendental, by definition, if it does not belong to $\overline{\mathbb{Q}}$.)

2.30 Let $\mathcal{B}$ be a basis of the vector space V over a field $\mathbb{K}$. Describe how a surjective map $\mathcal{P}_0(\mathbb{K} \times \mathcal{B}) \to V$ looks like and deduce that if the set $\mathbb{K} \times \mathcal{B}$ is infinite, then

$$|V| = |\mathbb{K} \times \mathcal{B}| = \max(|\mathbb{K}|, |\mathcal{B}|).$$

2.31 (☞, ♡) Let I be an infinite set, $\mathbb{K}$ a field and $\mathcal{B}$ a basis for the vector space $\mathbb{K}^I$ of maps $f : I \to \mathbb{K}$. Prove that the cardinality of $\mathcal{B}$ is strictly bigger than that of I. Deduce that every infinite-dimensional vector space cannot be isomorphic to its algebraic dual.

References

[Ke55] Kelley, J.L.: General Topology. D. Van Nostrand Company Inc., Toronto (1955)
[To03] Tourlakis, G.: Lectures in Logic and Set Theory, vol. 2. Cambridge University Press, Cambridge (2003)
[Wa93] Wagon, S.: The Banach-Tarski Paradox. Cambridge University Press, Cambridge (1993)

Chapter 3
Topological Structures

According to some psychologists, before the age of two-and-a-half children simply scribble. Between two-and-a-half and four they develop an understanding of sorts for topology, because they draw different pictures for open and closed figures. From four years on they can replicate all topological notions: a point inside a figure, outside it or on the border. Only between four and seven they start distinguishing simple figures (like squares, triangles) based on their size or angles.

A space is a set whose elements are called points. Putting a topological structure on a space X means being able to say, given a subset A, which are the interior points, which the exterior points and which the boundary points. Obviously this can't be done completely arbitrarily, and there are a few requirements dictated by common sense:

1. any point in X is an interior point of X;
2. if x is an interior point of A, then $x \in A$;
3. if x is an interior point of A and $A \subset B$, x is an interior point of B;
4. if x is an interior point of both A and B, then it is an interior point of $A \cap B$;
5. if $A°$ denotes the set of interior points of A, then any point of $A°$ is an interior point of $A°$.

Exterior points of a subset are those that are interior to the complementary set, and boundary points are neither interior nor exterior points.

The above five conditions can be adopted as axiomatic definition of a topological structure (see Exercise 3.14). Nowadays, though, the standard practice is to define topological structures by way of families of open sets, which is what we'll do next.

3.1 Topological Spaces

Definition 3.1 A **topology** on a set X is a family $\mathcal{T}$ of subsets of X, called **open sets**, satisfying the following requirements:

© Springer International Publishing Switzerland 2015
M. Manetti, *Topology*, UNITEXT - La Matematica per il 3+2 91,
DOI 10.1007/978-3-319-16958-3_3

(A1) $\emptyset$ and X are open sets;

(A2) the union of any number of open sets is an open set;

(A3) the intersection of two open sets is an open set.

A set equipped with a topology is called a **topological space**. Its elements are called **points**.

Condition A3 implies that any finite intersection of open sets is still an open set: in fact if $A_1, \ldots, A_n$ are open sets we can write $A_1 \cap \cdots \cap A_n = (A_1 \cap \cdots \cap A_{n-1}) \cap A_n$. By induction on n the set $A_1 \cap \cdots \cap A_{n-1}$ is open, so by A3 also $A_1 \cap \cdots \cap A_n$ is open.

Any set admits at least one topology, typically several. For instance the family $\mathcal{T} = \mathcal{P}(X)$ of subsets of X is a topology called the **discrete** topology, while the family $\mathcal{T}$ containing only the empty set and X is a topology called **trivial** or **indiscrete**.

Example 3.2 In the **Euclidean topology** on $\mathbb{R}$, a subset $U \subset \mathbb{R}$ is open if and only if it is the union of open intervals. Open sets thus defined satisfy A1, A2, A3 in Definition 3.1: for example, if

$$A = \bigcup_{i \in I} \,]a_i, b_i[\,, \qquad B = \bigcup_{h \in H} \,]c_h, d_h[\,,$$

are open sets, the distributive laws of union and intersection guarantee that

$$A \cap B = \left(\bigcup_{i \in I} \,]a_i, b_i[\, \right) \cap \left(\bigcup_{h \in H} \,]c_h, d_h[\, \right) = \bigcup_{i \in I, h \in H} \,]a_i, b_i[\, \cap\,]c_h, d_h[$$

is still a union of open intervals. The Euclidean topology is often called the **standard topology**.

Example 3.3 The **upper topology** on $\mathbb{R}$ is defined as the one having $]-\infty, a[$, with $a \in \mathbb{R} \cup \{+\infty\}$ as non-empty open sets.

Definition 3.4 Let X be a topological space. A subset $C \subset X$ is called **closed** if $X - C$ is open.

Since set-complementation swaps unions with intersections, the closed sets of a topology on X satisfy the following properties:

(C1) $\emptyset$ and X are closed;

(C2) any intersection of closed sets is closed;

(C3) the union of two closed sets is closed.

As for open sets, condition C3 implies that the finite union of closed sets is closed.

It is clearly possible to describe a topology by telling what are its closed sets. For example, the **cofinite topology** (or finite-complement topology) on X decrees a subset $C \subset X$ closed if and only if either $C = X$ or C is finite.

Definition 3.5 Let $\mathcal{T}$ be a topology on a set X. A subfamily $\mathcal{B} \subset \mathcal{T}$ is called a **basis** of $\mathcal{T}$ if every open set $A \in \mathcal{T}$ can be written as union of elements of $\mathcal{B}$.

Example 3.6 Open intervals form a basis for the Euclidean topology on the real line.

A basis determines uniquely the topology, because any union of basis elements is open, and conversely any open set has that form.

Theorem 3.7 *Let X be a set and $\mathcal{B} \subset \mathcal{P}(X)$ a family of subsets. There exists a topology on X for which $\mathcal{B}$ is a basis if and only if two conditions hold:*

1. $X = \cup\{B \mid B \in \mathcal{B}\}$;
2. *for any pair $A, B \in \mathcal{B}$ and any point $x \in A \cap B$ there exists $C \in \mathcal{B}$ such that $x \in C \subset A \cap B$.*

Proof The two conditions are clearly necessary, so let's prove they are also sufficient. Define open sets to be arbitrary unions of elements of $\mathcal{B}$. Then X is open (union of everything), $\emptyset$ is open (empty union), and the union of open sets is easily open. The second condition implies that for every $A, B \in \mathcal{B}$ we have

$$A \cap B = \cup\{C \mid C \in \mathcal{B}, C \subset A \cap B\}.$$

If $U = \cup A_i$ and $V = \cup B_j$ are unions of elements $A_i, B_j \in \mathcal{B}$, by the distributive laws we obtain

$$U \cap V = \underset{i,j}{\cup}(A_i \cap B_j) = \cup\{C \mid C \in \mathcal{B} \text{ and } \exists i, j \text{ such that } C \subset A_i \cap B_j\}.$$

$\square$

A common mistake the reader should prevent is to mix the notion of a topological basis with Theorem 3.7: in the definition the topology is given, and the criterion to be checked is not the one stated in the theorem, but that any open set can be written as union of open sets from the basis.

There is a natural order relation among the topologies of a given set.

Definition 3.8 Let $\mathcal{T}$ and $\mathcal{R}$ be topologies on the same set. One says that $\mathcal{T}$ is **finer** (or stronger, larger) than $\mathcal{R}$ (and that $\mathcal{R}$ is **coarser**, weaker, or smaller than $\mathcal{T}$), written $\mathcal{R} \subset \mathcal{T}$, when every open set of $\mathcal{R}$ is open in $\mathcal{T}$.

Example 3.9 The **lower-limit line** is defined as the set of real numbers with the topology whose basis consists of the family of half-open intervals $[a, b[$. As $]a, b[= \cup_{c>a}[c, b[$, this topology is finer that the Euclidean topology.[1]

Given an arbitrary collection $\{\mathcal{T}_i\}$ of topologies on a set X their intersection $\mathcal{T} = \cap_i \mathcal{T}_i$ is a topology. If the $\mathcal{T}_i$ are the only topologies containing a given family $\mathcal{S} \subset \mathcal{P}(X)$ of subsets of X, then $\mathcal{T}$ is the coarsest topology among those containing elements of $\mathcal{S}$ as open sets.

---

[1]Mathematicians do not invent strange-looking topologies just for fun: Robert Sorgenfrey, for instance, defined the lower-limit topology to answer the question of whether the product of paracompact spaces (Definition 7.14) is paracompact (it's not [So47]).

Example 3.10 Let $\{X_i \mid i \in I\}$ be a collection of topological spaces. Consider on their disjoint union $X = \coprod_i X_i$ the coarsest topology among those containing the topology of each X_i; here a subset $A \subset X$ is open iff $A \cap X_i$ is open in X_i for every i. The topological space defined in this way is called **disjoint union** of the topological spaces X_i.

Example 3.11 Let $\mathbb{K}$ be a field, $n > 0$ a given integer and $\mathbb{K}[x_1, \ldots, x_n]$ the polynomial ring in n variables and coefficients in $\mathbb{K}$. For any $f \in \mathbb{K}[x_1, \ldots, x_n]$ we define

$$D(f) = \{(a_1, \ldots, a_n) \in \mathbb{K}^n \mid f(a_1, \ldots, a_n) \neq 0\}.$$

Since $D(0) = \emptyset$, $D(1) = \mathbb{K}^n$ and $D(f) \cap D(g) = D(fg)$, the subsets $D(f)$ form a basis of open sets for a topology on $\mathbb{K}^n$ that is called **Zariski topology**. One indicates by $V(f)$ the complement of $D(f)$, and sets

$$V(E) = \bigcap_{f \in E} V(f) = \{(a_1, \ldots, a_n) \in \mathbb{K}^n \mid f(a_1, \ldots, a_n) = 0 \ \forall f \in E\}$$

for any subset $E \subset \mathbb{K}[x_1, \ldots, x_n]$. The subsets $V(E)$ are, as E varies, the only closed subsets in the Zariski topology. Given any subset $E \subset \mathbb{K}[x_1, \ldots, x_n]$ one writes $I = (E)$ for the ideal generated by E; as $E \subset I$ we have $V(I) \subset V(E)$. Conversely, given an element $f \in I$, there is an integer $n > 0$ and polynomials $f_1, \ldots, f_n \in E$ and $g_1, \ldots, g_n \in \mathbb{K}[x_1, \ldots, x_n]$ such that $f = f_1 g_1 + \cdots + f_n g_n$, and therefore

$$V(E) \subset V(f_1) \cap \cdots \cap V(f_n) \subset V(f).$$

This proves that

$$V(E) \subset \bigcap_{f \in I} V(f) = V(I),$$

whence $V(I) = V(E)$ and, further, that Zariski-closed subsets are precisely the subsets of type $V(I)$, as I varies among the ideals of the ring $\mathbb{K}[x_1, \ldots, x_n]$.

Exercises

3.1 ($\heartsuit$) True or false?

1. There is only one topological structure on the singleton.
2. A set with two elements admits exactly four different topological structures.
3. On a finite set any topology has an even number of open sets.
4. On an infinite set with the cofinite topology any pair of non-empty open sets has non-empty intersection.

3.2 Prove that closed intervals $[a, b] \subset \mathbb{R}$ are closed in the Euclidean topology.

3.3 Let $\infty \in X$ be a given element in a set. Check that

$$T = \{A \subset X \mid \infty \notin A \text{ or } X - A \text{ is finite}\}$$

defines a topology on X.

3.4 Let X be a set. Prove that

$$T = \{A \subset X \mid A = \emptyset \text{ or } X - A \text{ is finite or countable}\}$$

is a topology on X.

3.5 Let $(X, \leq)$ be an ordered set. Show that the subsets

$$M_x = \{y \in X \mid x \leq y\}$$

are, as $x \in X$ varies, a basis for a topology.

3.6 (There exist infinitely many primes, ♡) For any pair of integers $a, b \in \mathbb{Z}$, with $b > 0$, let's write $N_{a,b} = \{a + kb \mid k \in \mathbb{Z}\}$. Prove the following facts:

1. arithmetic progressions $\mathcal{B} = \{N_{a,b} \mid a, b \in \mathbb{Z}, b > 0\}$ form a basis for a topology T on $\mathbb{Z}$;
2. every $N_{a,b}$ is both open and closed in T;
3. call $P = \{2, 3, \ldots\} \subset \mathbb{N}$ the set of primes. Then

$$\mathbb{Z} - \{-1, 1\} = \cup\{N_{0,p} \mid p \in P\}.$$

Therefore if P were finite $\{-1, 1\}$ would be open in T.

3.2 Interior of a Set, Closure and Neighbourhoods

The concept of adherence defined in Sect. 1.3 generalises to arbitrary topological structures.

Definition 3.12 Given a subset B in a topological space X, one writes $\overline{B}$ for the intersection of all closed sets containing B:

$$\overline{B} = \bigcap \{C \mid B \subset C \subset X, \ C \text{ closed}\} .$$

Equivalently, $\overline{B}$ is the smallest closed set in X containing B, and is called the **closure** of B; its points are said to be **adherent** to B.

Example 3.13 If $A \subset B$, any closed set containing B also contains A, and so $\overline{A} \subset \overline{B}$. If A_i is a family of subsets in a topological space, then $\cup_i \overline{A_i} \subset \overline{\cup_i A_i}$: as $A_j \subset \cup_i A_i$, in fact, $\overline{A_j} \subset \overline{\cup_i A_i}$ for any j.

Definition 3.14 A subset A in a topological space X is said to be **dense** if $\overline{A} = X$ or, equivalently, if A meets every non-empty open subset of X. More generally, if $A \subset B$ are subsets, A is dense in B when $B \subset \overline{A}$.

Example 3.15   1. In a space with the trivial topology any non-empty subset is dense.
2. In a space with the discrete topology no proper subset is dense.
3. The set of rationals is dense in the space $\mathbb{R}$ equipped with the Euclidean topology.
4. In a space with the cofinite topology any infinite subset is dense.

Closely related to the closure are the notions of interior and boundary.

Definition 3.16 The **interior** B° of a subset B in X is the union of all open sets contained in B. Equivalently, B° is the largest open set contained in B; its points are called **interior points** of B.

Notice that a subset B of a topological space is open if and only if $B = B^\circ$ and closed if and only if $B = \overline{B}$. Passing to the complement we obtain the relationship $X - B^\circ = \overline{X - B}$.

Example 3.17 For the Euclidean topology on the real line $\mathbb{R}$ we have, for any $a < b$:

$$\overline{]a, b[} = \overline{[a, b[} = \overline{]a, b]} = [a, b], \qquad [a, b]^\circ = [a, b[^\circ =]a, b]^\circ =]a, b[\,.$$

Definition 3.18 The **boundary** of a subset B in a topological space is the closed set $\partial B = \overline{B} - B^\circ = \overline{B} \cap \overline{X - B}$. Hence boundary points are those adherent to B and to $X - B$.

For instance, in the Euclidean topology on $\mathbb{R}$ we have $\partial[a, b] = \{a, b\}$ and $\partial\,]a, b[= \{a, b\}$ for any $a < b$.

Definition 3.19 Let X be a topological space and $x \in X$ a point. A subset $U \subset X$ is called a **neighbourhood of** x if there is an open set V such that $x \in V$ and $V \subset U$, i.e. if x is an interior point of U.

Let's denote by $\mathcal{I}(x)$ the family of neighbourhoods of x. By definition if A is a subset in a topological space, $A^\circ = \{x \in A \mid A \in \mathcal{I}(x)\}$; in particular a subset is open if and only if it is a neighbourhood of each of its points.

Lemma 3.20 *The family $\mathcal{I}(x)$ of neighbourhoods of a point x is closed under extension and finite intersection, i.e.:*

1. if $U \in \mathcal{I}(x)$ and $U \subset V$, then $V \in \mathcal{I}(x)$;
2. if $U, V \in \mathcal{I}(x)$, then $U \cap V \in \mathcal{I}(x)$.

Proof If U is a neighbourhood of x there exists an open set A such that $x \in A \subset U$. If $U \subset V$ then *a fortiori* $x \in A \subset V$, and so V as well is a neighbourhood of x. If U, V are neighbourhoods of x there exist open sets A, B such that $x \in A \subset U$, $x \in B \subset V$. Hence $x \in A \cap B \subset U \cap V$. $\square$

Set-complementation allows to give a useful characterisation of the closure of a set.

Lemma 3.21 *Let B be a subset in a topological space X. A point $x \in X$ belongs to $\overline{B}$ if and only if $U \cap B \neq \emptyset$ for every neighbourhood $U \in \mathcal{I}(x)$.*

Proof By definition $x \notin \overline{B}$ if and only if x is an interior point of $X - B$; at the same time x is an interior point of $X - B$ if and only if there is a neighbourhood $U \in \mathcal{I}(x)$ such that $U \subset X - B$. $\square$

There is, for neighbourhoods, the analogue of the notion of a basis.

Definition 3.22 Let x be a point in a topological space X. A subfamily $\mathcal{J} \subset \mathcal{I}(x)$ is called a **local basis (of neighbourhoods)** at or around x, if for any $U \in \mathcal{I}(x)$ there exists $A \in \mathcal{J}$ such that $A \subset U$.

Example 3.23   1. Let $U \in \mathcal{I}(x)$ be a given neighbourhood. All neighbourhoods of x contained in U form a local basis at x.
2. If $\mathcal{B}$ is a basis for the topology, open sets in $\mathcal{B}$ containing x form a local basis at x.

A topology can be described entirely by listing the neighbourhoods of its points (see Exercise 3.14).

Exercises

3.7 ($\heartsuit$) Give an example of subsets $A, B \subset \mathbb{R}$ such that

$$A \cap B = \emptyset, \qquad \overline{A} \cap B \neq \emptyset, \qquad A \cap \overline{B} \neq \emptyset.$$

3.8 ($\heartsuit$) Let A, B be subsets in a topological space. Prove $\overline{A \cup B} = \overline{A} \cup \overline{B}$.

3.9 ($\heartsuit$) Let A be a dense subset of a topological space X. Prove $\overline{U \cap A} = \overline{U}$ for any open set $U \subset X$.

3.10 On the plane $\mathbb{R}^2$ consider the family $\mathcal{T}$ consisting of the empty set, $\mathbb{R}^2$ and all open discs $\{x^2 + y^2 < r^2\}$, $r > 0$. Prove that $\mathcal{T}$ defines a topology and determine the closure of the hyperbola $xy = 1$.

3.11 Show that the closed intervals $[-2^{-n}, 2^{-n}]$, $n \in \mathbb{N}$, form a local basis of neighbourhoods around 0 in the Euclidean topology on $\mathbb{R}$.

3.12 A topological space is called **T1** if every finite subset is closed. Prove that X is a T1 space if and only if

$$\{x\} = \bigcap_{U \in \mathcal{I}(x)} U$$

for every $x \in X$.

3.13 Let $\{U_n \mid n \in \mathbb{N}\}$ be a countable local basis of neighbourhoods at a point x in a topological space. Prove that the family

$$\{V_n = U_1 \cap \cdots \cap U_n \mid n \in \mathbb{N}\}$$

is still a local neighbourhood basis at x.

3.14 ($\heartsuit$) Let X be a set and suppose we are given, for any $x \in X$, a family $\mathcal{I}(x)$ of subsets in X so that the following five conditions hold:

1. $X \in \mathcal{I}(x)$ for every point $x \in X$;
2. $x \in U$ for any $U \in \mathcal{I}(x)$;
3. if $U \in \mathcal{I}(x)$ and $U \subset V$, then $V \in \mathcal{I}(x)$;
4. if $U, V \in \mathcal{I}(x)$, then $U \cap V \in \mathcal{I}(x)$;
5. if $U \in \mathcal{I}(x)$, there is a subset $V \subset U$ such that $x \in V$ and $V \in \mathcal{I}(y)$ for any $y \in V$.

Prove that there exists a unique topology on X for which $\mathcal{I}(x)$ is the family of neighbourhoods at x, for any $x \in X$.

3.15 A *closure operator* on a given set X is a map $C : \mathcal{P}(X) \to \mathcal{P}(X)$ obeying the four (*Kuratowski*) axioms:

1. $A \subset C(A)$ for every subset $A \subset X$;
2. $C(A) = C(C(A))$ for every $A \subset X$;
3. $C(\emptyset) = \emptyset$;
4. $C(A \cup B) = C(A) \cup C(B)$ for every $A, B \subset X$.

Prove that for any topological structure on X the map $A \mapsto \overline{A}$ is a closure operator, and conversely, that for any closure operator C on X there is a unique topological structure for which $C(A) = \overline{A}$.

3.16 Spot the mistake(s) in the following argument, allegedly proving the inclusion $\overline{\cup_i A_i} \subset \cup_i \overline{A_i}$. Let $x \in \overline{\cup_i A_i}$. Any neighbourhood of x meets $\cup_i A_i$, so any neighbourhood meets one of the A_i and therefore x belongs to the closure of A_i.

3.3 Continuous Maps

Definition 3.24 A map $f : X \to Y$ between two topological spaces is **continuous** if the pre-image

$$f^{-1}(A) = \{x \in X \mid f(x) \in A\}$$

of any open set $A \subset Y$ is open in X.

Before we continue let's remark that the operator $f^{-1} : \mathcal{P}(Y) \to \mathcal{P}(X)$ commutes with the operations of set-complementation and union:

$$f^{-1}(Y - A) = X - f^{-1}(A), \qquad f^{-1}(\cup_i A_i) = \cup_i f^{-1}(A_i).$$

As a result:

1. a map $f: X \to Y$ is continuous if and only if the pre-image $f^{-1}(C)$ of any closed set $C \subset Y$ is closed in X (complementation);
2. let $\mathcal{B}$ be a topological basis of Y. A map $f: X \to Y$ is continuous if and only if for any $B \in \mathcal{B}$ the set $f^{-1}(B)$ is open in X (each open set in Y is the union of elements of $\mathcal{B}$).

Lemma 3.25 *Let $f: X \to Y$ be a map between two topological spaces. Then f is continuous if and only if $f(\overline{A}) \subset \overline{f(A)}$ for any subset $A \subset X$.*

Proof Let's suppose f continuous; then for any $A \subset X$ the subset $f^{-1}(\overline{f(A)})$ is closed and contains A. Therefore $\overline{A} \subset f^{-1}(\overline{f(A)})$ and so $f(\overline{A}) \subset \overline{f(A)}$.

Now assume $f(\overline{A}) \subset \overline{f(A)}$ for any $A \subset X$. In particular for every closed set $C \subset Y$ (setting $A = f^{-1}(C)$)

$$f(\overline{f^{-1}(C)}) \subset \overline{f(f^{-1}(C))} = \overline{C} = C,$$

whence $\overline{f^{-1}(C)} \subset f^{-1}(C)$. But this means $f^{-1}(C)$ is closed. $\quad\square$

Lemma 3.25 implies that Definitions 1.4 and 3.24 are equivalent.

Theorem 3.26 *The composite of continuous maps is continuous.*

Proof Take continuous maps $f: X \to Y$ and $g: Y \to Z$ and let $A \subset Z$ be open. By the continuity of g we have that $g^{-1}(A)$ is open, and then $f^{-1}(g^{-1}(A))$ becomes open because f is continuous. To finish it suffices to note that $f^{-1}(g^{-1}(A)) = (gf)^{-1}(A)$. $\quad\square$

Definition 3.27 A map $f: X \to Y$ between topological spaces is **continuous at a point** $x \in X$ if for any neighbourhood U of $f(x)$ there is a neighbourhood V of x such that $f(V) \subset U$.

Theorem 3.28 *A map $f: X \to Y$ is continuous if and only if it is continuous at every point of X.*

Proof Suppose f continuous and let U be a neighbourhood of $f(x)$. By definition of neighbourhood there is an open set $A \subset Y$ such that $f(x) \in A \subset U$: the open set $V = f^{-1}(A)$ is a neighbourhood of x and $f(V) \subset U$.

Conversely, suppose f is continuous at every point and let A be an open set in Y. We claim $f^{-1}(A)$ is a neighbourhood of all of its points. If $x \in f^{-1}(A)$ then A is a neighbourhood of $f(x)$ and there is a neighbourhood V of x such that $f(V) \subset A$. This amounts to saying $V \subset f^{-1}(A)$, and therefore $f^{-1}(A)$ is a neighbourhood of x. $\quad\square$

Definition 3.29 A **homeomorphism** is a continuous bijective map with continuous inverse. More precisely, a continuous mapping $f: X \to Y$ is called a homeomorphism if there exists a continuous map $g: Y \to X$ such that the composites gf and fg are the identity maps on X and Y respectively.

Two topological spaces are said **homeomorphic** if there's a homeomorphism between them.

Definition 3.30 A map $f: X \to Y$ is called:

1. **open** if $f(A)$ is open in Y for any open set A of X;
2. **closed** if $f(C)$ is closed in Y for any closed set C of X.

Lemma 3.31 *Let* $f: X \to Y$ *be a continuous map. The following conditions are equivalent:*

1. *f is a homeomorphism;*
2. *f is closed and bijective;*
3. *f is open and bijective.*

Proof Let's prove (1) $\Rightarrow$ (2): every homeomorphism is bijective by very definition. If $g: Y \to X$ is the inverse of f, g is continuous and for any closed subset $C \subset X$, $f(C) = g^{-1}(C)$ is closed in Y.

As for (2) $\Rightarrow$ (3), let $A \subset X$ be open and set $C = X - A$. As f is bijective we have $f(A) = f(X - C) = Y - f(C)$, so $f(A)$ is the complement of the closed set $f(C)$.

At last, to show (3) $\Rightarrow$ (1) suppose $g: Y \to X$ is the inverse of f. Then for any open subset $A \subset X$, $g^{-1}(A) = f(A)$ is open and therefore g is continuous. $\square$

The term real(-valued) continuous map typically refers to a continuous map $f: X \to \mathbb{R}$ for the Euclidean topology on $\mathbb{R}$.

Exercises

3.17 Prove that constant maps are continuous, irrespective of the topologies considered.

3.18 Let T_1 and T_2 be given topologies on a set X. Prove that the identity map $(X, T_1) \to (X, T_2)$, $x \mapsto x$, is continuous if and only if T_1 is finer than T_2.

3.19 Given $f: \mathbb{R} \to \mathbb{R}$ and $k \in \mathbb{R}$ write

$$M(k) = \{x \in \mathbb{R} \mid f(x) > k\} \quad m(k) = \{x \in \mathbb{R} \mid f(x) < k\}.$$

Prove that f is continuous if and only if $M(k)$ and $m(k)$ are open for any k.

3.20 Let $f_i: Y_i \to X$ be a family of continuous maps with values in a given topological space X. Show that the induced map $f: \coprod_i Y_i \to X$ is continuous, where $\coprod_i Y_i$ is the disjoint union (Example 3.10) and $f(y) = f_i(y)$ if $y \in Y_i$.

3.21 Two subsets A, B of a topological space are called **adherent** to each other if

$$(A \cap \overline{B}) \cup (\overline{A} \cap B) \neq \emptyset.$$

Prove that continuous maps preserve the relation of adherence of subsets. Conversely if $f: X \to Y$ is a map between T1 spaces (see Exercise 3.12) that preserves adherence, f is continuous.

3.22 Prove that the composite of two homeomorphisms is a homeomorphism and the inverse of a homeomorphism is a homeomorphism, so that the set Homeo(X) of homeomorphisms from a topological space X to itself is a group. Introduce, on the category[2] of topological spaces, the relation '$X \sim Y \iff X$ is homeomorphic to Y'. Show that $\sim$ is an equivalence relation (called **topological equivalence**).

3.23 Tell, justifying the answer, whether the group of homeomorphisms of the interval $[0, 1]$ is Abelian.

3.24 ($\heartsuit$) Let X, Y be topological spaces and $\mathcal{B}$ a topological basis of X. Prove that $f: X \to Y$ is open if and only if $f(A)$ is open in Y for every $A \in \mathcal{B}$.

3.25 Let $f: X \to Y$ be a continuous mapping and $A \subset X$ a dense subset. Prove $f(A)$ is dense in $f(X)$.

3.26 Let $f: X \to Y$ be an open mapping and $D \subset Y$ a dense subset. Prove $f^{-1}(D)$ is dense in X.

3.27 Consider the following commutative diagram of continuous maps:

Show that h is closed provided f is onto and g closed.

3.28 Let X be a topological space. A map $f: X \to \mathbb{R}$ is *upper semi-continuous* if $f^{-1}(] - \infty, a[)$ is open in X for any $a \in \mathbb{R}$. Prove $f: X \to \mathbb{R}$ is upper semi-continuous if and only if, for any $x \in X$ and any $\varepsilon > 0$, there is a neighbourhood U of x such that $f(y) < f(x) + \varepsilon$ for every $y \in U$.

3.29 ($\heartsuit$) Find an example of a map $f: [0, 1] \to [0, 1]$ that is continuous only at the point 0.

3.30 ($\heartsuit$) Let $A \subset \mathbb{R}$ be a countable subset. Find a map $f: \mathbb{R} \to [0, 1]$ that is continuous at every point in $\mathbb{R} - A$, and only there.

---

[2]Rigorously speaking, the notion of a category is quite different from that of a set: while we wait for the correct definition (Sect. 10.4), the name 'category' can be used instead of 'collection'.

3.31 ($\bullet$, $\heartsuit$) Let X be a topological space and $f : X \to \mathbb{R}$ any map. Prove the subset of points of X at which f is continuous is the intersection of a countable family of open sets.

Remark 3.32 It can be proved (Exercise 6.25) that the set of rational numbers cannot be written as countable intersection of open sets in $\mathbb{R}$ in any way. A baffling consequence of Exercises 3.30 and 3.31 is that there do exist maps that are continuous only at irrational points, while there's no continuous mapping on $\mathbb{Q}$ only.

3.4 Metric Spaces

Definition 3.33 A **distance** on a set X is a function $d : X \times X \to \mathbb{R}$ meeting the following properties:

1. $d(x, y) \geq 0$ for any $x, y \in X$, and $d(x, y) = 0 \iff x = y$;
2. $d(x, y) = d(y, x)$ for every $x, y \in X$;
3. $d(x, y) \leq d(x, z) + d(z, y)$ for all $x, y, z \in X$.

Condition 3 is called **triangle inequality**.

Example 3.34 On an arbitrary set X the function

$$d : X \times X \to \mathbb{R}, \qquad d(x, y) = \begin{cases} 0 & \text{if } x = y, \\ 1 & \text{if } x \neq y, \end{cases}$$

is a distance.

Definition 3.35 A **metric space** is a pair (X, d) formed by a set X and a distance d on X.

Example 3.36 The line $\mathbb{R}$ with the Euclidean distance $d(x, y) = |x - y|$ is a metric space.

Example 3.37 The space $\mathbb{R}^n$, with the Euclidean distance

$$d(x, y) = \sqrt{(x_1 - y_1)^2 + \cdots + (x_n - y_n)^2},$$

is a metric space: the triangle inequality holds by virtue of Lemma 1.3.

Example 3.38 The identification $\mathbb{C} = \mathbb{R}^2$ of complex numbers with pairs of reals generalises to any Cartesian power: $\mathbb{C}^n = \mathbb{R}^{2n}$. The Euclidean distance on $\mathbb{R}^{2n}$ reads

$$d(x, y) = \sqrt{|x_1 - y_1|^2 + \cdots + |x_n - y_n|^2}, \qquad x, y \in \mathbb{C}^n$$

in the coordinates of $\mathbb{C}^n$. The function $d : \mathbb{C}^n \times \mathbb{C}^n \to \mathbb{R}$ is called **Euclidean distance on $\mathbb{C}^n$**.

Example 3.39 $\mathbb{R}^n$ admits other interesting distance functions, such as

$$d_1(x, y) = \sum_{i=1}^{n} |x_i - y_i| \quad \text{and} \quad d_\infty(x, y) = \max_i \{|x_i - y_i|\}.$$

We may compare these to the Euclidean distance d by the inequalities

$$d_\infty(x, y) \le d(x, y) \le d_1(x, y) \le n \cdot d_\infty(x, y).$$

The only non-trivial relation is the middle one, and to prove it it's enough to square it.

Example 3.40 Let d be a distance on a set X. The function

$$\overline{d} : X \times X \to \mathbb{R}, \qquad \overline{d}(x, y) = \min(1, d(x, y))$$

is still a distance, called **standard bound** of d. The only property that's not immediate is the triangle inequality $\overline{d}(x, y) \le \overline{d}(x, z) + \overline{d}(z, y)$. As $\overline{d} \le 1$, if $\overline{d}(x, z) + \overline{d}(z, y) \ge 1$ there is nothing to prove. If, instead, $\overline{d}(x, z) + \overline{d}(z, y) < 1$ then $\overline{d}(x, z) = d(x, z), \overline{d}(z, y) = d(z, y)$, so

$$\overline{d}(x, y) \le d(x, y) \le d(x, z) + d(y, z) = \overline{d}(x, z) + \overline{d}(z, y),$$

eventually showing that $\overline{d}$ is indeed a distance function.

Definition 3.41 Let (X, d) be a metric space. The subset

$$B(x, r) = \{y \in X \mid d(x, y) < r\}$$

is called **open ball** centred at x with radius r (for the distance d).

The name 'ball' stems from the Euclidean world. Balls for the two distances of Example 3.39 are actually hypercubes.

Any distance function induces in a natural way a topological structure.

Definition 3.42 (*Topology induced by a distance*) Let (X, d) be a metric space. The topology on X induced by d is called **metric topology**: $A \subset X$ is open in the metric topology if for any $x \in A$ there exists $r > 0$ such that $B(x, r) \subset A$.

Let's check that the family of open sets defined in 3.42 fulfils axioms A1, A2 and A3. If $A = \emptyset$ the condition is automatic. Since $x \in B(x, r)$ for any $r > 0$, we can write $X = \cup_{x \in X} B(x, 1)$. If $A = \cup_i A_i$ with A_i open for every i and $x \in A$, there is one index j such that $x \in A_j$, and we can find $r > 0$ such that $B(x, r) \subset A_j \subset A$. If A, B are open and $x \in A \cap B$ there exist $r, t > 0$ such that $B(x, r) \subset A, B(x, t) \subset B$; calling s the minimum between r and t, we have $B(x, s) \subset A \cap B$.

Example 3.43 The **Euclidean topology** (also known as **standard topology**) of the spaces $\mathbb{R}^n$ and $\mathbb{C}^n$ is by definition the one induced by the Euclidean distance. Unless we specify otherwise, the symbols $\mathbb{R}^n$ and $\mathbb{C}^n$ will henceforth always denote the standard topological structure on the respective space.

Lemma 3.44 *For the topology induced by a distance:*

1. *open balls are open;*
2. *a subset is open if and only if it is a union of open balls (therefore open balls form a topological basis);*
3. *a subset U is a neighbourhood of a point x if and only if it contains an open ball with centre x, i.e. iff there exists $r > 0$ such that $B(x, r) \subset U$.*

Proof Concerning (1), if $y \in B(x, r)$ we set $s = r - d(x, y) > 0$; if $d(z, y) < s$ then $d(x, z) \leq d(x, y) + d(y, z) < r$ and $B(y, s) \subset B(x, r)$ follows.

Now we prove (2). Since the union of open sets is open the 'if' statement is ok. For the converse, if A is open we can choose, for any $x \in A$, a real number $r(x) > 0$ such that $B(x, r(x)) \subset A$, whence $A = \cup_{x \in A} B(x, r(x))$.

As for (3), note that U is a neighbourhood of x if and only if there is an open A such that $x \in A \subset U$. If $B(x, r) \subset U$, since $B(x, r)$ is open, we have that U is a neighbourhood. *Vice versa*, if U is a neighbourhood there is an open set A such that $x \in A \subset U$, and therefore an $r > 0$ such that $B(x, r) \subset A \subset U$. $\qquad\square$

Example 3.45 Consider a metric space (X, d). The proof of Example 1.6 shows that for any $x \in X$ and any real number r, the subset $C = \{x' \in X \mid d(x, x') \leq r\}$ is closed in the metric topology. It is quite interesting to observe that in general such subset does not coincide with the closure of the open ball $B(x, r)$: just take open balls of radius 1 for the distance of Example 3.34.

For metric spaces we rediscover the classical definition of continuity given by the (in)famous couple 'epsilon-delta'.

Theorem 3.46 *Let $f : (X, d) \to (Y, h)$ be a map between two metric spaces and x a point of X. Then f is continuous at x if and only if for any $\varepsilon > 0$ there exists $\delta > 0$ such that $h(f(x), f(y)) < \varepsilon$ whenever $d(x, y) < \delta$.*

Proof Continuity at x means that for any neighbourhood V of $f(x)$ there is a neighbourhood U of x such that $f(U) \subset V$. It is enough to invoke the description of neighbourhoods in a metric space given in Lemma 3.44. $\qquad\square$

Example 3.47 (*Distance from a subset*) Let (X, d) be a metric space. For any non-empty subset $Z \subset X$ consider the *distance from Z*, the function defined by

$$d_Z : X \to \mathbb{R}, \qquad d_Z(x) = \inf_{z \in Z} d(x, z).$$

Notice $d_Z(x) = 0$ if and only if $x \in \overline{Z}$. Moreover, d_Z satisfies the triangle inequality $|d_Z(x) - d_Z(y)| \leq d(x, y)$, hence is continuous: in fact, because of the obvious symmetry it suffices to show that for any pair of points $x, y \in X$ we have $d_Z(y) - d_Z(x) \le d(x, y)$. From the definition there exists, for any $\varepsilon > 0, z \in Z$ such that $d_Z(x) + \varepsilon \ge d(x, z)$, and so

$$d_Z(y) \le d(z, y) \le d(z, x) + d(x, y) \le d_Z(x) + \varepsilon + d(x, y).$$

In particular $d_Z(y) - d_Z(x) \le \varepsilon + d(x, y)$, and everything follows when taking the limit as ε goes to 0.

Corollary 3.48 *Let d, h be distances on a set X. The topology induced by d is finer than the topology induced by h if and only if for any $x \in X$ and any $\varepsilon > 0$ there exists δ such that $d(x, y) < \delta \implies h(x, y) < \varepsilon$.*

Proof The d-topology is finer than the h-topology if and only if the identity map $(X, d) \to (X, h)$ is continuous. □

Definition 3.49 Two distances on a set X are said **equivalent** if they induce the same topology. A topological space X is **metrisable** if the topology is induced by a suitable distance.

For example if d is a distance, for any positive real number a the function ad is a distance with the same balls as d, and thus it is equivalent to d.

Corollary 3.50 *Let d, h be two distances on X and a, c positive real numbers. If*

$$d(x, y) \ge \frac{\min(a, h(x, y))}{c}$$

for any $x, y \in X$, the topology induced by d is finer than the one induced by h. In particular any distance is equivalent to its standard bound (Example 3.40).

Proof For the first statement it is enough to take, in Corollary 3.48, the number δ as the minimum of a/c and ε/c. Calling $\bar{d}(x, y) = \min(1, d(x, y))$ the standard bound of d, we have

$$d(x, y) \ge \frac{\min(1, \bar{d}(x, y))}{1}, \qquad \bar{d}(x, y) \ge \frac{\min(1, d(x, y))}{1}. \qquad \square$$

Example 3.51 Let (X, h) and (Y, k) be metric spaces and $f : [0, +\infty[^2 \to \mathbb{R}$ a map such that:

1. $0 \le f(c_1, c_2) \le f(a_1, a_2) + f(b_1, b_2)$ whenever $0 \le c_1 \le a_1 + b_1$ and $0 \le c_2 \le a_2 + b_2$;
2. $f(a, b) = 0$ if and only if $a = b = 0$.

Then $d : (X \times Y) \times (X \times Y) \to \mathbb{R}$

$$d((x_1, y_1), (x_2, y_2)) = f(h(x_1, x_2), k(y_1, y_2)),$$

is a distance on $X \times Y$. In particular, the functions

$$d_1, d_2, d_\infty : (X \times Y) \times (X \times Y) \to \mathbb{R},$$

$$d_1((x_1, y_1), (x_2, y_2)) = h(x_1, x_2) + k(y_1, y_2),$$

$$d_2((x_1, y_1), (x_2, y_2)) = \sqrt{h(x_1, x_2)^2 + k(y_1, y_2)^2},$$

$$d_\infty((x_1, y_1), (x_2, y_2)) = \max(h(x_1, x_2), k(y_1, y_2)),$$

are equivalent distances on $X \times Y$, as it follows from Corollary 3.50 and the inequalities $d_\infty \le d_2 \le d_1 \le 2d_\infty$.

Definition 3.52 Let (X, d) be a metric space. A subset $A \subset X$ is **bounded** if there is a real number M such that

$$d(a, b) \le M \text{ for any } a, b \in A.$$

A map $f : Z \to X$, with Z a set, is **bounded** if $f(Z)$ is a bounded set.

Corollary 3.50 shows, amongst other things, that boundedness is not a topological property: there exist pairs of equivalent distances on a set X such that X is bounded for one distance and unbounded for the other.

Exercises

3.32 (Quadrangle inequality, ♡) Let (X, d) be a metric space. Prove that any four points $x, y, z, w \in X$ satisfy

$$|d(x, y) - d(z, w)| \le d(x, z) + d(y, w).$$

When $w = z$ the quadrangle inequality reduces to the triangle inequality, while for $w = y$ it becomes $|d(x, y) - d(z, y)| \le d(x, z)$.

3.33 (♡) Let d be a distance on a finite set X. Prove that the induced topology is discrete.

3.34 Let X be a finite set, $Y = \mathcal{P}(X)$ the family of subsets of X and define $d : Y \times Y \to \mathbb{R}$ by

$$d(A, B) = |A| + |B| - 2|A \cap B| \quad (|A| = \text{cardinality of} A).$$

Prove that d is a distance on Y.

3.35 Prove that the metric topology is the coarsest among all topologies whose open balls are open.

3.36 Show that any set with the discrete topology is metrisable.

3.37 Show that any set (containing at least two points) with the trivial topology is not metrisable.

3.38 Prove that in a metric space an open ball of radius 1 cannot properly contain an open ball of radius 2. Find an example, or prove the non-existence, of a metric space where an open ball of radius 2 contains an open ball of radius 3 as a proper subset.

3.39 ($\heartsuit$) Let $f: [0, +\infty[\to [0, +\infty[$ be a continuous map with $f^{-1}(0) = 0$ and $f(c) \le f(a) + f(b)$ for any $c \le a + b$, and take a distance d on a set X. Prove that the function $h(x, y) = f(d(x, y))$ is a distance equivalent to d.

3.40 ($\heartsuit$) Let d be a distance on a set X. Prove that

$$\delta: X \times X \to \mathbb{R}, \qquad \delta(x, y) = \frac{d(x, y)}{1 + d(x, y)}$$

is a distance equivalent to d.

3.41 Let (X, d) be a metric space and $Z \subset X$ a non-empty subset. Prove that for any $x \in X$

$$d_Z(x) = \inf\{r \in \mathbb{R} \mid B(x, r) \cap Z \ne \emptyset\}.$$

3.42 ($\heartsuit$) Let (X, d) be a metric space and $A, B \subset X$ disjoint closed sets. Prove the existence of two disjoint open sets $U, V \subset X$ such that $A \subset U$ and $B \subset V$. (Hint: use the functions d_A, d_B from Example 3.47 in order to define the set $\{x \in X \mid d_A(x) < d_B(x)\}$.)

3.5 Subspaces and Immersions

Any subset Y in a topological space X inherits naturally a topological structure: one says that a subset $U \subset Y$ is **open in** Y if there is an open set V in X such that $U = Y \cap V$. It is straightforward that open sets thus defined satisfy conditions A1, A2 and A3 in Definition 3.1, and therefore define a topology on Y, called **subspace topology**.[3]

A **topological subspace** of X is a subset Y endowed with the topology of open sets in Y. As $Y - (V \cap Y) = Y \cap (X - V)$, it follows that $C \subset Y$ is closed in Y if and only if there is a closed set B in X such that $C = Y \cap B$. If $\mathcal{B}$ is a topological basis of X, $\{A \cap Y \mid A \in \mathcal{B}\}$ is a basis of the subspace topology on Y. The inclusion map $i: Y \to X$ is continuous, because for any open A of X we have $i^{-1}(A) = Y \cap A$, an open set in Y by definition. The subspace topology, moreover, is the coarsest among those that make the inclusion map continuous.

---

[3]In other books this sometimes is called **induced topology** on subsets. We shall shun the latter terminology to avoid confusion with what comes next.

Example 3.53 If (X, d) is a metric space and $Y \subset X$, the restriction of d to $Y \times Y$ turns Y into a metric space: the topology induced by the restricted distance coincides with the subspace topology: any open ball in Y arises as intersection with an open ball in X.

A topological subspace is called discrete if the subspace topology is discrete. Equivalently, a subspace $Z \subset X$ is discrete if and only if for any $z \in Z$ there is an open set $U \subset X$ such that $U \cap Z = \{z\}$. Integer numbers are a discrete subspace of $\mathbb{R}$, for example, whereas $\mathbb{Q}$ is not discrete in $\mathbb{R}$.

Proposition 3.54 *Consider X, Z topological spaces, Y a subspace in X, $f : Z \to Y$ a map and $if : Z \to X$ the composite of f with the inclusion of Y in X. If if is continuous then also f is, and conversely.*

Proof If f is continuous then if is continuous as composite of continuous maps.

Now suppose if continuous and take $A \subset Y$ open in Y. There is an open set U in X such that $A = Y \cap U$, so $f^{-1}(A) = (if)^{-1}(U)$ is open in Z. $\square$

Lemma 3.55 *Let Y be a subspace in a topological space X and A a subset of Y. The closure of A in Y coincides with the intersection of Y with the closure of A in X.*

Proof Let $\mathbb{C}$ be the family of closed sets in X containing A. For $C \in \mathbb{C}$ the subsets $C \cap Y$ are the only closed sets in Y containing A, so the closure of A in Y (by definition $\bigcap_{C \in \mathbb{C}} (Y \cap C)$, coincides with $Y \cap (\bigcap_{C \in \mathbb{C}} C)$. But the latter is the restriction to Y of the closure of A in X. $\square$

Lemma 3.56 *Let X be a topological space, $Y \subset X$ a subspace and $Z \subset Y$ a subset.*

1. *If Y is open in X, then Z is open in Y iff Z is open in X.*
2. *If Y is closed in X, then Z is closed in Y iff Z is closed in X.*
3. *If Y is a neighbourhood of y, Z is a neighbourhood of y in Y iff Z is a neighbourhood of y in X.*

Proof Only the last claim is non-trivial. Let U be an open subset in X such that $y \in U \subset Y$. If Z is a neighbourhood of y in Y there is an open set V in Y with $x \in V \subset Z$. The intersection $V \cap U$ is open in U, so it is open in X as well, and $y \in V \cap U \subset Z$.

Conversely: if there is V open in X with $x \in V \subset Z$ then $V = V \cap Y$ is open in Y, too. $\square$

Definition 3.57 A continuous, one-to-one map $f : X \to Y$ is called a (**topological**) **immersion** if every open set in X is of the form $f^{-1}(A)$ for some open set A in Y.

Put differently, $f : X \to Y$ is an immersion if and only if it induces a homeomorphism from X to $f(X)$.

Not all continuous injections are immersions. For example the identity map

$$Id : (\mathbb{R}, \text{ Euclidean topology}) \to (\mathbb{R}, \text{ trivial topology})$$

is continuous and 1-1 but no immersion.

Definition 3.58 A **closed immersion** is both an immersion and a closed map. An **open immersion** is an immersion and an open map.

Lemma 3.59 *Let* $f : X \to Y$ *be continuous.*

1. *If* f *is closed and one-to-one,* f *is a closed immersion.*
2. *If* f *is open and one-to-one,* f *is an open immersion.*

Proof (1) Assume f is injective, continuous and closed. For any closed set $C \subset X$, the image $f(C) \subset f(X)$ is closed in Y and hence closed in the subspace topology of $f(X)$. Therefore the restriction $f : X \to f(X)$ is continuous, bijective and closed, whence a homeomorphism.

(2) The argument is analogous and left as an exercise. $\qquad\qquad\qquad\square$

There exist immersions that are neither closed nor open; an immersion $f : X \to Y$ is closed if and only if $f(X)$ is closed, and open if and only if $f(X)$ is open.

Exercises

3.43 ($\heartsuit$) True or false?

1. The closure of a discrete subspace is discrete.
2. Every discrete subspace of a metric space is closed.

3.44 ($\heartsuit$) Let $B \subset X$ be a subset, A open in X and $Z \subset A$ the closure of $A \cap B$ in the subspace A. Prove $Z = A \cap \overline{B}$, where $\overline{B}$ is the closure of B in X. Compare this fact with Lemma 3.55 and Exercise 3.7.

3.45 ($\heartsuit$) Let X be a topological space: a subset $Z \subset X$ is said **locally closed** if for any $z \in Z$ there is an open subset $U \subset X$ containing z such that $Z \cap U$ is closed in U.

Prove that the following statements are equivalent:

1. Z is locally closed;
2. Z is open in $\overline{Z}$ (in the subspace topology);
3. Z is the intersection of a closed and an open subset of X.

3.46 Two subsets A and B of a topological space X are called **separated** if they are not adherent to each other: $\overline{A} \cap B = A \cap \overline{B} = \emptyset$. Prove that:

1. if $F, G \subset X$ are both open or both closed, $A = F - G$ and $B = G - F$ are separated;
2. if $A, B \subset X$ are separated, A and B are open and closed in $A \cup B$.

3.47 Consider $X = \mathbb{R} - \{1\}$, $Y = \{y^2 = x^2 + x^3\} \subset \mathbb{R}^2$ and $f : X \to Y$, $f(t) = (t^2 - 1, t^3 - t)$. Show that f is continuous and bijective, that any $x \in X$ has a neighbourhood U such that $f : U \to f(U)$ is a homeomorphism and that f is not a closed map.

3.48 ($\heartsuit$) Prove that any discrete subspace of $\mathbb{R}$ is countable.

3.6 Topological Products

If P, Q are topological spaces we write $P \times Q$ for their Cartesian product and call $p \colon P \times Q \to P, q \colon P \times Q \to Q$ the projections onto the factors. The collection $\mathfrak{T}$ of topologies on $P \times Q$ for which p, q are continuous mappings isn't empty because of the discrete topology, and the intersection of all topologies in $\mathfrak{T}$ is the coarsest among those rendering the projections continuous.

Definition 3.60 The **product topology** on $P \times Q$ is the coarsest of all those making the projections continuous.

Theorem 3.61 *With the previous notations:*

1. *subsets of the type $U \times V$, for any $U \subset P$ and $V \subset Q$ open, form a basis, that we call the **canonical basis** of the product topology;*
2. *the projections p, q are open maps, and for any $(x, y) \in P \times Q$ the restrictions $p \colon P \times \{y\} \to P, q \colon \{x\} \times Q \to Q$ are homeomorphisms;*
3. *a map $f \colon X \to P \times Q$ is continuous if and only if its components $f_1 = pf$ and $f_2 = qf$ are continuous.*

Proof (1) For the time being let's indicate with $\mathcal{P}$ the product topology. As $(U_1 \times V_1) \cap (U_2 \times V_2) = (U_1 \cap U_2) \times (V_1 \cap V_2)$, by Theorem 3.7 the subsets $U \times V$, with U and V open in P and Q respectively, form a basis of a topology $\mathcal{T}$. The projection $p \colon P \times Q \to P$ is continuous for $\mathcal{T}$ because for any open $U \subset P$ the set $p^{-1}(U) = U \times Q$ is a basis element, and hence it belongs to $\mathcal{T}$. One proves similarly that q is continuous for $\mathcal{T}$, hence $\mathcal{T}$ is finer than $\mathcal{P}$. On the other hand if $U \subset P$ and $V \subset Q$ are open, $U \times V = p^{-1}(U) \cap q^{-1}(V) \in \mathcal{P}$. Therefore any open set of $\mathcal{T}$ is union of open sets in the product topology, implying that $\mathcal{T}$ is coarser than $\mathcal{P}$. Altogether $\mathcal{T}$ coincides with the product topology.

(2) Take $y \in Q$, and then for any open pair $U \subset P$, $V \subset Q$

$$(U \times V) \cap (P \times \{y\}) = \begin{cases} U \times \{y\} & \text{if } y \in V \\ \emptyset & \text{if } y \notin V \end{cases},$$

and the open sets of $P \times \{y\}$ are precisely the possible unions of the products $U \times \{y\}$ as U varies among open sets in P. Consequently $p \colon P \times \{y\} \to P$ is a homeomorphism. Let A be open in $P \times Q$: to show that $p(A)$ is open we need to write $p(A) = \cup_{y \in Q} p(A \cap P \times \{y\})$ and note that any $p(A \cap P \times \{y\})$ is open in P.

As far as (3) is concerned, f is continuous if and only if the pre-image of any open set of a basis is open. Hence f is continuous if and only if for any open pair $U \subset P, V \subset Q$ the subset $f^{-1}(U \times V) = f_1^{-1}(U) \cap f_2^{-1}(V)$ is open in X. $\qquad \square$

Example 3.62 The equivalent metrics d_1, d_2 and d_∞ (Example 3.51) induce on the Cartesian product $(X, h) \times (Y, k)$ the product topology.

The previous construction generalised to the product of any *finite* number $P_1, \ldots, P_n$ of topological spaces. The product topology on $P_1 \times \cdots \times P_n$ is the coarsest among those rendering the projections continuous. The canonical basis of the product topology is given by subsets $U_1 \times \cdots \times U_n$, with U_i open in P_i.

Example 3.63 The Euclidean topology on $\mathbb{R}^n$ coincides with the product topology of n copies of $\mathbb{R}$.

Example 3.64 The projection on the first factor $p \colon \mathbb{R}^2 \to \mathbb{R}$ is not a closed map. The hyperbola $C = \{xy = 1\} \subset \mathbb{R}^2$ is the zero set of the continuous map $f(x, y) = xy - 1$, whence a closed set. On the other hand $p(C) = \mathbb{R} - \{0\}$ is definitely not a closed set.

Exercises

3.49 Given maps $f \colon X \to Y$ and $g \colon Z \to W$, define

$$f \times g \colon X \times Z \to Y \times W, \qquad (f \times g)(x, z) = (f(x), g(z)).$$

1. Prove that if f and g are continuous, $f \times g$ is continuous.
2. Show that $f \times g$ is open if f and g are open.
3. Prove by an example that if f and g are closed, $f \times g$ might not be closed.

3.50 Show that the product of topological spaces is associative, i.e. that $(X \times Y) \times Z$, $X \times (Y \times Z)$ and $X \times Y \times Z$ have the same topology for any X, Y and Z.

3.51 Let X, Y be topological spaces, $A \subset X$, $B \subset Y$ subsets. Prove that $\overline{A \times B} = \overline{A} \times \overline{B}$. In particular if A and B are closed, $A \times B$ is closed in the product.

3.52 Let (X, d) be a metric space. Show that $d \colon X \times X \to \mathbb{R}$ is continuous for the product topology.

3.53 Show that if Y is discrete, the product topology on $X \times Y$ coincides with the topology of the disjoint union $\coprod_{y \in Y} X \times \{y\}$.

3.54 Let $g \colon \mathbb{R} \to [0, 1]$ be continuous and such that $g(1) = 1$, $g(t) = 0$ if $|t - 1| \geq 1/2$. Consider $f \colon \mathbb{R}^2 \to [0, 1]$,

$$f(x, 0) = 0, \qquad f(x, y) = g(xy^{-1}) \quad \text{if} \quad y \neq 0.$$

Prove that f is not continuous, although for any given x, y both restrictions $t \mapsto f(x, t), t \mapsto f(t, y)$ are continuous.

3.7 Hausdorff Spaces

Definition 3.65 A topological space is called a **Hausdorff space**, or **T2** space, if any two distinct points admit disjoint neighbourhoods.

In other words, given distinct points x, y in a Hausdorff space X there exist neighbourhoods $U \in \mathcal{I}(x)$ and $V \in \mathcal{I}(y)$ such that $U \cap V = \emptyset$.

Not all spaces are Hausdorff: the trivial topology is not T2, except for the irrelevant cases where the space is empty or has only one element.

Example 3.66 Any metric space is Hausdorff. If d is the distance and $x \neq y$, then $d(x, y) > 0$. When $0 < r < \dfrac{d(x, y)}{2}$ the balls $B(x, r)$ and $B(y, r)$ are disjoint, simply because if there was a $z \in B(x, r) \cap B(y, r)$, the triangle inequality would imply $d(x, y) \leq d(x, z) + d(z, y) < 2r < d(x, y)$.

Non-metrisable Hausdorff spaces do exist: Exercises 3.61 and 3.62 are but two examples.

Lemma 3.67 *In a Hausdorff space finite subsets are closed.*

Proof It suffices to prove that points are closed, i.e. if $x \in X$ Hausdorff, then $X - \{x\}$ is open. Take $y \in X - \{x\}$, so that there are disjoint neighbourhoods $U \in \mathcal{I}(x)$, $V \in \mathcal{I}(y)$; *a fortiori* $V \subset X - \{x\}$, so $X - \{x\}$ is a neighbourhood of any of its points.   □

Proposition 3.68 *Subspaces and products of Hausdorff spaces are Hausdorff.*

Proof Let X be Hausdorff, $Y \subset X$ a subspace and $x, y \in Y$ distinct. There are open disjoint sets $U, V \subset X$ such that $x \in U$ and $y \in V$. Then $U \cap Y$ and $V \cap Y$ are open and disjoint in Y.

Let X, Y be Hausdorff and $(x, y), (z, w) \in X \times Y$ distinct; to fix ideas suppose $x \neq z$. Then there exist disjoint open sets $U, V \subset X$ such that $x \in U$ and $z \in V$. It follows that $(x, y) \in U \times Y$, $(z, w) \in V \times Y$ and $U \times Y \cap V \times Y = (U \cap V) \times Y = \emptyset$.   □

Theorem 3.69 *A topological space is Hausdorff if and only if the diagonal is closed in the product.*

Proof The diagonal of a topological space X is the subset

$$\Delta = \{(x, x) \mid x \in X\} \subset X \times X.$$

Suppose X is Hausdorff space and take $(x, y) \in X \times X - \Delta$. By definition $x \neq y$, so there exist open sets U, V of X such that $x \in U$, $y \in V$ and $U \cap V = \emptyset$. Hence $(x, y) \in U \times V \subset X \times X - \Delta$, proving that $X \times X - \Delta$ is a neighbourhood of any of its points and that the diagonal is closed.

If Δ is closed in $X \times X$ and $x \neq y$, there exist open sets $U, V \subset X$ such that $(x, y) \in U \times V \subset X \times X - \Delta$, whence $x \in U$, $y \in V$ and $U \cap V = \emptyset$.   □

Corollary 3.70 *Let $f, g: X \to Y$ be continuous, with Y Hausdorff. The set $C = \{x \in X \mid f(x) = g(x)\}$ is closed in X.*

Proof The map

$$(f, g): X \to Y \times Y, \quad (f, g)(x) = (f(x), g(x))$$

is continuous and $C = (f, g)^{-1} \Delta$, where Δ denotes the diagonal of Y. ☐

Example 3.71 If $f: X \to X$ is a continuous map of X, one says $x \in X$ is a **fixed point** of f if $f(x) = x$. If X is Hausdorff the set of fixed points for f is a closed subset: just apply Corollary 3.70 taking $X = Y$ and g equal the identity.

Lemma 3.72 *Let X be a Hausdorff space and $p_1, \ldots, p_n$ finitely many distinct points in X. Then there exists a finite sequence $U_1, \ldots, U_n$ of open sets in X such that $p_i \in U_i$ for any i and $U_i \cap U_j = \emptyset$ for every $i \neq j$.*

Proof For every pair $1 \leq i < j \leq n$ we can find disjoint open sets U_{ij} and U_{ji} such that $p_i \in U_{ij}$ and $p_j \in U_{ji}$. Now consider, for any given i, the open set $U_i = \cap \{U_{ij} \mid j \neq i\}$. ☐

Exercises

3.55 Tell whether an infinite set with the cofinite topology is Hausdorff. Explain the answer.

3.56 Let X, Y be topological spaces, Y Hausdorff. Prove that if there is a 1-1 continuous map $f: X \to Y$ then also X is Hausdorff.

3.57 Let $f, g: X \to Y$ be continuous, Y Hausdorff and $A \subset X$ a dense subset. Prove that $f(x) = g(x)$ for every $x \in A$ implies $f = g$.

3.58 Determine the cardinality of the set of continuous maps $f: \mathbb{R} \to \mathbb{R}$. (Hint: continuous real-valued maps are determined by their values on $\mathbb{Q}$.)

3.59 Prove that a topological space is Hausdorff if and only if for any point x we have

$$\{x\} = \bigcap_{U \in \mathcal{I}(x)} \overline{U} .$$

3.60 Let $f: X \to Y$ be continuous with Y Hausdorff. Prove that the graph $\Gamma = \{(x, f(x)) \in X \times Y \mid x \in X\}$ is closed in the product.

3.61 (Golomb [Go59], ✋, ♡) For any pair of coprime integers $a, b \in \mathbb{N}$ define $N_{a,b} = \{a + kb \mid k \in \mathbb{N}_0\} \subset \mathbb{N}$. Prove that:

1. the family $\mathcal{B} = \{N_{a,b} \mid a, b \in \mathbb{N}, \ gcd(a, b) = 1\}$ is a basis of a topology $\mathcal{T}$ on $\mathbb{N}$;
2. $\mathcal{T}$ is Hausdorff;
3. any multiple of b belongs to the closure of $N_{a,b}$;
4. for any pair of non-empty open sets $A, B \in \mathcal{T}$ we have $\overline{A} \cap \overline{B} \neq \emptyset$;

5. $\mathcal{T}$ is non-metrisable.

3.62 (☙, ♡) Denote by $\mathbb{R}_l$ the lower-limit line (Example 3.9). Prove that $\mathbb{R}_l$ and $X = \mathbb{R}_l \times \mathbb{R}_l$ are Hausdorff spaces, and the subspace

$$Y = \{(x, y) \in \mathbb{R}_l \times \mathbb{R}_l \mid x + y = 0\}$$

is closed and discrete in X. Suppose (by contradiction) that there exists a distance d on X. Prove that there exists a map $f : \mathbb{R} \to \,]0, +\infty[$ such that

$$B((t, -t), f(t)) \cap B((s, -s), f(s)) = \emptyset$$

for all $t \neq s$. Then deduce there is a map $h : \mathbb{R} \to \,]0, +\infty[$ such that

$$\left([t, t + h(t)[\times [-t, -t + h(t)[\right) \cap \left([s, s + h(s)[\times [-s, -s + h(s)[\right) = \emptyset$$

for every $t \neq s$, and this leads to a contradiction.

References

[Go59] Golomb, S.W.: A connected topology for the integers. Amer. Math. Monthly **66**, 663–665 (1959)
[So47] Sorgenfrey, R.H.: On the topological product of paracompact spaces. Bull. AMS **53**, 631– 632 (1947)

Chapter 4
Connectedness and Compactness

The concept of a topological space is way too general to allow to prove interesting and wide ranging results. When a mathematician faces useful notions that are too general, there is only one thing he does: **classify**.

Without going into a matter that, albeit compelling, concerns more the methodology of science than mathematics, we will show with the aid of a simple example how the idea of classifying suits topology—always bearing in mind, though, that the same reasoning is common to mathematics as a whole.

Suppose we have two spaces that our intuition suggests aren't topologically equivalent, and we would like to prove rigorously that they're not homeomorphic. First of all we look at their cardinalities: if different we are done, but if they're equal we cannot deduce much. For obvious reasons of space, and time, the thought of writing down all possible bijective maps between the two spaces, and verify if each one is a homeomorphism, is ill-advised, and quickly abandoned. The traditional way a mathematician chooses to follow is this:

Step 1. Introduce topological features that are invariant under homeomorphism. The properties of being metrisable or Hausdorff are of this sort, for example. In some situations homeomorphism-invariance is clear from the definitions, in others it's not at all transparent (a typical case is the dimension of a topological manifold) and a lot of theoretical work is needed to sort the issue out.

Step 2. Test the properties of Step 1 on the two spaces: if we are smart enough to find—and here's where a mathematician's intuition enters the game—a property that's easy to establish and with respect to which the spaces behave differently, then we can deduce the spaces are not homeomorphic. At this stage the theory allows to spare time by means of statements of the kind: 'if a space has property $x_1, x_2, \ldots$, then it also has property $y_1, y_2, \ldots$.'

This chapter is devoted to the study of two pivotal topological proprieties invariant under homeomorphisms: connectedness and compactness. More will follow in the ensuing chapters.

© Springer International Publishing Switzerland 2015

M. Manetti, *Topology*, UNITEXT - La Matematica per il 3+2 91,
DOI 10.1007/978-3-319-16958-3_4

4.1 Connectedness

It's part of the human intuition to answer 'two' when asked 'How many pieces make up the space $X = \mathbb{R} - \{0\}$?' This happens because we are able to distinguish two parts in X, namely $X_- = \{x < 0\}$ and $X_+ = \{x > 0\}$, that in our mind match the word 'piece' of the question.

Topological structures allow to define in a mathematically precise way the notions of **connected space** and **connected component**, that correspond to the naïve concepts of 'one single piece' and 'piece of cake' if the cake is already sliced.

Definition 4.1 A topological space X is called **connected** if the only subsets that are both open and closed are $\emptyset$ and X. A non-connected topological space is called **disconnected**.

Lemma 4.2 *On a topological space X the following properties are equivalent:*

1. *X is disconnected;*
2. *X is the disjoint union of two open, proper subsets;*
3. *X is the disjoint union of two closed, proper subsets.*

Proof (1) $\Rightarrow$ (2), (3). Let $A \subset X$ be open, closed and non-empty. If $A \neq X$ the complement $B = X - A$ is open, closed and non-empty and X is the disjoint union of A and B.

(2) $\Rightarrow$ (1). Suppose $A_1 \cup A_2 = X$ with A_1, A_2 open, non-empty and disjoint. Then $A_1 = X - A_2$ is closed, too.

(3) $\Rightarrow$ (1). If $C_1 \cup C_2 = X$, where C_1, C_2 are closed, non-empty and disjoint, $C_1 = X - C_2$ is open. $\square$

Example 4.3 The space $X = \{x \in \mathbb{R} \mid x \neq 0\}$ is disconnected. The non-empty subsets

$$X_- = X \cap \,] - \infty, 0[, \qquad X_+ = X \cap \,]0, +\infty[$$

are open in the subspace topology, disjoint and their union is X.

One says a topological subspace is connected if it is so for the induced topology.

Lemma 4.4 *Let X be a topological space and $A \subset X$ an open and closed subset. For any connected subspace $Y \subset X$ either $Y \subset A$ or $Y \cap A = \emptyset$.*

Proof The intersection $Y \cap A$ is open and closed in Y. As Y is connected, necessarily $Y \cap A = Y$ (hence $Y \subset A$) or $Y \cap A = \emptyset$. $\square$

Example 4.5 Let X be a topological space, $K \subset X$ a closed subset and $U \subset X$ an open subset containing K. If X and $U - K$ are connected, also $X - K$ is connected. In fact, suppose by contradiction $X - K = A \cup B$, with A, B open, disjoint and non-empty. Since $U - K$ is a connected subspace of $X - K$, by Lemma 4.4 either $U - K \subset A$ or $U - K \subset B$; to fix ideas suppose $U - K \subset A$ and $U - K \cap B = \emptyset$. Then $A \cup K = A \cup U$, $U \cap B = \emptyset$ and hence $X = (A \cup K) \cup B = (A \cup U) \cup B$ is the union of open, disjoint, non-empty sets, against the assumption.

Theorem 4.6 *The interval* $[0, 1]$ *is connected in the Euclidean topology.*

Proof Take C and D non-empty closed subspaces of $[0, 1]$ such that $C \cup D = [0, 1]$; we claim $C \cap D \neq \emptyset$. Suppose, to fix ideas, $0 \in C$, and write $d \in [0, 1]$ for the infimum of D. It will be enough to prove $d \in C \cap D$. As D is closed, d must belong in D, so if $d = 0$ we are done. If $d > 0$ we set $E = C \cap [0, d]$. As E is closed and contains $[0, d[$ it follows $d \in E$, so $d \in C$. □

Theorem 4.7 *Let* $f \colon X \to Y$ *be a continuous map. Then* X *connected implies* $f(X)$ *connected.*

Proof Let $Z \subset f(X)$ be a non-empty, open and closed subset in $f(X)$. By definition of subspace topology there exist an open $A \subset Y$ and a closed $C \subset Y$ such that $Z = f(X) \cap A = f(X) \cap C$. But f is continuous, so $f^{-1}(Z) = f^{-1}(A)$ is open, and $f^{-1}(Z) = f^{-1}(C)$ is closed. As X is connected, $f^{-1}(Z) = X$ and therefore $Z = f(X)$. □

Definition 4.8 A topological space X is **path(wise)-connected** if, given any two points $x, y \in X$, there is a continuous mapping $\alpha \colon [0, 1] \to X$ such that $\alpha(0) = x$ and $\alpha(1) = y$. Such an α is called a **path** from x to y.

The continuous image of a path connected space is clearly path connected. The intuitive notion of connectedness of a graph seen in Chap. 1 corresponds to path connectedness.

Lemma 4.9 *Any path connected space is connected.*

Proof Let X be a path connected space and $A, B \subset X$ open, non-empty and such that $A \cup B = X$: we want to prove $A \cap B \neq \emptyset$. Choose $x \in A$, $y \in B$ and a path $\alpha \colon [0, 1] \to X$ such that $\alpha(0) = x$ and $\alpha(1) = y$. The open sets $\alpha^{-1}(A)$, $\alpha^{-1}(B)$ are non-empty, their union is $[0, 1]$ and so there exists $t \in \alpha^{-1}(A) \cap \alpha^{-1}(B)$. The point $\alpha(t)$ belongs to $A \cap B$, which is thus not empty. □

We shall prove later on, as consequence of Proposition 10.5, that open connected subsets in $\mathbb{R}^n$ are path connected. If $n \geq 2$, there exist closed, connected subsets in $\mathbb{R}^n$ that are not path connected (Exercise 4.20). The most wicked—at least in our view—example of connected but path disconnected subset is described in Exercise 8.6.

Lemma 4.10 *Let* A *and* B *be path connected subspaces in a topological space. If* $A \cap B \neq \emptyset$, *then* $A \cup B$ *is path connected.*

Proof It suffices to show that if $x \in A$ and $y \in B$ (or conversely) there is a path $\alpha \colon [0, 1] \to A \cup B$ from $\alpha(0) = x$ to $\alpha(1) = y$. Pick $z \in A \cap B$ and choose paths $\beta \colon [0, 1] \to A$ and $\gamma \colon [0, 1] \to B$ such that $\beta(0) = x$, $\beta(1) = \gamma(0) = z$, $\gamma(1) = y$. Now define α as follows:

$$\alpha(t) = \begin{cases} \beta(2t) & \text{if } 0 \leq t \leq 1/2. \\ \gamma(2t - 1) & \text{if } 1/2 \leq t \leq 1. \end{cases}$$
□

Example 4.11 The spaces $\mathbb{R}^n$ and the spheres S^n are path connected for every $n > 0$. Indeed, a possible path joining any two points $x, y \in \mathbb{R}^n$ is the standard parametrisation of the segment between x, y: $\alpha(t) = tx + (1 - t)y$. The sphere S^n, $n > 1$, is the union of two open sets homeomorphic to $\mathbb{R}^n$ under stereographic projection, and the claim follows now from Lemma 4.10.

We recall that a subset $A \subset \mathbb{R}^n$ is **convex** if $tx + (1 - t)y \in A$ for every $x, y \in A$ and any $t \in [0, 1]$. Intervals, for instance, are precisely the convex subsets of $\mathbb{R}$. Another example is the open ball

$$B(p, r) = \{x \in \mathbb{R}^n \mid \|p - x\| < r\},$$

convex for any $p \in \mathbb{R}^n$ and any $r > 0$. In fact if $x, y \in B(p, r)$ and $t \in [0, 1]$, the triangle inequality gives

$$\|p - (tx + (1 - t)y)\| = \|t(p - x) + (1 - t)(p - y)\|$$
$$\leq \|t(p - x)\| + \|(1 - t)(p - y)\| = t\|p - x\| + (1 - t)\|p - y\|$$
$$< tr + (1 - t)r = r$$

so $tx + (1 - t)y \in B(p, r)$.

Corollary 4.12 *Any convex subset of $\mathbb{R}^n$ is path connected and hence connected.*

Proof Let A be convex in $\mathbb{R}^n$; for any $x, y \in A$ the map $\alpha: [0, 1] \to A$, $\alpha(t) = tx + (1 - t)y$ is continuous.   □

Corollary 4.13 *For a subset $I \subset \mathbb{R}$ the following are equivalent conditions:*

1. *I is an interval, i.e. a convex subset;*
2. *I is path connected;*
3. *I is connected.*

Proof Only (3) $\Rightarrow$ (1) is non-trivial. If $I \subset \mathbb{R}$ is not an interval, there exist $a < b < c$ such that $a, c \in I$ but $b \notin I$. The open disjoint sets $I \cap] - \infty, b[$ and $I \cap]b, +\infty[$ aren't empty and disconnect I.   □

We are now in the position of showing the first applications of connectedness.

Example 4.14 Let $f: [0, 1] \to [0, +\infty[$ be continuous and such that $f(1) = 0$. There exists an $x \in [0, 1]$ such that $f(x) = x$. In fact the range of $[0, 1] \to \mathbb{R}$, $x \mapsto f(x) - x$, is a connected set containing $f(1) - 1 = -1$ and $f(0) - 0 \geq 0$.

Example 4.15 The interval $]0, 1[$ is not homeomorphic to $[0, 1[$. Suppose the contrary, and that $f: [0, 1[\to]0, 1[$ was a homeomorphism. Then f would induce, by restriction, a homeomorphism $f:]0, 1[\to]0, 1[- \{f(0)\}$. The contradiction emerges from the observation that $]0, 1[$ is connected whereas $]0, 1[- \{f(0)\}$ is disconnected.

To prevent a common mistake let us point out that connectedness doesn't forbid the existence of open, non-empty subspaces homeomorphic to closed subspaces. Here's an example: in the Euclidean topology the intervals $[0, 1[$ and $[1, 2[$ are homeomorphic; yet the former is open and the latter closed as subspaces of the connected space $[0, 2[$.

Lemma 4.16 *Take $n > 0$ and a continuous map $f : S^n \to \mathbb{R}$. There exists $x \in S^n$ such that $f(x) = f(-x)$. In particular f cannot be one-to-one.*

Proof Consider the continuous map

$$g : S^n \to \mathbb{R}, \qquad g(x) = f(x) - f(-x).$$

As $n > 0$, the sphere S^n is connected, the image $g(S^n)$ is connected in $\mathbb{R}$ and hence convex. Choose an arbitrary point $y \in S^n$. The interval $g(S^n)$ contains $g(y), g(-y)$ and therefore also the convex combination

$$\frac{1}{2}g(y) + \frac{1}{2}g(-y) = \frac{1}{2}(f(y) - f(-y)) + \frac{1}{2}(f(-y) - f(y)) = 0.$$

Hence $0 \in g(S^n)$ and an $x \in S^n$ must exist such that $g(x) = 0$. $\qquad\square$

Example 4.17 No subset in $\mathbb{R}$ can be homeomorphic to an open set in $\mathbb{R}^n$, for any $n > 1$. In fact every open subset in $\mathbb{R}^n$ contains a subset homeomorphic to the sphere S^{n-1}, so Lemma 4.16 applies.

Lemma 4.18 *Let Y be a connected topological space and $f : X \to Y$ a continuous, onto map such that $f^{-1}(y)$ is connected for every $y \in Y$. If f is open, or closed, then X is connected.*

Proof Suppose f is open and let $A_1, A_2 \subset X$ be open, non-empty and such that $X = A_1 \cup A_2$. As $Y = f(A_1) \cup f(A_2)$ and Y is connected, there exists $y \in f(A_1) \cap f(A_2)$, so $f^{-1}(y) \cap A_i \neq \emptyset, i = 1, 2$. But $f^{-1}(y)$ is connected, so $f^{-1}(y) \cap A_1 \cap A_2 \neq \emptyset$, and *a fortiori* $A_1 \cap A_2 \neq \emptyset$.

If f is closed the same proof works with the proviso of taking A_1, A_2 closed. $\qquad\square$

Theorem 4.19 *The product of two connected spaces is connected.*

Proof We have to show that if X and Y are connected, so is $X \times Y$. For that it's enough to note that the projection $X \times Y \to Y$ is continuous, onto, open and its fibres are homeomorphic to X. The claim descends from Lemma 4.18. $\qquad\square$

Using induction we immediately obtain that finite products of connected spaces are connected.

Exercises

4.1 ($\heartsuit$) Let $p, q \in \mathbb{R}^2$ denote the points $(1, 0)$ and $(-1, 0)$. Which of the following subspaces in $\mathbb{R}^2$ are connected?

$$A = \{x \in \mathbb{R}^2 \mid \|x - p\| < 1 \text{ or } \|x - q\| < 1\};$$
$$B = \{x \in \mathbb{R}^2 \mid \|x - p\| < 1 \text{ or } \|x - q\| \le 1\};$$
$$C = \{x \in \mathbb{R}^2 \mid \|x - p\| \le 1 \text{ or } \|x - q\| \le 1\}.$$

4.2 Which are the connected subspaces in a discrete topological space?

4.3 Let A, B be subspaces such that $A \cup B$ and $A \cap B$ are connected. Prove that if A, B are both closed or both open, A and B are connected.

4.4 For any triple $p, q, v \in \mathbb{R}^n$, the path

$$\alpha_{p,q,v} \colon [0, 1] \to \mathbb{R}^n, \qquad \alpha_{p,q,v}(t) = (1 - t)p + tq + t(1 - t)v,$$

is a parabolic arc (possibly degenerate) with endpoints p and q.

Prove that if $p \ne q$, for any $x \in \mathbb{R}^n - \{p, q\}$ there exists at most one vector v orthogonal to $p - q$ and such that x belongs to the range of $\alpha_{p,q,v}$.

4.5 ($\heartsuit$) Prove that if $n \ge 2$ the complement of a countable subset in $\mathbb{R}^n$ is path connected.

4.6 Take $n \ge 2$ and $f \colon S^n \to \mathbb{R}$ a continuous map. Call A the set of points $t \in f(S^n)$ such that the fibre $f^{-1}(t)$ has finite cardinality.

Show that A contains at most two points. Give examples of maps f where A has cardinality 0, 1 and 2.

4.7 Prove that the product of two path connected spaces is path connected.

4.8 A subset A in $\mathbb{R}^n$ is **star-shaped** with respect to the point $a \in A$ if for every $b \in A$ the segment joining a with b is contained in A. Prove that star-shaped sets are path connected.

4.9 Let $X \subset \mathbb{R}^2$ be a bounded and convex subset. Show $\mathbb{R}^2 - X$ is path connected.

4.10 ($\clubsuit$, $\heartsuit$) Let A be an open convex set in $\mathbb{R}^2$ and p a point not in A. Prove that there exists a straight line through p that does not meet A.

4.11 ($\clubsuit$) Let X be a Hausdorff space with the property that $X - A$ is connected for any finite subset $A \subset X$. Prove that the space of configurations

$$\mathrm{Conf}_n(X) = \{(x_1, \dots, x_n) \in X^n \mid x_i \ne x_j \text{ for every } i \ne j\}$$

is connected for every $n > 0$.

4.2 Connected Components

Definition 4.20 Let X be a topological space. A subspace $C \subset X$ is called a **connected component** of X (sometimes just 'component') if it satisfies two properties:

1. C is connected;
2. $C \subset A$ and A connected imply $C = A$.

In other terms a connected component is a maximal element in the family of connected subspaces ordered by inclusion.

Example 4.21 Let X be a topological space and $C \subset X$ open, closed, connected and non-empty. Then C is a connected component of X. In fact C is connected by hypothesis, and if $C \subset A$, then C is open, closed and non-empty in A. Therefore if A is connected it follows $C = A$.

Lemma 4.22 *Let Y be a connected subspace in a topological space X and suppose $Y \subset W \subset \overline{Y}$. Then W is connected. In particular the closure of a connected subspace is connected.*

Proof Let $Z \subset W$ be non-empty, open and closed in W. Then $Z \cap Y$ is open and closed in Y. As Y is dense and Z is open in W, we have $Z \cap Y \neq \emptyset$. But Y is connected, so $Y \subset Z$. Now, Y is dense and Z is closed in W, hence $Z = W$. $\square$

Lemma 4.23 *Let x be a point in X and $\{Z_i \mid i \in I\}$ a family of connected subspaces in X all containing x. Then $W = \bigcup_i Z_i$ is connected.*

Proof Take $A \subset W$ simultaneously open, closed and non-empty. By Lemma 4.4, for every i we have $Z_i \subset A$ or $Z_i \cap A = \emptyset$. Since $A \neq \emptyset$ there's an index i such that $A \cap Z_i \neq \emptyset$, so $Z_i \subset A$ and in particular $x \in A$. But x belongs to all subspaces Z_j, so $Z_j \cap A$ isn't empty for every j, whence $Z_j \subset A$ for all j and therefore $W = A$. $\square$

Corollary 4.24 *Let A, B be connected. If $A \cap B \neq \emptyset$ then $A \cup B$ is connected.*

Proof Let $x \in A \cap B$ and apply Lemma 4.23 to the family $\{A, B\}$. $\square$

Lemma 4.25 *Let x be a point in the topological space X. Denote by $C(x)$ the union of all connected subspaces containing x:*

$$C(x) = \bigcup \{Y \mid x \in Y \subset X, \ Y \ connected\}.$$

Then $C(x)$ is a connected component of X that contains x.

Proof First, note $\{x\}$ is connected, so $x \in C(x)$. Applying Lemma 4.23 to the family of all connected subspaces containing x we infer that $C(x)$ is connected. Now let $A \subset X$ be a connected subspace containing $C(x)$. In particular $x \in A$, and by definition of $C(x)$ we have $A \subset C(x)$, so $C(x) = A$. $\square$

Definition 4.26 Retaining the notation of Lemma 4.25 one calls $C(x)$ the **connected component of** x in X.

Theorem 4.27 *Every topological space is the union of its connected components. Each connected component is closed and any point is contained in one, and only one, connected component.*

Proof Lemma 4.25 tells that every point belongs in some connected component. If C, D are connected components with $C \cap D \neq \emptyset$, by Corollary 4.24 the subspace $C \cup D$ is connected and the obvious inclusions $C \subset C \cup D$, $D \subset C \cup D$ imply $C = C \cup D = D$. At last, if C is a connected component, by Lemma 4.22 $\bar{C}$ is connected and contains C, so $C = \bar{C}$. $\qquad\square$

Connected components are in general not open (see Exercise 4.12). In many concrete cases, though, they will be open because of the following result.

Lemma 4.28 *If every point in a topological space X has a connected neighbourhood, the connected components of X are open.*

Proof Let $C \subset X$ denote a connected component, take $x \in C$ and U a connected neighbourhood of x. Since $x \in C \cap U$, the union $C \cup U$ is connected and hence $C \cup U \subset C$, which means that $U \subset C$ and C is a neighbourhood of x. $\qquad\square$

Any homeomorphism transforms connected components into connected components, with the effect that two homeomorphic spaces must have the same number of components.

Example 4.29 $X = \mathbb{R} - \{0\}$ is not homeomorphic to $Y = \mathbb{R} - \{0, 1\}$, because X has two connected components $] - \infty, 0[$ and $]0, +\infty[$, while Y has three: $] - \infty, 0[$, $]0, 1[$ and $]1, +\infty[$.

Exercises

4.12 ($\heartsuit$) Describe the connected components of the subspace $\mathbb{Q} \subset \mathbb{R}$ of rational numbers. Deduce that, in general, connected components are not open.

4.13 ($\heartsuit$) Let $Y \subset X$ be a connected subspace and $\{Z_i \mid i \in I\}$ a family of connected subspaces in X that meet Y. Prove that $W = Y \cup \bigcup_i Z_i$ is connected.

4.14 ($\heartsuit$) Let X, Y be connected spaces and $A \subset X$, $B \subset Y$ proper subsets. Prove $X \times Y - A \times B$ is connected.

4.15 Show that the two subspaces of $\mathbb{R}^2$ (see Fig. 4.1):

$$A = \{(x, y) \in \mathbb{R}^2 \mid y = 0\} \cup \{(x, y) \in \mathbb{R}^2 \mid x^2 + y^2 = 1 \text{ and } y \geq 0\};$$
$$R = \{(x, y) \in \mathbb{R}^2 \mid y = 0\} \cup \{(x, y) \in \mathbb{R}^2 \mid x^2 + (y - 1)^2 = 1\},$$

are not homeomorphic. (Hint: understand how many connected components one obtains by removing one point from either space.)

Fig. 4.1 Non-homeomorphic subspaces in the plane

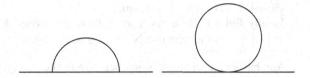

4.16 Let $\{A_n \mid n \in \mathbb{Z}\}$ be a countable family of connected subspaces of X such that $A_n \cap A_{n+1} \neq \emptyset$ for every n. Prove $A = \cup_n A_n$ is connected.

4.17 Consider a family $\{A_i \mid i \in I\}$ of non-empty connected subspaces in some X. Prove:

1. the family $\mathcal{A} \subset \mathcal{P}(I)$, formed by subsets $J \subset I$ for which the union $\cup\{A_j \mid j \in J\}$ is connected, contains maximal elements (with respect to the inclusion);
2. suppose that for any $i, j \in I$ there is a sequence $i_1, i_2, \ldots, i_n \in I$ such that $i_1 = i$, $i_n = j$ and $A_{i_k} \cap A_{i_{k+1}} \neq \emptyset$ for every $k = 1, \ldots, n-1$. Prove that the union $\cup\{A_i \mid i \in I\}$ is connected.

4.18 ($\heartsuit$) Let $\{A_n \mid n \in \mathbb{N}\}$ be a countable family of connected subspaces in X such that $A_{n+1} \subset A_n$ for every n. Is it true that $\cap_n A_n$ is necessarily connected?

4.19 Show that any homeomorphism $f : \mathbb{R} \to \mathbb{R}$ of finite order, i.e. such that $f^p = Id$ for some $p > 0$, has a fixed point. (Hint: call $b = \max(0, f(0), f^2(0), \ldots, f^{p-1}(0))$ and $a = f^{-1}(b)$, then prove that f fixes a point in the interval $[a, b]$.)

4.20 (☙, $\heartsuit$) Let $X \subset \mathbb{R}^2$ denote the union of the segment $\{x = 0, |y| \leq 1\}$ with the range of $f :]0, +\infty[\to \mathbb{R}^2$, $f(t) = (t^{-1}, \cos(t))$. Prove that X is closed in $\mathbb{R}^2$, connected and not path connected.

4.3 Covers

Definition 4.30 A **cover** of a set X is a family $\mathcal{A}$ of subsets such that $X = \cup\{A \mid A \in \mathcal{A}\}$. A cover is called finite if $\mathcal{A}$ is a finite family, countable if $\mathcal{A}$ is countable. If $\mathcal{A}, \mathcal{B}$ are covers of X and $\mathcal{A} \subset \mathcal{B}$, $\mathcal{A}$ is said a **subcover** of $\mathcal{B}$.

Quite frequently it's convenient to work with covers $\mathcal{A} = \{U_i \mid i \in I\}$ defined through indexed families. In such case, and unless we say so explicitly, the onto map $I \to \mathcal{A}, i \mapsto U_i$, is not assumed to be 1-1.

A number of important features of topological spaces regard the behaviour with respect to covers.

Definition 4.31 A cover $\mathcal{A}$ of a space X is said:

1. **open** if every $A \in \mathcal{A}$ is open;

2. **closed** if every $A \in \mathcal{A}$ is closed;
3. **locally finite** if for every $x \in X$ there is an open $V \subset X$ with $x \in V$ and $V \cap A \neq \emptyset$ for a finite number of $A \in \mathcal{A}$ at most.

Any basis of a topology is an open cover. The family $\{[n, n + 1] \mid n \in \mathbb{Z}\}$ is a locally finite closed cover of $\mathbb{R}$.

Definition 4.32 Let $\mathcal{A}$ be a cover of X. We call $\mathcal{A}$ an **identification cover** in case $U \subset X$ is open if and only if $U \cap A$ is open in A for every $A \in \mathcal{A}$.

Not every cover has the above property: $\{\{x\} \mid x \in \mathbb{R}\}$, for instance, is a closed cover of $\mathbb{R}$ but not an identification cover.

Proposition 4.33 *Let $\mathcal{A}$ be an identification cover of X. A map $f \colon X \to Y$ is continuous if and only if for any $A \in \mathcal{A}$ the restriction $f_{|A} \colon A \to Y$ is continuous.*

Proof Let $U \subset Y$ be open; if the restriction $f_{|A} \colon A \to Y$ is continuous for every $A \in \mathcal{A}$, then $f^{-1}(U) \cap A = f_{|A}^{-1}(U)$ is open in A and $f^{-1}(U)$ is open in X. □

Theorem 4.34 *Open covers and locally finite closed covers are identification covers.*

Proof Let $\mathcal{A}$ be an open cover of X, and $U \subset X$. If $U \cap A$ is open in A for any $A \in \mathcal{A}$, then $U \cap A$ is open in X as well, so $U = \cup\{A \cap U \mid A \in \mathcal{A}\}$ is open.

Consider now a finite, closed cover $X = C_1 \cup \cdots \cup C_n$ and take $U \subset X$ such that $U \cap C_i$ is open in C_i for every i. Write $B = X - U$, so $B \cap C_i = C_i - (U \cap C_i)$ is closed in C_i, and hence in X, for every i. Then $B = \cup_{i=1}^{n}(B \cap C_i)$ is closed.

If $\{C_i \mid i \in I\}$ is a locally finite, closed cover, we can find an open cover $\mathcal{A}$ in such a way that $\{C_i \cap A \mid i \in I\}$ is a closed, finite cover of A, for any $A \in \mathcal{A}$. Therefore, if $U \cap C_i$ is open in C_i for every i, $U \cap C_i \cap A$ is open in $A \cap C_i$ for all $A \in \mathcal{A}$, and then $U \cap A$ is open in A for all $A \in \mathcal{A}$. The conclusion is that U is open in X. □

Exercises

4.21 Let $\mathcal{A}$ be an open cover of X and $\mathcal{B}$ the family of open sets in X that are contained in some element of $\mathcal{A}$. Prove that $\mathcal{B}$ is a basis of the topology.

4.22 Prove that $\{[-n, n] \mid n \in \mathbb{N}\}$ is an identification cover of $\mathbb{R}$.

4.23 Let $\mathcal{A}$ be a cover of X such that $\cup\{A^{\circ} \mid A \in \mathcal{A}\} = X$. Show that $\mathcal{A}$ is an identification cover.

4.4 Compact Spaces

Definition 4.35 A topological space is said to be **compact** if any open cover admits a finite subcover. A subspace in a topological space is compact if it is compact for the induced (subspace) topology.

Equivalently, a subspace K of X is compact iff for any family $\mathcal{A}$ of open sets in X with $K \subset \cup\{A \mid A \in \mathcal{A}\}$, there exist finitely many $A_1, \ldots, A_n \in \mathcal{A}$ such that $K \subset A_1 \cup \cdots \cup A_n$.

Example 4.36 The space $\mathbb{R}^n$ is not compact, for any $n > 0$; to show this it's enough to find an open cover without finite subcovers. Consider the open balls with centre 0 and radius a natural number

$$\mathbb{R}^n = \cup_{m=1}^{+\infty} B(0, m).$$

If such cover had a finite subcover, there would exist $m_1, \ldots, m_k \in \mathbb{N}$ with

$$\mathbb{R}^n = \cup_{i=1}^{k} B(0, m_i) = B(0, \max(m_1, \ldots, m_k)),$$

which is patently false.

Example 4.37 Any finite set is compact, irrespective of the topology we put on it: for any open cover we can define a finite subcover by choosing one open set containing each point.

A discrete topological space is compact if and only if it is finite. In a discrete space X, in fact, the family of singletons $\{\{x\} \mid x \in X\}$ is an open cover. This has a finite subcover iff X is finite.

Theorem 4.38 *Let* $f : X \to Y$ *be a continuous map. If* X *is compact the range* $f(X)$ *is a compact subspace in* Y.

Proof Let $\mathcal{A}$ be a family of open sets in Y covering $f(X)$. Then the family $\{f^{-1}(A) \mid A \in \mathcal{A}\}$ is an open cover for X, so there are finitely many $A_1, \ldots, A_n \in \mathcal{A}$ with $X = f^{-1}(A_1) \cup \cdots \cup f^{-1}(A_n)$. This implies that $f(X) \subset A_1 \cup \cdots \cup A_n$. □

Theorem 4.39 *The closed interval* $[0, 1]$ *is compact in* $\mathbb{R}$.

Proof Let $\mathcal{A}$ be a family of open subsets in $\mathbb{R}$ that covers $[0, 1]$:

$$[0, 1] \subset \cup\{A \mid A \in \mathcal{A}\}.$$

Write $X \subset [0, +\infty[$ for the set of points t such that the closed interval $[0, t]$ is contained in a finite union of open subsets in $\mathcal{A}$.

The interval $[0, 0]$ is contained in an open set of $\mathcal{A}$, so $0 \in X$. Let's call b the least upper bound of X. If $b > 1$ there exists $t \in X$ such that $1 \leq t \leq b$, whence $[0, 1] \subset [0, t]$ is contained in a finite union of open subsets in $\mathcal{A}$. Now suppose $b \leq 1$, and let's prove this leads to a contradiction. If $b \leq 1$ there's an $A \in \mathcal{A}$ such that $b \in A$, and A being open implies the existence of $\delta > 0$ such that

$$]b - \delta, b + \delta[\subset A.$$

On the other hand, by the properties of the supremum there's a $t \in X$ such that $b - \delta < t \leq b$. Hence $[0, t]$ is contained in a finite union of covering sets, say

$$[0, t] \subset A_1 \cup \cdots \cup A_n.$$

Therefore if $0 \leq h < \delta$,

$$[0, b + h] = [0, t] \cup [t, b + h] \subset [0, t] \cup\,]b - \delta, b + \delta[\subset A \cup A_1 \cup \cdots \cup A_n,$$

and then $b + h \in X$ for any real number $0 \leq h < \delta$.   □

Example 4.40 The real line $\mathbb{R}$ isn't homeomorphic to the interval $[0, 1]$, for $[0, 1]$ is compact, whereas $\mathbb{R}$ is not.

Proposition 4.41 *1. Any closed subspace in a compact space is compact.*
 2. Finite unions of compact subspaces are compact.

Proof Let's begin with (1) and take Y closed in the compact space X. We need to prove that for any family $\mathcal{A}$ of open subsets such that $Y \subset \cup\{A \mid A \in \mathcal{A}\}$, there are finitely many sets $A_1, \ldots, A_n \in \mathcal{A}$ such that $Y \subset A_1 \cup \cdots \cup A_n$. The open family $\mathcal{A} \cup \{X - Y\}$ is a cover of the compact space X, so we can pick $A_1, \ldots, A_n \in \mathcal{A}$ such that $X = (X - Y) \cup A_1 \cup \cdots \cup A_n$. Immediately, then, $Y \subset A_1 \cup \cdots \cup A_n$.

Now let us see to (2). Let $K_1, \ldots, K_n$ be compact subspaces in some X, and $\mathcal{A}$ a family of open subsets in X covering $K = K_1 \cup \cdots \cup K_n$. For any $h = 1, \ldots, n$ there is a finite number of $\mathcal{A}_h \subset \mathcal{A}$ that covers the compact set K_h, so K is contained in the finite union $\cup\{A \mid A \in \mathcal{A}_1 \cup \cdots \cup \mathcal{A}_n\}$.   □

Corollary 4.42 *A subspace in $\mathbb{R}$ is compact if and only if it is closed and bounded.*

Proof Note preliminarily that the definition of a compact set involves only the notions of openness and cover, so it must be invariant under homeomorphism; that is, any space homeomorphic to a compact space is itself compact.

Let $A \subset \mathbb{R}$ be closed and bounded. Then $A \subset [-a, a]$ for some $a > 0$. The interval $[-a, a]$ is homeomorphic to $[0, 1]$ and hence compact. Therefore A is closed in a compact set, whence compact.

Conversely, if $A \subset \mathbb{R}$ is compact, the family of open intervals $\{\,] - n, n[\mid n \in \mathbb{N}\}$ covers A and so we can find a finite subcover:

$$A \subset\,] - n_1, n_1[\cup \cdots \cup\,] - n_s, n_s[\,.$$

Therefore $A \subset\,] - N, N[$, where N is the largest among $n_1, \ldots, n_s$. This implies A is bounded. Now for any $p \notin A$ the map $f(x) = 1/(x - p)$ is continuous and defined on $\mathbb{R} - \{p\}$. Its range $f(A)$ is compact, hence bounded, so $p \notin \bar{A}$ and then A is closed.   □

Corollary 4.43 *Let X be a compact topological space. Any continuous map $f : X \to \mathbb{R}$ has a maximum and a minimum.*

Proof The range $f(X)$ is compact in $\mathbb{R}$, so closed and bounded; but every closed bounded subset of $\mathbb{R}$ has maximum and minimum points.   □

Theorem 4.44 *Consider a closed map* $f: X \to Y$ *with* Y *compact. If* $f^{-1}(y)$ *is compact for every* $y \in Y$, *then* X *is compact.*

Proof Without loss of generality we may assume f is onto (otherwise we can always replace Y with $f(X)$). For any subset $A \subset X$ define

$$A' = \{y \in Y \mid f^{-1}(y) \subset A\}.$$

As $Y - A' = f(X - A)$ and f is closed, it follows that A' is open provided A is open. Take an open cover $\mathcal{A}$ of X and call $\mathcal{B}$ the family of finite unions of elements of $\mathcal{A}$. The family $\mathcal{B}' = \{B' \mid B \in \mathcal{B}\}$ is an open cover of Y: if $y \in Y$, the fibre $f^{-1}(y)$ is compact, so there are $A_1, \ldots, A_m \in \mathcal{A}$ such that $f^{-1}(y) \subset A_1 \cup \cdots \cup A_m$, and then $y \in B'$, where $B = A_1 \cup \cdots \cup A_m$.

But as Y is compact, we have a finite sequence $B_1, \ldots, B_n \in \mathcal{B}$ such that $Y = B_1' \cup \cdots \cup B_n'$. Then $X = B_1 \cup \cdots \cup B_n$, and since every B_i is a finite union in $\mathcal{A}$, we've found a finite subcover for $\mathcal{A}$. $\square$

Proposition 4.45 *Let* $\mathcal{B}$ *be a basis of the space* X. *If every cover of* X *made by elements of* $\mathcal{B}$ *has a finite subcover,* X *is compact.*

Proof For any open $U \subset X$ we set $\mathcal{B}_U = \{B \in \mathcal{B} \mid B \subset U\}$. By assumption $\mathcal{B}$ is a basis, so for any open set U we have $U = \cup\{B \mid B \in \mathcal{B}_U\}$.

Let $\mathcal{A}$ be an open cover of X and consider the open family $\mathcal{C} = \cup_{A \in \mathcal{A}} \mathcal{B}_A$. It's clear that $\mathcal{C}$ covers X by basis elements, and because of the assumptions we've made $\mathcal{C}$ has a finite subcover $\mathcal{F}$. For any open set $U \in \mathcal{F}$ let's choose an open set $A(U) \in \mathcal{A}$ such that $U \in \mathcal{B}_{A(U)}$, i.e. $U \subset A(U)$. Then $\{A(U) \mid U \in \mathcal{F}\}$ is a finite subcover of $\mathcal{A}$. $\square$

Proposition 4.46 *Let* $K_1 \supset K_2 \supset \cdots$ *be a countable descending chain of non-empty, closed and compact sets. Then*

$$\bigcap\{K_n \mid n \in \mathbb{N}\} \neq \emptyset.$$

Proof Given $n \in \mathbb{N}$, the set $K_1 - K_n$ is open in K_1. Now observe that the intersection of the closed sets K_n is empty precisely when the sets $K_1 - K_n$ cover K_1. $\square$

Exercises

4.24 Consider the metric space $\mathbb{Q}$ of rational numbers with the Euclidean distance. Prove that

$$K = \left\{x \in \mathbb{Q} \mid 0 \leq x \leq \sqrt{2}\right\}$$

is closed and bounded, yet not compact.

4.25 ($\heartsuit$) A family of subsets $\mathcal{A}$ in a set X is said to enjoy the **finite-intersection property** if for every (non-empty) finite subfamily $\mathcal{F} \subset \mathcal{A}$ we have $\cap\{A \mid A \in \mathcal{F}\} \neq \emptyset$.

Prove that a topological space is compact if and only if every closed family with the finite-intersection property has non-empty intersection. Convince yourself that Proposition 4.46 is a special case of this.

4.26 Let X be a compact space, $U \subset X$ open and $\{C_i \mid i \in I\}$ a closed family in X with

$$\bigcap_{i \in I} C_i \subset U .$$

Prove that there exist finitely many indices $i_1, \ldots, i_n \in I$ such that

$$C_{i_1} \cap \cdots \cap C_{i_n} \subset U .$$

4.27 Prove that the space of Exercise 3.3 is compact and Hausdorff. Deduce that any set admits a topology that renders it a compact T2 topological space.

4.28 Denote by $\mathbb{R}_{ut}$ the real line with the upper topology (Example 3.3). Prove that every compact, non-empty set in $\mathbb{R}_{ut}$ has a maximum. Conclude that if X is compact, every upper semi-continuous map $f : X \to \mathbb{R}$ (Exercise 3.28) has a maximum.

4.29 ($\heartsuit$) Let X be a compact space and $f : \mathbb{R} \to X$ a continuous and closed map. Show that there exists $x \in X$ such that $f^{-1}(x)$ is infinite.

4.30 ($\clubsuit$, $\heartsuit$) Let X be compact and $\{f_i : X \to [0, +\infty[\mid i \in I\}$ a non-empty family of continuous maps. Prove that

$$f : X \to [0, +\infty[, \qquad f(x) = \inf\{f_i(x) \mid i \in I\},$$

has a maximum point in X. Find an example showing that there is no minimum, in general.

4.5 Wallace's Theorem

Let $A \subset X$, $B \subset Y$ be subsets in two topological spaces. The product $A \times B$ is a subset of $X \times Y$ and we can introduce a topology on $A \times B$ in two different ways. We can consider $A \times B$ as a topological subspace of $X \times Y$, or we may view it as the topological product of A and B. It's plain to see that the restriction of the canonical basis of $X \times Y$ to $A \times B$ coincides with the canonical basis of $A \times B$, so that the two procedures eventually give rise to the same topological structure.

Theorem 4.47 (Wallace) *Let X, Y be topological spaces, $A \subset X$ and $B \subset Y$ compact subspaces and $W \subset X \times Y$ an open set such that $A \times B \subset W$. Then there exist open sets $U \subset X$ and $V \subset Y$ such that*

$$A \subset U, \quad B \subset V \quad and \quad U \times V \subset W.$$

Proof Consider first the special case in which A is a point, $A = \{a\}$. For any $b \in B$ there exist two open sets $U_b \subset X$, $V_b \subset Y$ such that

$$(a, b) \in U_b \times V_b \subset W.$$

The open family $\{V_b \mid b \in B\}$ covers B and by compactness we have $b_1, \ldots, b_n \in B$ such that $B \subset V_{b_1} \cup \cdots \cup V_{b_n}$. Define the open sets $U = U_{b_1} \cap \cdots \cap U_{b_n}$ and $V = V_{b_1} \cup \cdots \cup V_{b_n}$: then

$$\{a\} \times B \subset U \times V \subset \cup_i U_{b_i} \times V_{b_i} \subset W.$$

In case A is an arbitrary compact set, we've shown that for any $a \in A$ there are open sets U_a, V_a with

$$\{a\} \times B \subset U_a \times V_a \subset W.$$

The family $\{U_a \mid a \in A\}$ is an open cover of A; by compactness there are $a_1, \ldots, a_m \in A$ such that $A \subset U_{a_1} \cup \cdots \cup U_{a_m}$, so the open sets $U = U_{a_1} \cup \cdots \cup U_{a_m}$ and $V = V_{a_1} \cap \cdots \cap V_{a_m}$ satisfy $A \times B \subset U \times V \subset W$. $\square$

Wallace's theorem is particularly useful in view of its many interesting consequences.

Corollary 4.48 *Any compact subspace in a Hausdorff space is closed.*

Proof Let X be Hausdorff and $K \subset X$ compact. To show K is closed we'll prove that if $x \notin K$ there is an open set $U \subset X$ such that $x \in U$ and $U \cap K = \emptyset$.

The product $\{x\} \times K$ does not intersect the diagonal $\Delta \subset X \times X$, and since X is Hausdorff, $\{x\} \times K$ is contained in the open set $W = X \times X - \Delta$. By Wallace's theorem we have open sets $U, V \subset X$ with $\{x\} \times K \subset U \times V \subset W$. In particular $x \in U$ and $U \cap K = \emptyset$. $\square$

Corollary 4.49 *Let X, Y be topological spaces:*

1. *if X is compact, the projection $p \colon X \times Y \to Y$ is a closed map;*
2. *if X and Y are compact, $X \times Y$ is compact.*

Proof To prove (1) let $C \subset X \times Y$ be a given closed set. If $p(C) = Y$ there is nothing to prove; otherwise, take $y \notin p(C)$ and let us prove y has a neighbourhood disjoint from $p(C)$. As $X \times \{y\} \subset X \times Y - C$, by Wallace's theorem there is an open neighbourhood V of y such that $X \times V \cap C = \emptyset$, so $V \cap p(C) = \emptyset$.

Concerning (2) we may just apply Theorem 4.44 to the second projection $X \times Y \to Y$. $\square$

An induction argument guarantees that a finite product of compact spaces is compact.

Corollary 4.50 *A subset $A \subset \mathbb{R}^n$ is compact, in the Euclidean topology, if and only if A is closed and bounded.*

Proof If $A \subset \mathbb{R}^n$ is closed and bounded it must be contained in $[-a, a]^n$ for some $a > 0$. The interval $[-a, a]$ is homeomorphic to $[0, 1]$, whence compact. By Corollary 4.49 the Cartesian product $[-a, a]^n$ is compact, and so is any closed subspace in it.

Conversely, if $A \subset \mathbb{R}^n$ is compact, the map $d_0 \colon A \to \mathbb{R}$, $d_0(x) = \|x\|$, is continuous and so has a maximum. This means A is bounded. Moreover $\mathbb{R}^n$ is Hausdorff, so A is also closed. $\qquad\square$

Example 4.51 Spheres S^n and closed balls D^n are closed and bounded in Euclidean space, hence compact.

Corollary 4.52 *Let $f \colon X \to Y$ be continuous, X compact and Y Hausdorff. Then f is closed. If f is additionally bijective, then it is a homeomorphism.*

Proof Let $A \subset X$ be closed. Then it is compact, $f(A)$ is compact and hence closed in Y. Every continuous and closed bijection is a homeomorphism. $\qquad\square$

Exercises

4.31 Let $K \subset \mathbb{R}^n$ be a compact subspace and $U \subset \mathbb{R}^n$ open, non-empty and containing K. Prove $U \cap (\mathbb{R}^n - K) \neq \emptyset$.

4.32 Consider the closed unit ball $D^n \subset \mathbb{R}^n$ and an open set $U \subset D^n$ containing the boundary $\partial D^n = S^{n-1}$. Prove there exists a real number $r < 1$ such that $D^n = B(0, r) \cup U$.

4.33 ($\heartsuit$) Prove that a continuous and closed map $f \colon \mathbb{R} \to S^n$ cannot be 1-1.

4.34 Let $K \subset \mathbb{R}^n$ be a compact subset. Prove that

$$X = \{(a_0, \ldots, a_n) \in S^n \mid \exists\, (x_1, \ldots, x_n) \in K,\ a_0 = a_1 x_1 + \cdots + a_n x_n\}$$

is a closed subset of the unit sphere.

4.35 Show that a map between compact Hausdorff spaces is continuous iff its graph is closed in the product.

4.36 (Compactly-generated spaces) A topological space is called **compactly generated** if it is Hausdorff and the family of its compact subspaces is an identification cover.[1] Prove that if every point in a Hausdorff space X has a compact neighbourhood then X is compactly generated.

4.37 (Proper maps) A continuous map $f \colon Y \to Z$ is called **proper** if the pre-image $f^{-1}(K)$ of any compact set $K \subset Z$ is compact.

Prove that a Hausdorff space X is compactly generated (Exercise 4.36) iff every proper map $f \colon Y \to X$ is closed.

---

[1]For more on the topic of compactly-generated spaces and their usefulness in topology we refer to Steenrod's article [Ste67].

4.38 Let Z be a compact Hausdorff space. Prove that a subset $Y \subset Z$ is compact if and only if the projection $X \times Y \to X$ is closed, for any space X.

4.39 ($\heartsuit$) Let (X, d) be a metric space and ε a positive real number. Consider a sequence $x_0, x_1, \ldots, x_n \in X$ such that $d(x_{i-1}, x_i) < 2\varepsilon$ for every $i = 1, \ldots, n$; the endpoints x_0 and x_n are called ε-apart. We may think of such points as the footprints left by a man with legs long ε that wanders through X. Prove that:

1. if X is connected, then any two points of X are ε-apart, whichever $\varepsilon > 0$;
2. give an example of disconnected space in which any two points are ε-apart, whichever $\varepsilon > 0$;
3. if X is compact and two arbitrary points are ε-apart for any $\varepsilon > 0$, then X is connected.

4.40 ($\spadesuit$) Let $(X, \mathcal{T})$ be a Hausdorff space and $\mathcal{K}$ the family of subsets $A \subset X$ such that $A \cap K$ is open in K for any compact set $K \subset X$. Prove that:

1. $\mathcal{K}$ is a finer topology than $\mathcal{T}$; furthermore, $\mathcal{K} = \mathcal{T}$ precisely when $(X, \mathcal{T})$ is compactly generated (Exercise 4.36);
2. the space $(X, \mathcal{K})$ is Hausdorff and compactly generated;
3. let Y be a compactly generated space and $f : Y \to (X, \mathcal{T})$ a continuous map. Then $f : Y \to (X, \mathcal{K})$ is continuous.

The topology $\mathcal{K}$ is called **Kelley extension** of $\mathcal{T}$ and $(X, \mathcal{K})$ the '**Kelleyfication**' of $(X, \mathcal{T})$.

4.6 Topological Groups

A **topological group** is a set G on which a group structure and a topological structure coexist: this means that the product

$$G \times G \to G, \qquad (g, h) \mapsto gh,$$

and inversion

$$G \to G, \qquad g \mapsto g^{-1}$$

are continuous operations. For any given $h \in G$, the map

$$R_h : G \to G \quad \text{defined as} \quad R_h(g) = gh$$

is the composite of the product in G with the inclusion $G \to G \times G$, $g \mapsto (g, h)$. Hence R_h, called **right multiplication by** h, is continuous as composite of continuous functions. Furthermore, R_h is bijective with inverse $R_{h^{-1}}$, hence a homeomorphism. In a similar manner the **left multiplication by** h

$$L_h \colon G \to G \quad \text{defined by} \quad L_h(g) = hg$$

is a homeomorphism with inverse $L_{h^{-1}}$.

Lemma 4.53 *For any g and h in a topological group G there exists a homeomorphism φ from G to G such that $\varphi(g) = h$.*

Proof Consider $\varphi = R_{g^{-1}h}$, or $\varphi = L_{hg^{-1}}$. $\square$

Example 4.54 Every group endowed with the discrete topology becomes a topological group.

Example 4.55 The additive groups $(\mathbb{R}^n, +)$ and $(\mathbb{C}^n, +)$, with the Euclidean topology, are topological groups.

Example 4.56 The group $\mathrm{GL}(n, \mathbb{R})$ is open in the space $\mathbb{R}^{n^2}$ and as such it admits a natural topological structure, with which it becomes a topological group. The entries of the product of two matrices A, B depend in a continuous way on the entries of A and B, and the entries of A^{-1} depend continuously on those of A.

Lemma 4.57 *Let G be a topological group with neutral element e. Then G is Hausdorff if and only if $\{e\}$ is closed.*

Proof One implication is clear, since points are closed in Hausdorff spaces.

Now suppose $\{e\}$ is closed in G and consider the map $\phi \colon G \times G \to G, \phi(g, h) = gh^{-1}$, composite of continuous maps hence itself continuous. Since $\phi(g, h) = e$ iff $g = h$, the pre-image of $\{e\}$ is precisely the diagonal, which is closed. To finish we invoke Theorem 3.69. $\square$

Let's now overview some features of the main linear groups:

$$\mathrm{GL}(n, \mathbb{R}) = \{A \in M_{n,n}(\mathbb{R}) \mid \det(A) \neq 0\}.$$
$$\mathrm{SL}(n, \mathbb{R}) = \{A \in M_{n,n}(\mathbb{R}) \mid \det(A) = 1\}.$$
$$\mathrm{SO}(n, \mathbb{R}) = \{A \in M_{n,n}(\mathbb{R}) \mid AA^T = I, \det(A) = 1\}.$$
$$\mathrm{GL}(n, \mathbb{C}) = \{A \in M_{n,n}(\mathbb{C}) \mid \det(A) \neq 0\}.$$
$$\mathrm{SL}(n, \mathbb{C}) = \{A \in M_{n,n}(\mathbb{C}) \mid \det(A) = 1\}.$$
$$\mathrm{U}(n, \mathbb{C}) = \{A \in M_{n,n}(\mathbb{C}) \mid \overline{A}^T A = I\}.$$
$$\mathrm{SU}(n, \mathbb{C}) = \mathrm{U}(n, \mathbb{C}) \cap \mathrm{SL}(n, \mathbb{C}).$$

They're all subspaces of $\mathbb{R}^{n^2}$ or $\mathbb{C}^{n^2}$, in particular they are metrisable and so Hausdorff.

Proposition 4.58 *The groups $\mathrm{GL}^+(n, \mathbb{R}) = \{A \in \mathrm{GL}(n, \mathbb{R}) \mid \det(A) > 0\}$ and $\mathrm{GL}(n, \mathbb{C})$ are connected.*

Proof We shall prove only that $GL^+(n, \mathbb{R})$ is connected, for $GL(n, \mathbb{C})$ is completely analogous.

The group $GL^+(1, \mathbb{R})$ coincides with the interval $]0, +\infty[$ and is connected; by induction we can suppose $n > 1$ and $GL^+(n-1, \mathbb{R})$ connected. Consider the map $p: M_{n,n}(\mathbb{R}) \to \mathbb{R}^n$ sending a matrix to its first column vector. If we isolate the first column from the others we may write $M_{n,n}(\mathbb{R}) = \mathbb{R}^n \times M_{n,n-1}(\mathbb{R})$, and then p is the projection onto the first factor of the product. This makes p continuous and open, and so its restriction to the open set $GL^+(n, \mathbb{R}) \subset M_{n,n}(\mathbb{R})$ is open; moreover, $p(GL^+(n, \mathbb{R})) = \mathbb{R}^n - \{0\}$.

We claim that the fibres of $p: GL^+(n, \mathbb{R}) \to \mathbb{R}^n - \{0\}$ are connected: from this plus Lemma 4.18 the connectedness of $GL^+(n, \mathbb{R})$ will follow. The fibre $p^{-1}(1, 0, \ldots, 0)$ is the product $\mathbb{R}^{n-1} \times GL^+(n-1, \mathbb{R})$, so connected. Given any $y \in \mathbb{R}^n - \{0\}$ the fibre $p^{-1}(y)$ is non-empty, so let $A \in GL^+(n, \mathbb{R})$ be such that $p(A) = y$. Left-multiplication by A, $L_A(B) = AB$, is a homeomorphism of $GL^+(n, \mathbb{R})$. From the relation $p(AB) = Ap(B)$ follows $L_A\left(p^{-1}(1, 0, \ldots, 0)\right) = p^{-1}(y)$, so the fibres of p are all homeomorphic. $\square$

Corollary 4.59 *The topological groups* $SL(n, \mathbb{R})$ *and* $SL(n, \mathbb{C})$ *are connected, while* $GL(n, \mathbb{R})$ *has two connected components.*

Proof We just saw that $GL(n, \mathbb{C})$ and $GL^+(n, \mathbb{R})$ are connected. Dividing the first column vector by the determinant yields maps $GL(n, \mathbb{C}) \to SL(n, \mathbb{C})$ and $GL^+(n, \mathbb{R}) \to SL(n, \mathbb{R})$, both continuous and onto.

The group $GL(n, \mathbb{R})$ is the disjoint union of the open sets $GL^+(n, \mathbb{R})$ and $GL^-(n, \mathbb{R}) = \det^{-1}(]-\infty, 0[)$, and multiplying by any matrix with negative determinant induces a homeomorphism from $GL^+(n, \mathbb{R})$ to $GL^-(n, \mathbb{R})$. $\square$

For the study of the orthogonal and unitary groups we'll need a preliminary result.

Lemma 4.60 *Let* $f: X \to Y$ *be an onto, continuous map from* X *compact to* Y *connected and Hausdorff. If all fibres* $f^{-1}(y)$ *are connected,* $y \in Y$, *then* X *is connected.*

Proof The map f is closed so we can apply Lemma 4.18. $\square$

Proposition 4.61 *The topological groups* $SO(n, \mathbb{R}), U(n, \mathbb{C}), SU(n, \mathbb{C})$ *are compact and connected.*

Proof We will only discuss $SO(n, \mathbb{R})$, because the cases $U(n, \mathbb{C})$ and $SU(n, \mathbb{C})$ are entirely similar. Consider the continuous mapping

$$M_{n,n}(\mathbb{R}) \to M_{n,n}(\mathbb{R}) \times \mathbb{R}, \qquad A \mapsto (AA^T, \det(A));$$

$SO(n, \mathbb{R})$ is the pre-image of $(I, 1)$ and hence closed in $M_{n,n}(\mathbb{R})$. Since the column vectors of an orthogonal matrix have norm 1, $SO(n, \mathbb{R})$ is contained in the bounded set $\{(a_{ij}) \mid \sum_{i,j} a_{ij}^2 = n\}$, proving its compactness.

The map $p\colon \mathrm{SO}(n, \mathbb{R}) \to S^{n-1}$ sending a matrix to its first column vector is continuous. Moreover, $p(AB) = Ap(B)$ for any $A, B \in \mathrm{SO}(n, \mathbb{R})$, and $L_A(p^{-1}(v)) = p^{-1}(Av)$, which implies that the fibres are all homeomorphic. The fibre over $(1, 0, \ldots, 0)$ is $\mathrm{SO}(n - 1, \mathbb{R})$. Since $\mathrm{SO}(n, \mathbb{R})$ is compact and S^{n-1} is connected Hausdorff, connectedness is proven by induction on n, keeping into account Lemma 4.60 and the fact that $\mathrm{SO}(1, \mathbb{R}) = \{1\}$. $\square$

Exercises

4.41 Prove that

$$\mathrm{SU}(2, \mathbb{C}) \times \mathrm{U}(1, \mathbb{C}) \to \mathrm{U}(2, \mathbb{C}), \qquad (A, z) \mapsto A \cdot \begin{pmatrix} z & 0 \\ 0 & 1 \end{pmatrix}$$

is a homeomorphism.

4.42 Prove that $\mathrm{SL}(n, \mathbb{R})$ and $\mathrm{SL}(n, \mathbb{C})$ are not compact, for any $n > 1$.

4.43 ($\heartsuit$) Prove that for any $n, m \geq 2$ the space

$$X = \{A \in M_{n,n}(\mathbb{R}) \mid A^m = I\}$$

is disconnected.

4.44 Let $A, B \subset \mathbb{R}^n$ be closed sets and call

$$W(A, B) = \{g \in \mathrm{GL}(n, \mathbb{R}) \mid g(A) \cap B \neq \emptyset\}.$$

Suppose A is compact and prove that $W(A, B)$ is closed in $\mathrm{GL}(n, \mathbb{R})$. Show—by finding an example—that $W(A, B)$ might not be closed if neither A nor B are compact.

4.45 Let G be a topological group and $H \subset G$ a subgroup. Prove that H is open and closed in G provided the interior H° is not empty. (Hint: write H and its complement as union of open sets of the form $R_g(H^\circ)$ for suitable $g \in G$.)

4.46 Let G be a topological group with neutral element e. Prove the following facts:

1. if $H \subset G$ is a subgroup, then $\overline{H}$ is a subgroup;
2. the connected component of e in G is a closed subgroup.

4.47 Let H be a closed, connected subgroup in $(\mathbb{R}^n, +)$. Prove H is a vector subspace of $\mathbb{R}^n$. (Hint: show $\overline{H} = \cap_n H_n$, where H_n is the subspace generated by $B(0, 1/n) \cap H$.)

4.48 Let G be a topological group. For subsets $A, B \subset G$ define

$$A^{-1} = \{a^{-1} \mid a \in A\}, \qquad AB = \{ab \mid a \in A, b \in B\}.$$

Prove that $\overline{A} = \cap AU = \cap AU^{-1}$, where U varies among all neighbourhoods of the neutral element.

4.49 Let G be a Hausdorff topological group and $H \subset G$ a discrete subgroup (the points in H are open in the subspace topology). Prove that H is closed in G and compare this fact with Exercise 3.43. (Hint: use Exercise 4.48; show that there's a neighbourhood U of the neutral element e such that $UU^{-1} \cap H = \{e\}$, and also that for every $x \in HU$ there's a unique $h \in H$ such that $x \in hU$.)

4.50 Prove that:

1. any discrete subgroup of the real additive group $(\mathbb{R}, +)$ has the form $\mathbb{Z}a$ for some $a \in \mathbb{R}$;
2. any discrete subgroup of $U(1, \mathbb{C})$ is finite cyclic.

4.7 Exhaustions by Compact Sets

Definition 4.62 An **exhaustion by compact sets** of a topological space X is a sequence of compact subspaces $\{K_n \mid n \in \mathbb{N}\}$ such that:

1. $K_n \subset K_{n+1}^\circ$ for any n;
2. $\cup_n K_n = X$.

For an arbitrary exhaustion by compact sets $\{K_n\}$ the family of interiors $\{K_n^\circ\}$ is an open cover of X, so for any compact $H \subset X$ there's an n such that $H \subset K_n^\circ \subset K_n$.

Example 4.63 For any $n \in \mathbb{N}$ define $K_n = \{(x, y) \mid x^2 + y^2 \leq n^2\} \subset \mathbb{R}^2$. The sequence $\{K_n\}$ is an exhaustion of $\mathbb{R}^2$ by compact sets.

Example 4.64 Let's use the exhaustion of Example 4.63 to show that $\mathbb{R}^2$ isn't homeomorphic to $Y = \mathbb{R}^2 - \{0\}$. We argue by contradiction and assume there is a homeomorphism $f : \mathbb{R}^2 \to Y$. Then the sequence $D_n = f(K_n)$ is an exhaustion of Y by compact sets, and what we saw earlier tells that there's an integer N such that the compact set $S^1 = \{(x, y) \mid x^2 + y^2 = 1\}$ lies inside D_N. The function

$$f : Y \to]0, +\infty[, \qquad f(x, y) = x^2 + y^2,$$

has a maximum $M > 1$ and a minimum $1 > m > 0$ on the compact set D_N. Hence

$$\{(x, y) \in Y \mid x^2 + y^2 < m\} \subset \{(x, y) \in Y - D_N \mid x^2 + y^2 < 1\},$$

$$\{(x, y) \in Y \mid x^2 + y^2 > M\} \subset \{(x, y) \in Y - D_N \mid x^2 + y^2 > 1\},$$

and $Y - D_N$ is the disjoint union of the open non-empty sets

$$\{(x, y) \in Y - D_N \mid x^2 + y^2 < 1\} \quad \text{and} \quad \{(x, y) \in Y - D_N \mid x^2 + y^2 > 1\}.$$

But this is absurd, since $Y - D_N = f(\mathbb{R}^2 - K_N)$ and $\mathbb{R}^2 - K_N$ is connected.

Using exhaustions by compact sets we can, in some sense, describe the 'behaviour at infinity' of topological spaces.

The routine method for capturing the behaviour at infinity of a space X is to look at its **one-point compactification** (or **Alexandrov compactification**) $\widehat{X}$, which is the following.

Take a topological space X and a point ∞ not belonging in X. Define $\widehat{X} = X \cup \{\infty\}$ and consider on $\widehat{X}$ the family

$$\mathcal{T} = \{A \mid A \text{ open in } X\} \cup \{\widehat{X} - K \mid K \text{ closed and compact in } X\}.$$

It's easy to see that $\mathcal{T}$ defines a topology and the natural inclusion $X \hookrightarrow \widehat{X}$ is an open immersion. We want to prove that $\widehat{X}$ *is a compact space*. If $\widehat{X} = \cup U_i$ is an open cover, there is an index, say 0, such that $\infty \in U_0$, and so $\cup_{i \neq 0} U_i$ covers the compact set $K = \widehat{X} - U_0$. If $U_1 \cup \cdots \cup U_n$ is a finite subcover of K then $\widehat{X} = U_0 \cup U_1 \cup \cdots \cup U_n$.

Proposition 4.65 *Retaining the previous notation, $\widehat{X}$ is Hausdorff if and only if X is Hausdorff and every point in X has a compact neighbourhood.*

Proof X is open in $\widehat{X}$, so any two points have disjoint neighbourhoods in X if and only if they have disjoint neighbourhoods in $\widehat{X}$. Take $x \in X$: then x, ∞ have disjoint neighbourhoods in $\widehat{X}$ iff there exist a closed, compact $K \subset X$ and an open U such that $x \in U$ and $U \cap (\widehat{X} - K) = \emptyset$. Fix K; an open U as above exists iff x belongs to the interior of K, so x, ∞ have disjoint neighbourhoods in $\widehat{X}$ iff there's a closed, compact $K \subset X$ such that $x \in K^\circ$. $\qquad\square$

Proposition 4.66 *Let $f : X \to Y$ be an open immersion of Hausdorff spaces. Then*

$$g : Y \to \widehat{X}, \qquad g(y) = \begin{cases} x & \text{if } y = f(x), \\ \infty & \text{if } y \notin f(X), \end{cases}$$

is continuous. In particular, a compact Hausdorff space Y coincides with the one-point compactification of $Y - \{y\}$, for any $y \in Y$.

Proof Take U open in $\widehat{X}$. If $U \subset X$ then $g^{-1}(U) = f(U)$. If, instead, $U = \widehat{X} - K$ for some compact set $K \subset X$, then $g^{-1}(U) = Y - f(K)$.

In the special situation where Y is compact and Hausdorff, $X = Y - \{y\}$ and f is the inclusion, the map g is a continuous bijection between compact Hausdorff spaces, hence a homeomorphism. $\qquad\square$

Exercises

4.51 ($\heartsuit$) Use the exhaustion of Example 4.63 to show that $\mathbb{R}^2$ is not homeomorphic to $\mathbb{R} \times [0, 1]$.

4.52 Prove that $\mathbb{R} \times [0, 1]$ is not homeomorphic to $\mathbb{R} \times [0, +\infty[$.

4.53 Prove that the open cylinder $\{x^2 + y^2 < 1\} \subset \mathbb{R}^3$ isn't homeomorphic to the closed cylinder $\{x^2 + y^2 \leq 1\} \subset \mathbb{R}^3$.

4.54 ($\heartsuit$) Prove the following subspaces of $\mathbb{R}^3$ are not homeomorphic:

$$X = \{(x, y, z) \mid z > 0\} \cup \{(x, y, z) \mid x^2 + y^2 < 1, \ z = 0\};$$

$$Y = \{(x, y, z) \mid z > 0\} \cup \{(x, y, z) \mid x^2 + y^2 > 1, \ z = 0\}.$$

4.55 Interpret the space of Exercise 3.3 as a one-point compactification.

4.56 Let $f: X \to Y$ be a proper map (Exercise 4.37). Prove that f extends naturally to a continuous map $\hat{f}: \widehat{X} \to \widehat{Y}$.

4.57 ($\clubsuit$, $\heartsuit$) Let $U \subset S^n$ be an open subset of the sphere homeomorphic to $\mathbb{R}^n$. Show that the complement $S^n - U$ is connected.

Reference

[Ste67] Steenrod, N.E.: A convenient category of topological spaces. Mich. Math. J. **14**, 133–152 (1967)

Chapter 5
Topological Quotients

5.1 Identifications

The dual notion to an immersion is that of an identification.[1]

Definition 5.1 A continuous and onto map $f \colon X \to Y$ is called an **identification** if the open sets of Y are precisely the subsets $A \subset Y$ such that $f^{-1}(A)$ is open in X.

As f^{-1} commutes with complementation of sets, a continuous onto map $f \colon X \to Y$ is an identification if and only if closed sets in Y are those of the form $C \subset Y$ for $f^{-1}(C)$ closed in X, and conversely.

Let $f \colon X \to Y$ be a continuous map; we remind that a subset $A \subset X$ is called saturated by f (or f-saturated) whenever $x \in A$, $y \in X$ and $f(x) = f(y)$ imply $y \in A$. Equivalently, f-saturated sets have the form $f^{-1}(B)$ for some $B \subset Y$.

Saying that f is an identification amounts to saying that open sets in Y are exactly the images $f(A)$ of saturated sets A. Warning: not every open set in X is saturated, so identifications may not be open maps.

Example 5.2 Let $\mathcal{A}$ be a cover of the space X and consider the disjoint union $Y = \coprod \{A \mid A \in \mathcal{A}\}$. The natural map $Y \to X$ is continuous, onto and an identification if and only if $\mathcal{A}$ is an identification cover.

Definition 5.3 A **closed identification** is an identification and a closed map. An **open identification** is an identification together with an open map.

An identification can be at the same time open and closed, like for instance homeomorphisms.

Lemma 5.4 *Let* $f \colon X \to Y$ *be continuous and onto. If* f *is closed, it is a closed identification. If* f *is open, it is an open identification.*

---

[1]The name *identification* is not used by everybody: *quotient map* and *proclusion* are employed at times to indicate the same.

© Springer International Publishing Switzerland 2015
M. Manetti, *Topology*, UNITEXT - La Matematica per il 3+2 91,
DOI 10.1007/978-3-319-16958-3_5

Proof f being onto implies $f(f^{-1}(A)) = A$ for any $A \subset Y$. Therefore if f is open and $A \subset Y$ is such that $f^{-1}(A)$ is open, then $A = f(f^{-1}(A))$ is open, making f an identification. The same argument goes through for closed maps. $\qquad\square$

Example 5.5 The mapping

$$f: [0, 2\pi] \to S^1 = \{(x, y) \mid x^2 + y^2 = 1\}, \qquad f(t) = (\cos(t), \sin(t)),$$

is a closed identification: it's onto and continuous, from a compact set to a Hausdorff space, hence closed. Had we wanted to prove directly that f is closed, we would've noted that the restrictions

$$f: [0, \pi] \to S^1 \cap \{y \geq 0\}, \qquad f: [\pi, 2\pi] \to S^1 \cap \{y \leq 0\},$$

are homeomorphisms, so that $C \subset [0, 2\pi]$ closed implies that both

$$f(C) \cap \{y \geq 0\} = f(C \cap [0, \pi]), \qquad f(C) \cap \{y \leq 0\} = f(C \cap [\pi, 2\pi]),$$

are closed; then $f(C)$ is closed, too. Notice that f isn't an open map, for $[0, 1[$ is open in $[0, 2\pi]$ but its image inside S^1 isn't.

Lemma 5.6 (Universal property of identifications) *Let $f: X \to Y$ be an identification and $g: X \to Z$ a continuous map. There exists a continuous mapping $h: Y \to Z$ such that $g = hf$ if and only if g is constant on the fibres of f.*

$$\begin{array}{ccc} X & \xrightarrow{\;g\;} & Z \\ {\scriptstyle f}\downarrow & \nearrow & \\ & {\scriptstyle h} & \\ Y & & \end{array}$$

Proof What 'g constant on the fibres of f' means is that $f(x) = f(y)$ implies $g(x) = g(y)$. This condition is clearly necessary for the existence of h; we'll prove its sufficiency.

As f is onto, we can define $h: Y \to Z$ by $h(y) = g(x)$, where $x \in X$ is any point such that $f(x) = y$: by assumption h is well defined, and $g = hf$. We claim h is continuous. Let $U \subset Z$ be open, then $f^{-1}(h^{-1}(U)) = g^{-1}(U)$ is open in X, and since f is an identification $h^{-1}(U)$ is open in Y. $\qquad\square$

Example 5.7 Let's describe the n-dimensional unit ball and sphere as

$$D^n = \{x \in \mathbb{R}^n \mid \|x\|^2 \leq 1\}, \qquad S^n = \{(x, y) \in \mathbb{R}^n \times \mathbb{R} \mid \|x\|^2 + y^2 = 1\},$$

and consider

$$f: D^n \to S^n, \qquad f(x) = (2x\sqrt{1 - \|x\|^2}, 2\|x\|^2 - 1).$$

Such a map is continuous and onto. Its restriction

$$f: \{x \in D^n \mid \|x\| < 1\} \to \{(x, y) \in S^n \mid y < 1\}$$

is a homeomorphism, and furthermore $f(\partial D^n) = (0, 1) \in S^n$; in practice f contracts the boundary of the ball to a single point on the sphere. Since D^n is compact and S^n Hausdorff, f is a closed identification.

Exercises

5.1 Prove that identifications are preserved under composition of maps.

5.2 Let $f: X \to Y$ be an identification and $g: Y \to Z$ a continuous map. If $gf: X \to Z$ is an identification, show that g is an identification as well.

5.3 ($\heartsuit$) Consider two open identifications $f: X \to Y$ and $g: Z \to W$. Show that

$$f \times g: X \times Z \to Y \times W, \qquad (f \times g)(x, z) = (f(x), g(z)),$$

is an open identification. (Warning: if f and g aren't both open, $f \times g$ might not be an identification, see Exercise 5.12.)

5.4 ($\heartsuit$) Let $f: X \to Y$ be an identification. Show that if the connected components of X are open, so are the connected components of Y.

5.5 Let $f: X \to Y$ be an identification all of whose fibres $f^{-1}(y)$ are connected. Prove that every open, closed and non-empty subset of X is saturated. Conclude that Y connected implies X connected.

5.6 ($\heartsuit$) Let $f: X \to Y$ be an open identification and $A \subset X$ a saturated set; prove that A^o and $\overline{A}$ are saturated. Find an example showing that this not always the case if f is a closed identification.

5.7 ($\maltese$) Let X, Y, Z be Hausdorff spaces, $f: X \to Y$ an open identification, $g: X \to Z$ a continuous map and $D \subset Y$ a dense subset. Suppose g is constant on $f^{-1}(y)$ for any $y \in D$. Prove that there exists $h: Y \to Z$ such that $g = hf$. (Hint: show that the map

$$\{(x_1, x_2) \in X \times X \mid f(x_1) = f(x_2)\} \to X, \qquad (x_1, x_2) \mapsto x_1,$$

is open and use the results of Exercises 3.26 and 3.57.)

5.2 Quotient Topology

Consider a space X, a set Y and a surjective map $f: X \to Y$. As f^{-1} commutes with unions and intersections the family of subsets $A \subset Y$ such that $f^{-1}(A)$ is open in X defines a topology on Y, called the **quotient topology** (or identification topology) with respect to f.

The quotient topology is the only topology on Y that makes f an identification, and the finest topology for which f is continuous.

Now imagine having an equivalence relation $\sim$ on X. Denote by $X/\sim$ the set of equivalence classes and by $\pi: X \to X/\sim$ the surjection mapping $x \in X$ to its class $\pi(x) = [x]$. The topological space $X/\sim$, equipped with the quotient topology of π, is called **quotient (space)**.

Proposition 5.8 *Let $f: X \to Y$ be a continuous map, $\sim$ an equivalence relation on X and $\pi: X \to X/\sim$ the canonical quotient map. There exists a continuous mapping $g: X/\sim \to Y$ such that $g\pi = f$ if and only if f is constant on equivalence classes.*

Proof Immediate consequence of Lemma 5.6.   □

Example 5.9 Let $f: X \to Y$ be continuous, and for any $x, y \in X$ define $x \sim y$ if $f(x) = f(y)$. The relation $\sim$ is an equivalence relation, and Proposition 5.8 guarantees f induces a continuous 1-1 map $\bar{f}: X/\sim \to Y$ on the quotient. From the definition of quotient topology it follows immediately that f is an identification if and only if $\bar{f}$ is a homeomorphism.

A particularly interesting class of quotients is produced by so-called **contractions**: let X be a space, $A \subset X$ a subset and consider the smallest equivalence relation $\sim$ on X having A as equivalence class. Precisely: for any $x, y \in X$

$$x \sim y \quad \text{if and only if} \quad x = y \text{ or } x, y \in A.$$

It is common to denote with X/A the quotient of X by the relation $\sim$; the idea is that the canonical quotient map $X \to X/A$ contracts, or shrinks, A to a point.

Example 5.10 The mapping

$$f: S^{n-1} \times [0, 1] \to D^n, \qquad f(x, t) = tx,$$

is continuous and onto. As $S^{n-1} \times [0, 1]$ is compact and D^n Hausdorff, f is closed and hence an identification. Now, $f(x) = f(y)$ iff $x = y$ or $x, y \in S^{n-1} \times \{0\}$. By Example 5.9 f induces a homeomorphism between the quotient $S^{n-1} \times [0, 1]/S^{n-1} \times \{0\}$ and the ball D^n.

Example 5.11 Look at the hypercube $I^n = [0, 1]^n \subset \mathbb{R}^n$ and contract its boundary $\partial I^n = [0, 1]^n - \,]0, 1[^n$ to a point. The quotient $I^n/\partial I^n$ is homeomorphic to the sphere S^n, because I^n is homeomorphic to D^n and we saw in Example 5.7 that there is an identification $D^n \to S^n$ contracting the ball's boundary to a point.

Example 5.12 The Möbius strip can be physically constructed by taking a rectangular strip of flexible material (such as paper, leather, thin plate etc.) and glueing two opposite sides after twisting one.

Topologically speaking, the Möbius strip M arises from the (closed) square $[0, 1] \times [0, 1] \subset \mathbb{R}^2$ by identifying $(0, y)$ with $(1, 1 - y)$, for every $y \in [0, 1]$: to visualise the identification better you might want to look at the sewing pattern of Fig. 1.5.

Example 5.13 The Klein bottle may be defined as the quotient of the square $[0, 1] \times [0, 1] \subset \mathbb{R}^2$ by the relation that identifies $(0, y) \sim (1, 1 - y)$ and $(x, 0) \sim (x, 1)$, for any $x, y \in [0, 1]$.

Let us finally remark that the quotient of a compact space is compact and the quotient of a connected space is connected.

This is no longer true for Hausdorff spaces (see Exercise 5.8). If $f : X \to Z$ is an identification, Z is Hausdorff precisely when, for any $x, y \in X$ with $f(x) \neq f(y)$, there exist open, saturated and disjoint sets $A, B \subset X$ with $x \in A$ and $y \in B$. By Exercise 5.6 if f is open the requirement that A and B be open may be replaced by that of being neighbourhoods of x and y respectively. The picture in the compact case is even simpler.

Theorem 5.14 *Let $f : X \to Y$ be an identification, where X is compact and Hausdorff. The following statements are equivalent:*

1. *Y is Hausdorff;*
2. *f is a closed identification;*
3. *$K = \{(x_1, x_2) \in X \times X \mid f(x_1) = f(x_2)\}$ is closed in the product.*

Proof The implication $(1) \Rightarrow (3)$ follows because K is the pre-image of the diagonal under the continuous map $f \times f : X \times X \to Y \times Y$.

Now let's show that (3) implies (2), i.e. that $f^{-1}(f(A))$ is closed for any closed $A \subset X$. As X is compact the projections $p_1, p_2 : X \times X \to X$ are closed, and the formula

$$f^{-1}(f(A)) = p_1(K \cap p_2^{-1}(A))$$

implies that $f^{-1}(f(A))$ is closed provided A and K are.

To finish we need to show $(2) \Rightarrow (1)$: f is closed and onto and points in X are closed, so the points of Y are closed. Take $a \neq b \in Y$. Then the closed sets $A = f^{-1}(a)$, $B = f^{-1}(b)$ are compact and disjoint. But X is Hausdorff, so Wallace's theorem gives open sets $U, V \subset X$ with

$$A \times B \subset U \times V \subset X \times X - \Delta,$$

i.e. $A \subset U, B \subset V, U \cap V = \emptyset, (X - U) \cup (X - V) = X$. By assumption f is closed, so $U' = Y - f(X - U), V' = Y - f(X - V)$ are open disjoint neighbourhoods of a and b respectively. $\qquad\square$

Exercises

5.8 ($\heartsuit$) Consider the equivalence relation

$$x \sim y \text{ if } x = y \text{ or if } |x| = |y| > 1$$

defined on $\mathbb{R}$ with the Euclidean topology. Prove that the quotient $\mathbb{R}/\sim$ is not Hausdorff, and any of its points has an open neighbourhood homeomorphic to the interval $]-1, 1[$.

5.9 Prove, as A varies among the subsets of $[0, 1]$ made by two distinct points, that the quotient space $[0, 1]/A$ can be of three distinct homeomorphism-types.

5.10 Let $D^2 = \{x^2 + y^2 \le 1\} \subset \mathbb{R}^2$. Prove that $\mathbb{R}^2/D^2$ is homeomorphic to $\mathbb{R}^2$.

5.11 Let X be Hausdorff, $K \subset X$ compact and X/K the contraction of K to a point. Prove the following facts:

1. X/K is Hausdorff;
2. (Excision property.) Let A be a open set of X contained in K. Then the natural map $(X - A)/(K - A) \to X/K$ is a homeomorphism;
3. if X is compact, X/K coincides with the one-point compactification of $X - K$.

5.12 ($\heartsuit$) Consider an identification $f : \mathbb{R} \to X$ such that $f(x) = f(y)$ if and only if $x = y$ or $x, y \in \mathbb{Z}$. Call $C \subset \mathbb{R} \times \mathbb{Q}$ the union of the lines $x + y = n + \dfrac{\sqrt{2}}{n}$, as $n \in \mathbb{N}$ varies. Prove that:

1. f is a closed identification;
2. C is closed and saturated by

$$f \times Id : \mathbb{R} \times \mathbb{Q} \to X \times \mathbb{Q};$$

3. $(f \times Id)(C)$ is not closed in the product topology; more precisely, the point $(f(0), 0)$ belongs to the closure of $(f \times Id)(C)$;
4. $f \times Id$ is not an identification.

5.3 Quotients by Groups of Homeomorphisms

We indicate with $\text{Homeo}(X)$ the set of homeomorphisms of a topological space X to itself. The composition of maps induces on $\text{Homeo}(X)$ a group structure, where the identity map is the neutral element.

Let $G \subset \text{Homeo}(X)$ be a subgroup. Then

$$x \sim y \quad \text{if there exists } g \in G \text{ such that } y = g(x),$$

for $x, y \in X$, defines an equivalence relation on X, whose cosets are called **G-orbits**, and whose quotient space is written X/G.

Proposition 5.15 *Let* $G \subset \text{Homeo}(X)$ *be a group of homeomorphisms of a space* X, *and* $\pi \colon X \to X/G$ *the canonical quotient map. Then* π *is an open map. If* G *is a finite group,* π *is closed as well.*

Proof For any subset $A \subset X$

$$\pi^{-1}(\pi(A)) = \cup\{g(A) \mid g \in G\}.$$

If A is open, $g(A)$ is open for any $g \in G$, so $\pi^{-1}(\pi(A))$ is a union of open sets, hence open; by definition of quotient topology $\pi(A)$ is open.

In case G is finite the same argument with A closed remains valid. □

Although even the quotient of a Hausdorff space by some group of homeomorphisms may not be Hausdorff (cf. Exercise 5.13), we have a better arsenal to decide whether a given quotient $X/\sim$ is Hausdorff in case we mod out by $G \subset \text{Homeo}(X)$.

Proposition 5.16 *Let* G *be a group of homeomorphisms of* X. *The quotient* X/G *is Hausdorff if and only if*

$$K = \{(x, g(x)) \in X \times X \mid x \in X, \, g \in G\}$$

is closed in the product.

Proof The canonical map $\pi \colon X \to X/G$ is open and surjective. Hence

$$p \colon X \times X \to X/G \times X/G, \qquad p(x, y) = (\pi(x), \pi(y)),$$

is open and onto, and so an identification.

Observe that $p(x, y)$ belongs in the diagonal Δ of $X/G \times X/G$ iff x and y live in the same G-orbit, i.e. $(x, y) \in K$. This proves that $p^{-1}(\Delta) = K$, and because p is an identification, Δ is closed if and only if K is closed. □

Proposition 5.17 *Let* G *be a group of homeomorphisms of a Hausdorff space* X *and* $\pi \colon X \to X/G$ *the canonical map. Suppose there is an open subset* $A \subset X$ *such that:*

1. *the quotient map* $\pi \colon A \to X/G$ *is onto;*
2. $\{g \in G \mid g(A) \cap A \neq \emptyset\}$ *is finite.*

Then X/G *is Hausdorff.*

Proof Denote by $g_1, \ldots, g_n$ the elements of G such that $g(A) \cap A \neq \emptyset$. Fix $p, q \in X/G$ distinct and choose two points $x, y \in A$ in their orbits: $\pi(x) = p$, $\pi(y) = q$. As X is Hausdorff, for any $i = 1, \ldots, n$ we can find an open pair (U_i, V_i) in X with $x \in U_i$, $g_i(y) \in V_i$ and $U_i \cap V_i = \emptyset$. Consider now the open sets

$$U = A \cap \bigcap_{i=1}^{n} U_i, \qquad V = A \cap \bigcap_{i=1}^{n} g_i^{-1}(V_i)$$

and we will show $U \cap g(V) = \emptyset$ for any $g \in G$. If $g(A) \cap A = \emptyset$, then by $U, V \subset A$ we have $U \cap g(V) \subset A \cap g(A) = \emptyset$. Instead if $g = g_i$ for some i, from $U \subset U_i$ and $V \subset g^{-1}(V_i)$ we infer $U \cap g(V) \subset U_i \cap V_i = \emptyset$. But $x \in U$ and $y \in V$, so to show that x, y belong to disjoint, open, saturated sets we might as well prove

$$\left(\bigcup_{g \in G} g(U) \right) \cap \left(\bigcup_{h \in G} h(V) \right) = \bigcup_{g,h \in G} (g(U) \cap h(V)) = \emptyset.$$

If we had $g(U) \cap h(V) \neq \emptyset$ for some $g, h \in G$ then

$$U \cap g^{-1}h(V) = g^{-1}(g(U) \cap h(V)) \neq \emptyset,$$

contradicting what was proved above. $\qquad\square$

Example 5.18 Let $G \subset \text{Homeo}(\mathbb{R}^n)$ be the group of translations by vectors with integer components, i.e. $g \in G$ iff there exists $a \in \mathbb{Z}^n$ with $g(x) = x + a$ for every $x \in \mathbb{R}^n$. The quotient $\mathbb{R}^n / G$ is Hausdorff: take

$$A = \{(x_1, \ldots, x_n) \in \mathbb{R}^n \mid |x_i| < 1 \text{ for any } i\},$$

then $A \to \mathbb{R}^n / G$ is onto and, if $a = (a_1, \ldots, a_n) \in \mathbb{Z}^n$, then $A \cap (A + a) \neq \emptyset$ only if $|a_i| \leq 1$ for any i. Now apply Proposition 5.17.

Corollary 5.19 *Let G be a finite group of homeomorphisms of a Hausdorff space X. Then X/G is Hausdorff.*

Proof Apply Proposition 5.17 to the open set $A = X$. $\qquad\square$

Exercises

5.13 Prove that the quotient $\mathbb{R}^n / \text{GL}(n, \mathbb{R})$ is not Hausdorff, for any integer $n > 0$.

5.14 Let X be Hausdorff, $G \subset \text{Homeo}(X)$. Suppose that for any $x, y \in X$ there exist neighbourhoods U of x and V of y such that $g(U) \cap V \neq \emptyset$ for at most finitely many $g \in G$. Prove that X/G is Hausdorff. (Hint: mimic the proof of Proposition 5.17.)

5.15 ($\heartsuit$) Prove that

$$e^{2\pi i -} : \mathbb{R} \to S^1, \qquad x \mapsto e^{2\pi i x} = (\cos(2\pi x), \sin(2\pi x))$$

is an identification that induces a homeomorphism $\mathbb{R}/\mathbb{Z} \cong S^1$. Here $\mathbb{Z}$ is understood as the subgroup of homeomorphisms of $\mathbb{R}$ spanned by the translation $x \mapsto x + 1$. Show that the map $e^{2\pi i -}$ is open but not closed.

5.16 Take X Hausdorff and $g \colon X \to X$ a homeomorphism. Let $G \simeq \mathbb{Z}$ indicate the group of homeomorphisms of $\mathbb{R} \times X$ generated by $(t, x) \mapsto (t + 1, g(x))$.
   Show that the quotient $(\mathbb{R} \times X)/G$ is a Hausdorff space.

5.4 Projective Spaces

Amongst the foremost examples of quotient spaces by groups of homeomorphisms we must include the real projective spaces $\mathbb{P}^n(\mathbb{R})$. By definition $\mathbb{P}^n(\mathbb{R})$ is the quotient of $\mathbb{R}^{n+1} - \{0\}$ by the group of homotheties (dilations): in other words the quotient under the equivalence relation

$$x \sim y \iff x = \lambda y \text{ for some } \lambda \in \mathbb{R} - \{0\}.$$

Note that there's a natural 1-1 correspondence between $\mathbb{P}^n(\mathbb{R})$ and the set of 1-dimensional vector subspaces (lines) of $\mathbb{R}^{n+1}$ through the origin. Given a vector $(x_0, \ldots, x_n) \in \mathbb{R}^{n+1} - \{0\}$, the equivalence class in $\mathbb{P}^n(\mathbb{R})$ is usually written $[x_0, \ldots, x_n]$.

We put on $\mathbb{P}^n(\mathbb{R})$ the quotient topology; by Proposition 5.15 the canonical map $\pi \colon \mathbb{R}^{n+1} - \{0\} \to \mathbb{P}^n(\mathbb{R})$ is an open identification.

Composing the inclusion $i \colon S^n \to \mathbb{R}^{n+1} - \{0\}$ and π produces a continuous, onto map that factorises through a continuous bijection $f \colon S^n / \sim \to \mathbb{P}^n(\mathbb{R})$, where for $x, y \in S^n$ we set $x \sim y$ iff $x = \pm y$. We claim that f is a homeomorphism; to this end let's consider the commutative diagram

$$
\begin{array}{ccccc}
S^n & \xrightarrow{i} & \mathbb{R}^{n+1} - \{0\} & \xrightarrow{r} & S^n \\
\downarrow{\pi'} & & \downarrow{\pi} & & \downarrow{\pi'} \\
S^n/\sim & \xrightarrow{f} & \mathbb{P}^n(\mathbb{R}) & \xrightarrow{f^{-1}} & S^n/\sim
\end{array}
\qquad r(x) = \frac{x}{\|x\|}.
$$

Both the inclusion i and the map r are continuous, so also πi and $\pi' r$ are continuous. By Proposition 5.8, then, f and f^{-1} are continuous.

Because $S^n / \sim = S^n / \pm Id$, $\pm Id \subset \mathrm{Homeo}(S^n)$ being the subgroup of order two formed by the identity and the antipodal map, Corollary 5.19 implies that the real projective spaces are Hausdorff.

Example 5.20 Consider the space $X = D^2 / \sim$ obtained from the two-dimensional disc $D^2 = \{x \in \mathbb{R}^2 \mid \|x\| \leq 1\}$ by modding out the relation that identifies antipodal points on the boundary:

$$x \sim y \iff x = y \quad \text{or} \quad \|x\| = \|y\| = 1, \ x = -y.$$

The aim is to show that X is homeomorphic to the real projective plane $\mathbb{P}^2(\mathbb{R})$. For starters, we identify the disc D^2 with the upper hemisphere

$$S_+^2 = \{(x, y, z) \in \mathbb{R}^3 \mid x^2 + y^2 + z^2 = 1, \ z \geq 0\} \subset S^2$$

using the homeomorphism $S_+^2 \to D^2$, $(x, y, z) \mapsto (x, y)$. Thus X is homeomorphic to $S_+^2 / \sim$, where $\sim$ is the antipodal identification of $(x, y, 0) \in S_+^2$ with $(-x, -y, 0)$. Call $i \colon S_+^2 \to S^2$ the inclusion and $\pi \colon S^2 \to \mathbb{P}^2(\mathbb{R})$, $p \colon S_+^2 \to X$ the canonical quotients. Then we have a commutative diagram of continuous maps

$$
\begin{array}{ccc}
S^2_+ & \xrightarrow{\ i\ } & S^2 \\
\downarrow p & & \downarrow \pi \\
X & \xrightarrow{\ j\ } & \mathbb{P}^2(\mathbb{R})
\end{array}
$$

Our j is bijective and maps a compact space to a Hausdorff space, so j is a homeomorphism.

Example 5.21 For every index $i = 0, \ldots, n$ the subset

$$A_i = \{[x_0, \ldots, x_n] \in \mathbb{P}^n(\mathbb{R}) \mid x_i \neq 0\}$$

is an open set homeomorphic to $\mathbb{R}^n$ and therefore $\{A_0, \ldots, A_n\}$ is an identification cover. In fact, if we denote by $\pi \colon \mathbb{R}^{n+1} - \{0\} \to \mathbb{P}^n(\mathbb{R})$ the quotient map,

$$\pi^{-1}(A_i) = \{(x_0, \ldots, x_n) \in \mathbb{R}^{n+1} - \{0\} \mid x_i \neq 0\}$$

is open. The map

$$f \colon \mathbb{R}^n \to A_0, \qquad f(y_1, \ldots, y_n) = [1, y_1, \ldots, y_n]$$

is continuous and bijective. In addition,

$$f^{-1}\pi \colon \pi^{-1}(A_0) \to \mathbb{R}^n, \qquad f^{-1}\pi(x_0, x_1, \ldots, x_n) = \left(\frac{x_1}{x_0}, \ldots, \frac{x_n}{x_0}\right),$$

is continuous, and by the universality of identifications f^{-1} is continuous. Eventually, a simple permutation of the indices gives, for any i, a homeomorphism $A_0 \simeq A_i$.

Complex projective spaces are defined in a similar way. $\mathbb{P}^n(\mathbb{C})$ is the quotient of $\mathbb{C}^{n+1} - \{0\}$ by the equivalence relation

$$x \sim y \quad \Longleftrightarrow \quad \text{there exists } \lambda \in \mathbb{C} - \{0\} \text{ such that } x = \lambda y,$$

so $\mathbb{P}^n(\mathbb{C}) = (\mathbb{C}^{n+1} - \{0\})/G$, where G is the group of complex homotheties.

We endow $\mathbb{P}^n(\mathbb{C})$ with the quotient topology.

Proposition 5.22 *Projective spaces, both complex and real, are connected, compact and Hausdorff.*

Proof The projections

$$S^n \to \mathbb{P}^n(\mathbb{R}), \qquad S^{2n+1} = \{z \in \mathbb{C}^{n+1} \mid \|z\| = 1\} \to \mathbb{P}^n(\mathbb{C})$$

are continuous and onto, proving that the spaces are compact and connected.

We already know real projective spaces are Hausdorff, so let us consider the complex case. By Proposition 5.16 $\mathbb{P}^n(\mathbb{C})$ is Hausdorff if and only if

$$K = \{(x, y) \in (\mathbb{C}^{n+1} - \{0\}) \times (\mathbb{C}^{n+1} - \{0\}) \mid x = \lambda y\}$$

is closed. Viewing any point in $\mathbb{C}^{n+1}$ as a column vector we end up identifying $(\mathbb{C}^{n+1} - \{0\}) \times (\mathbb{C}^{n+1} - \{0\})$ with the space of $(n + 1) \times 2$-matrices with non-zero columns. Therefore K is the intersection of $(\mathbb{C}^{n+1} - \{0\}) \times (\mathbb{C}^{n+1} - \{0\})$ with matrices of rank one. This set is closed because it's defined by the vanishing of all 2×2-minors. $\qquad\square$

Exercises

5.17 Prove that the quotients maps

$$S^n \to \mathbb{P}^n(\mathbb{R}), \qquad S^{2n+1} = \{z \in \mathbb{C}^{n+1} \mid \|z\| = 1\} \to \mathbb{P}^n(\mathbb{C}),$$

are closed identifications.

5.18 Verify that $\mathbb{P}^1(\mathbb{R}) \to S^1$,

$$[x_0, x_1] \mapsto \left(\frac{x_0^2 - x_1^2}{x_0^2 + x_1^2}, \frac{2x_0 x_1}{x_0^2 + x_1^2} \right),$$

is a homeomorphism.

5.19 Verify that $\mathbb{P}^1(\mathbb{C}) \to S^2$,

$$[z_0, z_1] \mapsto \left(\frac{|z_0|^2 - |z_1|^2}{|z_0|^2 + |z_1|^2}, \frac{\bar{z}_0 z_1 - \bar{z}_1 z_0}{|z_0|^2 + |z_1|^2}, \frac{\bar{z}_0 z_1 + \bar{z}_1 z_0}{|z_0|^2 + |z_1|^2} \right),$$

is a homeomorphism.

5.20 A projective transformation of $\mathbb{P}^n(\mathbb{R})$ is a map $\mathbb{P}^n(\mathbb{R}) \to \mathbb{P}^n(\mathbb{R})$ induced by a 1-1 linear map $\mathbb{R}^{n+1} \to \mathbb{R}^{n+1}$ when passing to the quotient. Prove that projective transformations are homeomorphisms.

5.21 Consider the subset $Z \subset \mathbb{P}^1(\mathbb{R}) \times \mathbb{R}^4$ of elements $([x_0, x_1], (a_1, a_2, a_3, a_4))$ with

$$x_0^4 + a_1 x_0^3 x_1 + a_2 x_0^2 x_1^2 + a_3 x_0 x_1^3 + a_4 x_1^4 = 0$$

Prove that Z is closed in the product and conclude that the set of vectors $(a_1, a_2, a_3, a_4) \in \mathbb{R}^4$ for which the monic polynomial $t^4 + a_1 t^3 + a_2 t^2 + a_3 t + a_4$ has at least one real root is closed in $\mathbb{R}^4$.

5.22 (Graßmannians, ☙) Given integers $0 < k < n$ we denote with $G(k, \mathbb{R}^n)$ the set of linear subspaces in $\mathbb{R}^n$ of dimension k, and with

$$X_{n,k} = \{A \in M_{n,k}(\mathbb{R}) \mid \text{rank}(A) = k\}$$

the set of $n \times k$-matrices with maximal rank. There is an obvious onto map $\pi \colon X_{n,k} \to G(k, \mathbb{R}^n)$ that sends a matrix to the subspace spanned by its column vectors. Let's put on $X_{n,k}$ the subspace topology, and on $G(k, \mathbb{R}^n)$ the quotient topology. Prove that:

1. the map $\text{GL}^+(n, \mathbb{R}) \to X_{n,k}$ giving the first k columns of a matrix is open and onto. Deduce that $G(k, \mathbb{R}^n)$ is a connected space;
2. $\pi(A) = \pi(B)$ if and only if $A = BC$ for some $C \in \text{GL}(k, \mathbb{R})$. Conclude that π is an open identification;
3. the obvious composite

$$\text{SO}(n, \mathbb{R}) \to \text{GL}^+(n, \mathbb{R}) \to X_{n,k} \to G(k, \mathbb{R}^n)$$

   is onto; deduce that $G(k, \mathbb{R}^n)$ is compact;
4. the space $G(k, \mathbb{R}^n)$ is Hausdorff (Hint: for a given pair $(A, B) \in X_{n,k} \times X_{n,k}$ we have $\pi(A) = \pi(B)$ iff the matrix $(A, B) \in M_{n,2k}$ has rank $\leq k$);
5. given a subspace $H \subset \mathbb{R}^n$ of dimension $n - k$, the set $V \in G(k, \mathbb{R}^n)$ of subspaces such that $V \cap H = 0$ is open and dense in $G(k, \mathbb{R}^n)$.

Eventually, use minors to define a continuous map $X_{n,k} \to \mathbb{R}^N - \{0\}$, with $N = \binom{n}{k}$, that factorises through a continuous 1-1 map $p \colon G(k, \mathbb{R}^n) \to \mathbb{P}^{N-1}(\mathbb{R})$.

5.23 (♣) Let X be a space, n a fixed integer and consider, for any point $x \in X$, a projective subspace $L_x \subset \mathbb{P}^n(\mathbb{R})$. Assume that the subset

$$Z = \{(x, p) \in X \times \mathbb{P}^n(\mathbb{R}) \mid p \in L_x\}$$

is closed. Prove that for any $k \leq n$

$$U_k = \{x \in X \mid \dim L_x < k\}$$

is open in X (Hint: $x \in U_k$ iff there exists a subspace $H \subset \mathbb{P}^n(\mathbb{R})$ of dimension $n - k$ such that $L \cap L_x = \emptyset$). How can we use this result to study the map that sends a matrix $A \in M_{n,n}(\mathbb{R})$ to the vector subspace L_A of matrices $B \in M_{n,n}(\mathbb{R})$ satisfying $AB = BA$?

5.5 Locally Compact Spaces

Definition 5.23 A topological space is **locally compact** if every point has a compact neighbourhood.

Example 5.24 Open sets in $\mathbb{R}^n$ are locally compact: if $U \subset \mathbb{R}^n$ is open and $x \in U$ we can find $r > 0$ satisfying $B(x, r) \subset U$; then the closed ball $\overline{B(x, R)}$, for any $0 < R < r$, is a compact neighbourhood of x in U.

Straight from the definition we have that any compact space is locally compact. Furthermore,

Proposition 5.25 *Every closed subspace in a locally compact space is locally compact. The product of two locally compact spaces is locally compact.*

Proof Let Y be closed in X locally compact. Any $y \in Y$ has a compact neighbourhood $U \subset X$. The intersection $Y \cap U$ is a neighbourhood of y in Y and is also compact because closed in U.

If X, Y are locally compact and $(x, y) \in X \times Y$, the product $U \times V$ of two compact neighbourhoods of $x \in U \subset X$ and $y \in V \subset Y$ is a compact neighbourhood of (x, y). $\qquad\square$

On a par with compactness, local compactness is especially useful side by side with the Hausdorff property.

Theorem 5.26 *Every point in a locally compact Hausdorff space has a local basis of compact neighbourhoods.*

Proof Pick x in a Hausdorff space X and suppose K is a compact neighbourhood of x. We'll prove that the family of compact neighbourhoods of x inside K is a basis. Take U open in X containing x and consider the compact set $A = K - U$ and the inclusions

$$\{x\} \times A \subset K \times K - \Delta \subset K \times K,$$

where Δ is the diagonal. Wallace's theorem gives two disjoint open sets $V, W \subset K$ with $x \in V$ and $A \subset W$. The subset $V \cap K^o$ is open in K^o and therefore open in X. As K is closed in X, also $K - W$ is closed in X, and so the closure of V in X is contained in $K - W$. Hence

$$x \in V \cap K^o \subset \overline{V} \subset K - W \subset K - A = K \cap U,$$

and $\overline{V}$ is thus a compact neighbourhood of x inside $K \cap U$. $\qquad\square$

Corollary 5.27 *Consider an identification $f : X \to Y$ and a locally compact Hausdorff space Z. Then*

$$f \times Id : X \times Z \to Y \times Z, \quad (x, z) \mapsto (f(x), z),$$

is an identification, too.

Proof To simplify the notation we'll write $F = f \times Id$. The key observation is that if $A \subset X \times Z$ is F-saturated, then $(A : B)$ is f-saturated for any $B \subset Z$. Recall that

$$(A:B) = \{x \in X \mid \{x\} \times B \subset A\}.$$

We have to show that $x \in (A : B)$, $x' \in X$ and $f(x) = f(x')$ imply $x' \in (A:B)$: this is clear because for any $b \in B$ we have $(x, b) \in A$, $F(x, b) = F(x', b)$, and since A is saturated it follows that $(x', b) \in A$.

Let $U \subset Y \times Z$ be such that $A = F^{-1}(U)$ is open in $X \times Z$ and take $(y, z) \in U$; we want to find neighbourhoods $V \subset Y$, $W \subset Z$ such that $(y, z) \in V \times W \subset U$. Choose $x \in X$ with $f(x) = y$. As Z is locally compact, by Theorem 5.26 there is a compact neighbourhood W of z such that $\{x\} \times W \subset A$. Now, the map $\pi : X \times W \to X$ is closed, so

$$(A:W) = X - \pi(X \times W - A)$$

is an open set that we know is f-saturated. By definition of identification $V = f((A : W))$ is open and $(y, z) \in V \times W \subset U$.

Eventually, if f is open then $f \times Id$ is open, irrespective of what Z is.   □

We saw in Exercise 5.12 that Corollary 5.27 does not hold without the local compactness of Z.

Exercises

5.24 Let $A \subset \mathbb{R}^2$ be the union of all straight lines $y = nx$, with n positive integer. Prove that A isn't locally compact, and the onto map $\mathbb{R} \times \mathbb{N} \to A$, $(t, n) \mapsto (t, nt)$, isn't an identification.

Show further that $\mathbb{R} \times \mathbb{N}$ is a locally compact Hausdorff space, and its quotient $\mathbb{R} \times \mathbb{N}/\{0\} \times \mathbb{N}$ is Hausdorff but not locally compact.

5.25 Let $Y = X/\sim$ be an arbitrary quotient; prove that if X is compactly generated (Exercise 4.36) and Y Hausdorff, then Y is compactly generated, too.

5.26 Let X be a locally compact space. Prove that every compact subspace has a compact neighbourhood (i.e., for any $K \subset X$ compact there exists $H \subset X$ compact such that $K \subset H^\circ$).

5.27 Let Y be dense in a Hausdorff space X. Prove that Y locally compact implies Y open.

5.28 (✿) Consider a continuous map $f : X \to Y$ between locally compact Hausdorff spaces. Suppose $K \subset Y$ is a compact set with compact pre-image $f^{-1}(K)$. Prove that there exist open sets $f^{-1}(K) \subset U$, $K \subset V$ such that $f(U) \subset V$ and the restriction $f : U \to V$ is proper (Exercise 4.37). (Hint: let W be a compact neighbourhood of $f^{-1}(K)$; notice that $K \cap f(\partial W) = \emptyset$ and consider $U = W - f^{-1}(f(\partial W))$.)

5.6 The Fundamental Theorem of Algebra $\curvearrowright$

We have all the ingredients to present the most classical proof of the fundamental theorem of algebra, which relies on results from point-set topology. Corollary 15.26 will offer a different proof involving homotopy theory.

Lemma 5.28 *Let $p(z)$ be a polynomial of positive degree with complex coefficients, and suppose $p(0) \neq 0$. Then there exists $z \in \mathbb{C}$ such that $|p(z)| < |p(0)|$.*

Proof Call $k > 0$ the multiplicity of 0 as root of the polynomial $p(z) - p(0)$, so that

$$p(z) = p(0) - z^k (b_0 + b_1 z + \cdots + b_r z^r), \qquad \text{with} \quad b_0 \neq 0.$$

Suppose c is a kth root of $\dfrac{p(0)}{b_0}$ and consider the continuous map

$$g : [0, 1] \to \mathbb{R}, \qquad g(t) = |p(ct)| = \left| p(0) - t^k p(0) - t^{k+1} \sum_{i=1}^{r} b_i c^{k+i} t^{i-1} \right|.$$

By the triangle inequality

$$g(t) \leq |p(0)|(1 - t^k) + t^{k+1} \left| \sum_{i=1}^{r} b_i c^{k+i} t^{i-1} \right|.$$

Since by assumption $|p(0)| > 0$, for any positive and sufficiently small t we have $g(t) < g(0)$. $\qquad\square$

Theorem 5.29 *Let $p(z)$ be a polynomial with complex coefficients of positive degree. There exists $z_0 \in \mathbb{C}$ such that $p(z_0) = 0$.*

Proof Without loss of generality we can assume $p(z)$ is monic:

$$p(z) = z^n + a_{n-1} z^{n-1} + \cdots + a_0, \qquad n > 0.$$

Consider the continuous map $f : \mathbb{C} \to [0, +\infty[$, $f(z) = |p(z)|$, and let's prove it has an absolute minimum. By the triangle inequality

$$f(z) = |p(z)| \geq |z^n| - \sum_{i=0}^{n-1} |a_i| |z|^i = |z^n| \left(1 - \sum_{i=0}^{n-1} \frac{|a_i|}{|z|^{n-i}} \right).$$

Choose a real number $R > 0$ so large that

$$\frac{|a_0|}{R^n} < \frac{1}{2}, \qquad \sum_{i=0}^{n-1} \frac{|a_i|}{R^{n-i}} < \frac{1}{2}.$$

For every $z \in \mathbb{C}$ with $|z| \geq R$ we have

$$f(z) \geq |z^n| \left(1 - \sum_{i=0}^{n-1} \frac{|a_i|}{R^{n-i}} \right) \geq \frac{|z|^n}{2} \geq \frac{R^n}{2} \geq |a_0| = f(0),$$

so the minimum of f restricted to the compact set $\{z \mid |z| \leq R\}$ is a minimum for f on the whole plane.

Call z_0 the absolute minimum point of $|p(z)|$. Then $g(z) = |p(z + z_0)|$ has an absolute minimum at 0. By Lemma 5.28 the polynomial $q(z) = p(z + z_0)$ vanishes at 0, whence $p(z_0) = 0$.   □

Definition 5.30 For any positive integer n the **elementary symmetric functions** $\sigma_1, \ldots, \sigma_n \colon \mathbb{C}^n \to \mathbb{C}$ are the polynomials defined through the relationship

$$t^n + \sigma_1(z_1, \ldots, z_n)t^{n-1} + \cdots + \sigma_n(z_1, \ldots, z_n) = \prod_{i=1}^{n}(t + z_i).$$

Put in a different way, the values of $\sigma_1, \ldots, \sigma_n$ on the complex n-tuple $(a_1, \ldots, a_n)$ are the coefficients of the monic polynomial of degree n having $-a_1, \ldots, -a_n$ as roots. When $n = 3$ the elementary symmetric polynomials are

$$\sigma_1 = z_1 + z_2 + z_3, \quad \sigma_2 = z_1 z_2 + z_2 z_3 + z_3 z_1, \quad \sigma_3 = z_1 z_2 z_3.$$

Lemma 5.31 *The map*

$$\sigma \colon \mathbb{C}^n \to \mathbb{C}^n, \quad {}'\sigma(z) = (\sigma_1(z), \ldots, \sigma_n(z)),$$

is a closed identification.

Proof By the fundamental theorem of algebra σ is onto; we claim it's actually a closed map. Denote by $V_n \cong \mathbb{C}^{n+1}$ the vector space of homogeneous polynomials of degree n with complex coefficients in the variables t_0, t_1, and let $\mathbb{P}(V_n) \cong \mathbb{P}^n(\mathbb{C})$ be its projectivisation. Then

$$g \colon \mathbb{C}^n \to \mathbb{P}(V_n), \quad g(a_1, \ldots, a_n) = [t_1^n + a_1 t_0 t_1^{n-1} + \cdots + a_n t_0^n],$$

induces a homeomorphism from $\mathbb{C}^n$ to the open set

$$U = \{[h(t_0, t_1)] \in \mathbb{P}(V_n) \mid h(0, 1) \neq 0\}.$$

Call $p \colon \mathbb{P}(V_1)^n \to \mathbb{P}(V_n)$ the map induced by multiplying the coordinates:

$$p([u_1 t_1 + z_1 t_0], \ldots, [u_n t_1 + z_n t_0]) = [(u_1 t_1 + z_1 t_0) \cdots (u_n t_1 + z_n t_0)].$$

The mapping p is closed because $\mathbb{P}(V_1)^n$ is compact and $\mathbb{P}(V_n)$ Hausdorff, and by the projection formula the restriction $p\colon p^{-1}(U) \to U$ is closed too. Now observe that

$$f\colon \mathbb{C}^n \to p^{-1}(U), \qquad f(z_1, \ldots, z_n) = ([t_1 + z_1 t_0], \ldots, [t_1 + z_n t_0]),$$

is a homeomorphism and

$$
\begin{array}{ccc}
\mathbb{C}^n & \xrightarrow{\sigma} & \mathbb{C}^n \\
\downarrow{f} & & \downarrow{g} \\
\mathbb{P}(V_1)^n & \xrightarrow{p} & \mathbb{P}(V_n)
\end{array}
$$

commutes. $\square$

Every permutation τ of $\{1, \ldots, n\}$ induces a homeomorphism $\tau\colon \mathbb{C}^n \to \mathbb{C}^n$ defined by

$$\tau^{-1}(z_1, \ldots, z_n) = (z_{\tau(1)}, \ldots, z_{\tau(n)})$$

(the permutation of indices). This defines a representation of the symmetric group Σ_n as a subgroup of $\mathrm{Homeo}(\mathbb{C}^n)$, and it makes sense to consider the quotient $\mathbb{C}^n / \Sigma_n$.

Theorem 5.32 *In the notations above the map σ is compatible with the Σ_n-action and induces a homeomorphism $\mathbb{C}^n / \Sigma_n \simeq \mathbb{C}^n$. In particular σ is an open and closed identification.*

Proof By definition σ is constant on the orbits of Σ_n, so Proposition 5.8 gives a commutative diagram of continuous maps.

We know from algebra courses that every polynomial of degree n has n uniquely defined roots (up to order), so h is bijective. By Lemma 5.31 σ is an identification and hence h is a homeomorphism. Now Proposition 5.15 tells that $\mathbb{C}^n \to \mathbb{C}^n / \Sigma_n$ is open and closed, making σ open and closed, too. $\square$

Exercises

5.29 Consider the set $Y \subset \mathbb{R}^n$ of vectors $(a_1, \ldots, a_n)$ such that the polynomial $t^n + a_1 t^{n-1} + \cdots + a_n$ has a multiple factor. Prove that Y is closed in $\mathbb{R}^n$. (Hint: call V_d the space of real homogeneous polynomials of degree d in t_0, t_1). Then

$$g\colon \mathbb{R}^n \to \mathbb{P}(V_n), \qquad g(a_1, \ldots, a_n) = [t_1^n + a_1 t_0 t_1^{n-1} + \cdots + a_n t_0^n]$$

is continuous and $Y = \cup_{2d \leq n} g^{-1}(Z_d)$, where Z_d is the image of the closed map

$$\mathbb{P}(V_d) \times \mathbb{P}(V_{n-2d}) \to \mathbb{P}(V_n), \qquad ([p], [q]) \mapsto [p^2 q].)$$

5.30 (♣) Prove that, for $n \geq 2$, there's no continuous map $s \colon \mathbb{C}^n \to \mathbb{C}^n$ such that $\sigma s = Id$. (Hint: it may be useful to restrict σ to configurations of n points, cf. Exercise 4.11).

5.31 (♣, ♡) Call $\mathbb{C}[t]_n$ the space of complex monic polynomials of degree n; the affine structure naturally turns $\mathbb{C}[t]_n$ into a topological space homeomorphic to $\mathbb{C}^n$. Prove that:

1. the subspace in $\mathbb{C}[t]_n$ of polynomials with multiple roots is closed and connected, its complement is dense and path connected;
2. the map
$$M_{n,n}(\mathbb{C}) \to \mathbb{C}[t]_n, \qquad A \mapsto \det(A + tI),$$

   is an open identification;
3. the subset in $M_{n,n}(\mathbb{C})$ of diagonalisable matrices contains a dense open set (NB: any neighbourhood of the zero matrix contains non-diagonalisable matrices).

Chapter 6
Sequences

6.1 Countability Axioms

Definition 6.1 A topological space is said to be **second countable** if the topology admits a countable basis of open sets.

A space with this property is also said to satisfy the **second countability axiom**.

Example 6.2 The Euclidean line $\mathbb{R}$ has a countable basis: the family of open intervals with rational endpoints

$$\{ \,]c, d[\mid c, d \in \mathbb{Q}\}$$

is countable and forms a basis for the Euclidean topology. Any open set in $\mathbb{R}$, in fact, is the union of open intervals, and for any $]a, b[$ we have

$$]a, b[= \bigcup \{ \,]c, d[\mid a \leq c < d \leq b, \ c, d \in \mathbb{Q}\}.$$

Example 6.3 Any subspace in a second-countable space is second countable: if $\mathcal{B}$ is a countable basis for X and $Y \subset X$, then $\{A \cap Y \mid A \in \mathcal{B}\}$ is a countable basis for Y.

Example 6.4 The product of two second-countable spaces is still second countable. It's easy show that if $\mathcal{A}$ and $\mathcal{B}$ are bases of X and Y respectively,

$$\mathcal{C} = \{A \times B \mid A \in \mathcal{A}, \ B \in \mathcal{B}\}$$

is a basis of $X \times Y$. If $\mathcal{A}$ and $\mathcal{B}$ are countable, so is $\mathcal{C}$.

It's also quite simple to find examples of quotients of second-countable spaces that do not inherit the property, like the one of Exercise 6.1.

We recall (Definition 3.14) that a subset is dense if it intersects every non-empty open set.

Definition 6.5 A topological space is called **separable** if it contains a countable dense subset.

© Springer International Publishing Switzerland 2015
M. Manetti, *Topology*, UNITEXT - La Matematica per il 3+2 91,
DOI 10.1007/978-3-319-16958-3_6

Lemma 6.6 *Every second-countable space is separable.*

Proof Let $\mathcal{B}$ be a countable basis of X; for any open set $U \in \mathcal{B}$ choose a point $p_U \in U$. The set of such points $E = \{p_U \mid U \in \mathcal{B}\}$ is countable, and dense because it meets every basis element.   □

Lemma 6.7 *Any separable metric space has a countable basis.*

Proof Let (X, d) be a separable metric space and choose a countable dense subset $E \subset X$. It is enough to show that the countable family of open balls

$$\mathcal{B} = \{B(e, 2^{-n}) \mid e \in E, n \in \mathbb{N}\}$$

is a basis. Take U open and $x \in U$, then choose an integer $n \in \mathbb{N}$ so that $B(x, 2^{1-n}) \subset U$. As E is dense, there exists an $e \in E \cap B(x, 2^{-n})$; symmetrically, x belongs to the ball $B(e, 2^{-n})$, so from the triangle inequality we have $B(e, 2^{-n}) \subset B(x, 2^{1-n}) \subset U$.   □

Example 6.8 Denote by $\ell^2(\mathbb{R})$ the set of real sequences $\{a_n\}$, $n \geq 1$, such that $\sum_n a_n^2 < +\infty$, and set

$$\|a\| = \sqrt{\sum_{n=1}^{\infty} a_n^2}$$

for any $a = \{a_n\} \in \ell^2(\mathbb{R})$. Since for every $a, b \in \mathbb{R}$ we have $(a + b)^2 + (a - b)^2 = 2(a^2 + b^2)$, immediately $\ell^2(\mathbb{R})$ is a vector subspace. Given sequences $\{a_n\}, \{b_n\} \in \ell^2(\mathbb{R})$, the triangle inequality

$$\sqrt{\sum_{n=1}^{N}(a_n - b_n)^2} \leq \sqrt{\sum_{n=1}^{N} a_n^2} + \sqrt{\sum_{n=1}^{N} b_n^2}$$

holds, for any $N > 0$. Passing to the limit for $N \to \infty$ gives

$$\|a - b\| \leq \|a\| + \|b\|,$$

hence $\ell^2(\mathbb{R})$ is metrised by the distance $d(a, b) = \|a - b\|$.

Now let's consider the countable set $E \subset \ell^2(\mathbb{R})$ of eventually null, rational sequences. Given any sequence $a = \{a_n\} \in \ell^2(\mathbb{R})$ and any real number $\varepsilon > 0$, we may find an integer $N > 0$ and N rational numbers $b_1, \ldots, b_N$ such that

$$\sum_{n>N} a_n^2 < \varepsilon, \qquad (a_n - b_n)^2 < \frac{\varepsilon}{N} \quad \text{for every} \ \ n \leq N.$$

Consider the sequence $b = (b_1, b_2, \ldots, b_N, 0, 0, \ldots)$: it satisfies

$$\|a - b\|^2 = \sum_{n \leq N} (a_n - b_n)^2 + \sum_{n > N} a_n^2 < 2\varepsilon .$$

This proves that E is dense, the metric space $\ell^2(\mathbb{R})$ is separable, and by Lemma 6.7 $\ell^2(\mathbb{R})$ has a countable basis.

Proposition 6.9 *In a second-countable space every open cover admits a countable subcover.*

Proof Let X have a countable basis $\mathcal{B}$ and consider an open cover $\mathcal{A}$. For every $x \in X$ choose an open set $U_x \in \mathcal{A}$ and an element $B_x \in \mathcal{B}$ such that $x \in B_x \subset U_x$. The open family $\mathcal{B}' = \{B_x \mid x \in X\}$ is a subcover of $\mathcal{B}$, and so it is countable. We can find a countable subset $E \subset X$ such that $\mathcal{B}' = \{B_x \mid x \in E\}$, whence $\{U_x \mid x \in E\}$ is a countable subcover of $\mathcal{A}$. $\qquad\square$

Definition 6.10 A topological space satisfies the **first countability axiom**, or is **first countable**, if every point admits a countable local basis of neighbourhoods.

If a space is second countable then it is first countable: the open sets in a countable basis that contain a given point forms a local basis at that point.

Lemma 6.11 *Metric spaces are first countable.*

Proof Let x be a point in a metric space. The open balls $\{B(x, 2^{-n}) \mid n \in \mathbb{N}\}$ are a local basis of neighbourhoods around x. $\qquad\square$

The next example shows that not all first-countable spaces are metrisable.

Example 6.12 The lower-limit line $\mathbb{R}_l$ (cf. Example 3.9) is separable and first countable, but doesn't have countable bases. In particular it can't be metrisable. The subset of rational numbers is dense in $\mathbb{R}_l$, making $\mathbb{R}_l$ separable. For any real a, the countable open family $\{[a, a + 2^{-n}[\mid n \in \mathbb{N}\}$ is a local basis around a. Take an open basis $\mathcal{B}$ of $\mathbb{R}_l$; for any $x \in \mathbb{R}_l$ the open set $[x, +\infty[$ contains x, and so there is $U(x) \in \mathcal{B}$ such that $x \in U(x) \subset [x, +\infty[$. In particular, if $x < y$ then $x \notin U(y)$ and hence $U(x) \neq U(y)$. Therefore the map $\mathbb{R}_l \to \mathcal{B}, x \mapsto U(x)$, is 1-1 and $\mathcal{B}$ is not countable.

Theorem 6.13 *Let X be a second-countable, locally compact Hausdorff space. There exists an exhaustion of X by compact sets, i.e. a sequence*

$$K_1 \subset K_2 \subset \cdots$$

of compact sets covering X and such that $K_n \subset K_{n+1}^\circ$ for any n.

Proof Let $\mathcal{B}$ be a countable basis, $\mathcal{B}_c \subset \mathcal{B}$ the subfamily of open sets with compact closure. Since X is locally compact and Hausdorff, $\mathcal{B}_c$ covers X. Call $\mathcal{A}$ the collection of finite unions of elements of $\mathcal{B}_c$. By construction $\mathcal{A}$ is at most countable, every

$A \in \mathcal{A}$ has compact closure and for any compact set $K \subset X$ there's an element $A \in \mathcal{A}$ such that $K \subset A$. Fix an onto map $g \colon \mathbb{N} \to \mathcal{B}_c$, then define recursively a sequence of compact sets $K_1 \subset K_2 \subset \cdots$ by

$$K_1 = \overline{g(1)}, \qquad K_n = \overline{A_n \cup g(n)},$$

where A_n is one of the elements in $\mathcal{A}$ that contain K_{n-1}. This makes the family $\{K_n\}$ an exhaustion by compact sets. □

Exercises

6.1 (♡) Given real numbers $x, y \in \mathbb{R}$, we define $x \sim y$ iff $x = y$ or $x, y \in \mathbb{Z}$. Prove that $\mathbb{R}/\sim$ doesn't satisfy the first axiom of countability.

6.2 Let $\mathcal{B}$ be a basis in a second-countable space. Prove that there exists a countable subfamily $\mathcal{C} \subset \mathcal{B}$ giving a basis for the topology.

6.3 Consider a continuous and onto map $f \colon X \to Y$. Show that if X is separable then also Y is separable. If, in addition, f is open and X second countable, then Y has a countable basis.

6.4 Prove that the product of two separable spaces is separable. Give an example of a separable space containing a non-separable subspace. (Hint: Exercise 3.62.)

6.5 Let $f \colon X \to Y$ be a continuous, closed and onto map such that $f^{-1}(y)$ is compact for every $y \in Y$. Prove that Y has a countable basis if X has a countable basis.

6.6 (✊) Take disjoint copies $\mathbb{R}_a$ of the real line, parametrised by a real number $a \in \,]-\pi/2, \pi/2[$, and define continuous maps

$$f_a \colon (\mathbb{R}_a - \{0\}) \times \,]-1, 1[\,\to \mathbb{R}^2,$$

$$f_a(x, t) = \left(\frac{\cos(a)}{x} - t\sin(a),\ \frac{\sin(a)}{x} + t\cos(a) \right).$$

Now consider the disjoint union of $\mathbb{R}^2$ with the strips $\mathbb{R}_a \times \,]-1, 1[$, and denote by

$$X = \frac{\mathbb{R}^2 \cup_a \mathbb{R}_a \times \,]-1, 1[}{\sim}$$

the quotient under the smallest equivalence relation $\sim$ such that

$$(x, t) \sim f_a(x, t), \qquad \text{for every}\quad (x, t) \in (\mathbb{R}_a - \{0\}) \times \,]-1, 1[.$$

Prove that:

1. X is connected, separable and Hausdorff; each point has an open neighbourhood homeomorphic to $\mathbb{R}^2$;
2. X contains a discrete subset whose cardinality is more than countable. In particular, X isn't second countable;
3. X can't be written as countable union of compact sets.

6.2 Sequences

A **sequence** in a topological space X is a map $a : \mathbb{N} \to X$; the domain is usually viewed as a set of indices, so one writes a_i rather than $a(i)$.

Definition 6.14 Let $\{a_n\}$ be a sequence in a space X.

1. The sequence **converges** to a point $p \in X$ if for any neighbourhood U of p there is an $N \in \mathbb{N}$ such that $a_n \in U$ for every $n \geq N$.
2. A point $p \in X$ is called a **limit point** of the sequence if for any neighbourhood U of p and for every $N \in \mathbb{N}$ there exist $n \geq N$ such that $a_n \in U$.

Saying that a sequence accumulates at p means that p is a limit point. If a sequence converges to p then it also accumulates to p, so p is a limit point; the converse is generally false.

Every sequence in a Hausdorff space converges at most to a point, if it converges at all. Suppose the sequence $\{a_n\}$ converges to distinct points p and q; if X is Hausdorff, there are disjoint neighbourhoods $p \in U, q \in V$ and integers N, M such that $a_n \in U$ for every $n \geq N$ and $a_n \in V$ for all $n \geq M$; therefore for any $n \geq \max(N, M)$ we have $a_n \in U \cap V$, contradicting the assumption $U \cap V = \emptyset$.

Definition 6.15 A sequence $\{a_n\}$ in X is said to be **convergent** if it converges to some point $p \in X$. When X is Hausdorff, one also says that p is the **limit** of $\{a_n\}$. In symbols

$$a_n \to p, \quad \text{or} \quad \lim_{n \to \infty} a_n = p.$$

Definition 6.16 A **subsequence** of a sequence $a : \mathbb{N} \to X$ is the composite of the mapping a with some strictly increasing function $k : \mathbb{N} \to \mathbb{N}$.

Lemma 6.17 *Let $\{a_n\}$ be a sequence in X. If there's a subsequence $\{a_{k(n)}\}$ converging to $p \in X$, then p is a limit point for $\{a_n\}$.*

Proof Let U be a neighbourhood of p. By definition of convergence there exists an integer N such that $a_{k(n)} \in U$ for any $n \geq N$. Given any integer M we can find an integer m such that $a_{k(m)} \in U$ and $k(m) \geq M$: therefore p is a limit point for $\{a_n\}$. $\qquad \square$

In first-countable spaces many topological properties can be defined by means of sequences and subsequences. In order not to bore the reader too much we shall discuss this matter only with regard to closure and compactness.

Proposition 6.18 *Let X be a first-countable space and $A \subset X$ a subset. For every $x \in X$ the following conditions are equivalent:*

1. *there exists a sequence with values in A that converges to x;*
2. *the point x is a limit point of an A-valued sequence;*
3. *the point x belongs in the closure of A.*

Proof The implication (1) $\Rightarrow$ (2) is clear; let us prove (2) $\Rightarrow$ (3). Suppose x is a limit point for $a \colon \mathbb{N} \to A$ and write $A' = a(\mathbb{N})$; for any neighbourhood U of x we have $U \cap A' \neq \emptyset$, so $x \in \overline{A'} \subset \overline{A}$.

As for (3) $\Rightarrow$ (1), let $\{U_n \mid n \in \mathbb{N}\}$ be a countable local basis of neighbourhoods around x. As $x \in \overline{A}$, for any n we know $U_1 \cap \cdots \cap U_n \cap A \neq \emptyset$, and we may pick a point $a_n \in U_1 \cap \cdots \cap U_n \cap A$. The sequence $\{a_n\}$ converges to x, for if U is a neighbourhood of x, then there's an integer N such that $U_N \subset U$, and therefore $a_n \in U_N \subset U$ for any $n \geq N$. □

Lemma 6.19 *In a compact space any sequence has limit points.*

Proof Let X be compact and $a \colon \mathbb{N} \to X$ a sequence. For every m we set $C_m = \overline{\{a_n \mid n \geq m\}} \subset X$. By definition $x \in X$ is a limit point for a if $x \in C_m$ for every m. Proposition 4.46 tells that the countable descending chain of non-empty closed sets C_m has non-empty intersection. □

Definition 6.20 A space X is **sequentially compact** if every sequence has a convergent subsequence.

Lemma 6.21 *A first-countable space is sequentially compact if and only if every sequence admits a limit point. In particular any first-countable compact space is also sequentially compact.*

Proof The 'only if' statement follows by the definitions and Lemma 6.17. For the other implication, assume every sequence has a limit point and let's show the space is sequentially compact.

Take a sequence $\{a_n\}$; we claim it contains a convergent subsequence. Choose a limit point p and call $\{U_m\}$ a countable local basis of neighbourhoods at p. Possibly to replacing U_m with $U_1 \cap \cdots \cap U_m$, we may assume that $U_{m+1} \subset U_m$ for every m. The fact that p is a limit point allows to define a sequence $k(m)$ of integers such that $k(m + 1) > k(m)$ and $a_{k(m)} \in U_m$ for any m. In particular the subsequence $a_{k(m)}$ converges to p. □

Proposition 6.22 *The following are equivalent in a second-countable space X:*

1. X is compact;
2. every sequence in X has a limit point;
3. X is sequentially compact.

Proof We already know (1) $\Rightarrow$ (2), and also that (2) $\Rightarrow$ (3) holds in first-countable spaces.

There remains to show that if X has a countable basis, then (3) $\Rightarrow$ (1). Suppose X isn't compact, and let's find a sequence with no convergent subsequences. Take an open cover $\mathcal{A}$ without finite subcovers; by Proposition 6.9 $\mathcal{A}$ has a countable subcover $\{A_n \mid n \in \mathbb{N}\}$. *A fortiori* $\{A_n \mid n \in \mathbb{N}\}$ doesn't admit finite subcovers, so for every $n \in \mathbb{N}$ there is one $a_n \in X - \cup_{j=1}^{n} A_j$. We shall prove that every subsequence of $\{a_n\}$ cannot converge. Given a strictly increasing map $k \colon \mathbb{N} \to \mathbb{N}$ and a point $p \in X$ there exists N such that $p \in A_N$, whence $a_{k(n)} \notin A_N$ for every $k(n) \ge N$; in particular $\{a_{k(n)}\}$ doesn't converge to (the arbitrary point) p. $\qquad\square$

Remark 6.23 Compact, but not sequentially compact spaces do exist (Exercise 7.7). There are also sequentially compact spaces that aren't compact (Exercise 7.8).

Exercises

6.7 Let $a \colon \mathbb{N} \to \mathbb{R}$ be a 1-1 and onto map between the natural numbers and the rationals. Find the limits points of the sequence a. (Hint: every non-empty open set in $\mathbb{R}$ contains infinitely many rational numbers.)

6.8 Let X and Y be first countable and Hausdorff. Prove that $f \colon X \to Y$ is continuous iff it maps convergent sequences to convergent sequences.

6.9 ($\heartsuit$) Prove that every first-countable Hausdorff space is compactly generated (Exercise 4.36).

6.10 A point x in a space X is called a **limit point of the subset** $A \subset X$ if every neighbourhood of x contains points of A other than x.

Prove that $x \in X$ is a limit point of $\{a_n\}$ if and only if $(x, 0) \in X \times [0, 1]$ is a limit point of the subset

$$\{(a_n, 2^{-n}) \mid n \in \mathbb{N}\} \subset X \times [0, 1].$$

6.11 ($\heartsuit$) If a space is not first countable the implication (2) $\Rightarrow$ (1) in proposition 6.18 usually fails: the following example is due to R. Arens (1950).

Take $X = \mathbb{N}_0 \times \mathbb{N}_0$ with the topology for which a set $C \subset X$ is closed iff $(0, 0) \in C$ or $C \cap (\{n\} \times \mathbb{N}_0)$ is infinite for at most finitely many n. Consider the subset $A = X - \{(0, 0)\}$ and prove:

1. $(0, 0)$ is a limit point for a sequence with values in A;
2. there's no sequence in A that converges to $(0, 0)$.

6.3 Cauchy Sequences

Definition 6.24 A sequence $\{a_n\}$ in a metric space (X, d) is a **Cauchy sequence** if for any $\varepsilon > 0$ there exists an integer N such that $d(a_n, a_m) < \varepsilon$ for any $n, m \geq N$.

Every convergent sequence is a Cauchy sequence: in fact, if p is the limit of $\{a_n\}$, then for every $\varepsilon > 0$ there is an N such that $a_n \in B(p, \varepsilon/2)$ for any $n \geq N$; the triangle inequality implies $d(a_n, a_m) < \varepsilon$ for any $n, m \geq N$.

Lemma 6.25 *A Cauchy sequence is convergent if and only if it has limit points. In particular, any Cauchy sequence in a sequentially compact metric space converges.*

Proof Let $\{a_n\}$ be a Cauchy sequence in a metric space (X, d), with limit point $p \in X$. We'll show that for every $\varepsilon > 0$ we can find $N \in \mathbb{N}$ such that $d(p, a_n) < \varepsilon$ for any $n \geq N$. The Cauchy property implies that there's an $M \in \mathbb{N}$ such that $d(a_n, a_m) < \varepsilon/2$ for any $n, m \geq M$; as p is a limit point, there also is an $N \geq M$ such that $d(p, a_N) < \varepsilon/2$. Now use the triangle inequality

$$d(p, a_n) \leq d(p, a_N) + d(a_N, a_n) < \varepsilon \text{ for every } n \geq M.$$

If a Cauchy sequence has a convergent subsequence, by Lemma 6.17 it has a limit point. $\qquad\square$

Definition 6.26 A metric space X is called **complete** if every Cauchy sequence converges to a point in X.

From Lemma 6.21 it follows that every compact metric space is sequentially compact; hence, by Lemma 6.25, it also complete.

Theorem 6.27 *The Euclidean spaces $\mathbb{R}^n$ and $\mathbb{C}^n$ are complete metric spaces.*

Proof As $\mathbb{C}^n = \mathbb{R}^{2n}$ we can deal with $\mathbb{R}^n$ only. Let $\{a_n\}$ be a Cauchy sequence in $\mathbb{R}^n$ and $N \in \mathbb{N}$ a number such that

$$\|a_n - a_N\| < 1 \text{ for any } n \geq N.$$

Writing

$$R = \max\{\|a_1\|, \|a_2\|, \ldots, \|a_N\|\},$$

by the triangle inequality the sequence $\{a_n\}$ is contained in the compact subspace

$$D = \{x \in \mathbb{R}^n \mid \|x\| \leq R + 1\}.$$

Lemma 6.25, then, guarantees that the sequence must converge. $\qquad\square$

Lemma 6.28 *Let $\{f_n : \mathbb{N} \to A_n \mid n \in \mathbb{N}\}$ be a countable family of maps. If every A_n is finite, there exists a strictly increasing map $g : \mathbb{N} \to \mathbb{N}$ such that $f_n(g(m)) = f_n(g(n))$ for any $m \geq n$.*

Proof A_1 is finite so at least one fibre of f_1 contains infinitely many elements. Lemma 2.3 warrants the existence of a strictly increasing map $g_1 : \mathbb{N} \to \mathbb{N}$ making $f_1 g_1$ constant. On the same grounds at least one of the fibres of $f_2 g_1$ is infinite, and we can find a strictly increasing $g_2 : \mathbb{N} \to \mathbb{N}$ so that $f_2 g_1 g_2$ is constant. Continuing in this way we can define a sequence of strictly increasing maps $g_n : \mathbb{N} \to \mathbb{N}$ such that, for any n, $f_n g_1 g_2 \cdots g_n$ is constant.

Now let's consider the 'diagonal' map

$$g : \mathbb{N} \to \mathbb{N}, \qquad g(n) = g_1 \cdots g_n(n),$$

and prove that it has the desired properties. Given $n, m \in \mathbb{N}$, $n < m$, we set $l = g_{n+1} \cdots g_m(m)$; then $l \geq m$ and

$$f_n(g(m)) = f_n g_1 \cdots g_n(l) = f_n g_1 \cdots g_n(n) = f_n(g(n)). \qquad \square$$

Theorem 6.27 yields that $\mathbb{R}$ is complete as a consequence of the sequential compactness of $[0, 1]$. It is instructive to show the opposite implication, i.e. prove that $[0, 1]$ is sequentially compact because $\mathbb{R}$ is complete. To do this we have to show that every sequence in $[0, 1]$ has a Cauchy subsequence.

Suppose we have a sequence $\{a_n\}$ in $[0, 1]$. Call A_n the set of the 2^n intervals

$$[0, 2^{-n}[\, , \ldots , [i2^{-n}, (i + 1)2^{-n}[\, , \ldots , [1 - 2^{-n}, 1], \quad 0 \leq i < 2^n,$$

and let $f_n : \mathbb{N} \to A_n$ be the map such that $a_m \in f_n(m)$ for any m. By Lemma 6.28 there's a strictly increasing map $g : \mathbb{N} \to \mathbb{N}$ such that $f_n(g(n)) = f_n(g(m))$ for any $n \leq m$. The subsequence $\{b_n = a_{g(n)}\}$ satisfies $|b_n - b_m| \leq 2^{-n}$ for any $n \leq m$, so it is Cauchy.

Proposition 6.29 *A subspace in a complete metric space is closed if and only if it is complete for the induced metric.*

Proof Let (X, d) be complete and $A \subset X$ closed. Every Cauchy sequence in A is Cauchy in X, so it converges to a limit $a \in \overline{A}$; therefore (A, d) is complete provided A is closed.

Vice versa, every sequence in A that converges to $a \in \overline{A}$ is Cauchy in (A, d), so if (A, d) is complete, $a \in A$. $\qquad \square$

Exercises

6.12 ($\heartsuit$) Find an example showing that completeness is not a homeomorphism-invariant property.

6.13 Define a distance d on $\mathbb{R} - \mathbb{Z}$ that renders $(\mathbb{R} - \mathbb{Z}, d)$ complete and homeomorphic to $\mathbb{R} \times \mathbb{Z}$.

6.14 Prove that the metric space $\ell^2(\mathbb{R})$ introduced in Example 6.8 is complete.

6.15 Let $\ell^\infty(\mathbb{R})$ be the space of bounded real sequences $\{a_n\}$, with distance

$$d(\{a_n\}, \{b_n\}) = \sup_n |a_n - b_n|.$$

Prove that $\ell^\infty(\mathbb{R})$ is a complete metric space but it isn't separable.

6.16 (Contraction theorem, $\heartsuit$) Let (X, d) be a metric space. A function $f: X \to X$ is called a **contraction (mapping)** if there exists a real number $\gamma < 1$ such that

$$d(f(x), f(y)) \leq \gamma d(x, y) \quad \text{for every} \quad x, y \in X.$$

Prove that if (X, d) is complete, non-empty, and $f: X \to X$ a contraction, there exists a unique point $z \in X$ such that $f(z) = z$. Show, moreover, that for every $x \in X$ the sequence $\{f^n(x)\}$ converges to z.

6.17 ($\heartsuit$) Find a map $f: \mathbb{R} \to \mathbb{R}$ without fixed points and such that

$$|f(x) - f(y)| < |x - y|$$

for every x, y.

6.18 ($\maltese$) Let A be an open set in a complete metric space (X, d). Prove there exists a distance h on A inducing the subspace topology and such that (A, h) is complete.

6.4 Compact Metric Spaces

Definition 6.30 A metric space is said to be **totally bounded** if it can be covered by a finite number of open balls of radius r, for any positive real number r.

Each totally bounded metric space is bounded: if X is the finite union of balls $B(x_1, 1), \ldots, B(x_n, 1)$ and M is the maximum distance between any two centres x_i, by the triangle inequality the distance of any two points of X is at most $M + 2$. Any discrete and infinite space, with the distance of Example 3.34, is bounded but not totally bounded.

Lemma 6.31 *Every sequentially compact metric space is totally bounded.*

Proof Take (X, d) sequentially compact and suppose, by contradiction, there is an $r > 0$ such that X cannot be covered by finitely many open balls of radius r. We construct recursively a sequence $\{a_n\}$: choose $a_1 \in X$ arbitrarily, and for any $n > 1$ let a_n be some element in the non-empty closed set

$$X - \bigcup_{i=1}^{n-1} B(a_i, r).$$

Since $d(a_n, a_m) \geq r$ for every $n > m$, no subsequence of $\{a_n\}$ can converge. $\qquad\square$

Lemma 6.32 *Every totally bounded metric space is second countable.*

Proof Let X be totally bounded. For any $n \in \mathbb{N}$ there's a finite subset $E_n \subset X$ such that $X = \cup\{B(e, 2^{-n}) \mid e \in E_n\}$. The countable set $E = \cup E_n$ is thus dense, making X separable. Invoking Lemma 6.7 allows to conclude. □

Theorem 6.33 *In a metric space X the following conditions are equivalent:*

1. *X is compact;*
2. *every sequence in X has a limit point;*
3. *X is sequentially compact;*
4. *X is complete and totally bounded.*

Furthermore, if any one of the above holds then X is second countable.

Proof Call d the distance. A metric space satisfies the first countability axiom, so $(1) \Rightarrow (2)$ and $(2) \Rightarrow (3)$ are proven.

As for $(3) \Rightarrow (1)$, the previous Lemmas 6.31 and 6.32 show that a sequentially compact metric space has a countable basis, and Proposition 6.22 applies.

We claim $(2) + (3) \Rightarrow (4)$: Lemma 6.31, in fact, says that the space is totally bounded, and Lemma 6.25 that it's complete.

In order to prove $(4) \Rightarrow (3)$ it suffices to show that in a totally bounded metric space we can extract a Cauchy subsequence from any sequence $\{a_n\}$. For any n call A_n a finite set of open balls of radius 2^{-n} covering X, and pick a map $f_n : \mathbb{N} \to A_n$ with the property that $f_n(i)$ is a ball containing a_i, for every i. By Lemma 6.28 there's a subsequence $\{a_{g(n)}\}$ such that, for any $n < m$, $a_{g(n)}$ and $a_{g(m)}$ belong in the same ball of radius 2^{-n}. Then $\{a_{g(n)}\}$ is a Cauchy sequence. □

Lemma 6.34 *A subspace A in a metric space is totally bounded if and only if $\overline{A}$ is totally bounded.*

Proof Suppose $\overline{A}$ is totally bounded and take $r > 0$; there exist $a_1, \ldots, a_n \in \overline{A}$ with

$$\overline{A} \subset \cup_{i=1}^{n} B(a_i, r/2).$$

Choosing, for any $i = 1, \ldots, n$, a point $b_i \in B(a_i, r/2) \cap A$, from the triangle inequality we have

$$A \subset \overline{A} \subset \cup_{i=1}^{n} B(b_i, r).$$

Suppose now A is totally bounded and take $r > 0$; there exist $a_1, \ldots, a_n \in A$ such that

$$A \subset \cup_{i=1}^{n} B(a_i, r/2),$$

so

$$\overline{A} \subset \cup_{i=1}^{n} \overline{B(a_i, r/2)} \subset \cup_{i=1}^{n} B(a_i, r). \qquad \square$$

Definition 6.35 A subspace A in a space X is **relatively compact** if it is contained in a compact subspace of X.

When X is Hausdorff, $A \subset X$ is relatively compact if and only if $\overline{A}$ is compact.

Corollary 6.36 *A subspace in a complete metric space is relatively compact if and only if it is totally bounded.*

Proof Let X be complete and $A \subset X$ totally bounded; by Proposition 6.29 and Lemma 6.34 $\overline{A}$ is complete and totally bounded, hence compact.

Conversely, if $\overline{A}$ is compact it is totally bounded, so A is totally bounded. □

Exercises

6.19 Show that $\mathbb{R}$, with the standard bound of the Euclidean distance (Example 3.40), is a bounded metric space, but not totally bounded.

6.20 (♡) Let (X, d) be a compact metric space and $f : X \to X$ an isometry, i.e. a distance-preserving map. Prove that f is a homeomorphism. (The only non-trivial fact is that f is onto.)

6.21 Prove that the image of a relatively compact space under a continuous mapping is relatively compact.

6.22 Let (X, d) be a metric space: for any $C \subset X$ and any real number $r > 0$ write

$$B(C, r) = \bigcup_{y \in C} B(y, r).$$

Prove $\cap_{r>0} B(C, r) = \overline{C}$ and $B(C, r) = \{x \in X \mid d_C(x) < r\}$, where d_C denotes the *distance from the set* C (Example 3.47).

6.23 (Hausdorff distance) Call $\mathcal{X}$ the family of non-empty, closed and bounded subspaces in a metric space (X, d). For every $C, D \in \mathcal{X}$ define, using the notation of Exercise 6.22,

$$h(C, D) = \inf\{r \mid C \subset B(D, r), \ D \subset B(C, r)\}.$$

Prove the following statements:

1. $h(C, D) < r$ if and only if for any $z \in C$ and any $w \in D$ there exist $x \in D, y \in C$ such that $d(z, x) < r$ and $d(w, y) < r$;
2. h is a distance on $\mathcal{X}$ (called the *Hausdorff distance*);
3. (X, d) is totally bounded if and only if it is bounded and there exists a finite set $S \subset X$ such that $h(X, S) < r$ for any $r > 0$;
4. let $C, D \in \mathcal{X}$. If $h(C, D) < r$ and $A \subset C$, then $h(A, D \cap B(A, r)) < r$;

5. if (X, d) is totally bounded, $(\mathcal{X}, h)$ is totally bounded;
6. let $\{A_n\}$ be a sequence in $\mathcal{X}$ such that $h(A_n, A_{n+1}) < 2^{-n}$ for any n. Then every $x_N \in A_N$ extends to a sequence $\{x_n\}$ such that $x_n \in A_n$ and $d(x_n, x_{n+1}) < 2^{-n}$ for any n;
7. if (X, d) is complete, $(\mathcal{X}, h)$ is complete; (Hint: take $\{A_n\}$ in $\mathcal{X}$ such that $h(A_n, A_{n+1}) < 2^{-n}$ and call A the set of limit points of sequences $\{x_n\}$ such that $x_n \in A_n$ and $d(x_n, x_{n+1}) < 2^{-n}$. Prove that A_n converges in the closure of A.)
8. if (X, d) is compact, $(\mathcal{X}, h)$ is compact.

6.5 Baire's Theorem

Definition 6.37 A subset in a topological space is said to be **nowhere dense** if its closure has empty interior, and **meagre** if it is contained in the union of countably many nowhere-dense subsets.

Being nowhere dense or meagre is not an intrinsic property, in other words it also depends on the ambient space X. For example, the point $\{0\}$ is nowhere dense in $\mathbb{R}$ but not in $\mathbb{Z}$ (it's not even meagre in the integers); so it makes sense to speak of nowhere-dense and meagre subspaces, whereas a nowhere-dense or meagre topological space alone is meaningless.

We may, rather punningly, distinguish meagre sets in two categories: truly thin and slender subsets, and 'false lean' ones. The former have empty interior, while the second sort do not albeit still being meagre. A Baire space is a space that has no subsets of the second type:

Definition 6.38 A topological space X is a **Baire space** if each meagre subset has empty interior.

To check such a property it obviously suffices to show that countable unions of nowhere-dense closed sets have non-empty interior, or equivalently, countable intersections of open dense sets are dense.

Example 6.39 The empty set is a Baire space. Any non-empty discrete set is a Baire space: the only nowhere-dense subset is $\emptyset$.

The space $\mathbb{Q}$ is not a Baire space: every point is nowhere dense and closed, and $\mathbb{Q}$ is the countable union of its elements.

Theorem 6.40 (Baire) *Complete metric spaces and locally compact Hausdorff spaces are Baire spaces.*

Proof The two situations require distinct proofs, albeit with a common underlying idea.

Start with X locally compact and Hausdorff. Let $\{C_n \mid n \in \mathbb{N}\}$ be a countable family of nowhere-dense closed sets in X and $U_0 \subset X$ a non-empty open set: we have to prove $U_0 \not\subset \cup_n C_n$. We define a sequence of open sets

$$U_0 \supset U_1 \supset U_2 \supset U_3 \supset \cdots$$

such that for any $n \geq 1$ the following conditions hold:

1. $U_n \neq \emptyset$;
2. $\overline{U_n}$ is compact and contained in $U_{n-1} - C_n$.

By assumption C_1 is nowhere dense and, in particular, $U_0 - C_1$ is non-empty and open; choose an arbitrary point $x \in U_0 - C_1$. By Theorem 5.26 x has a local basis of compact neighbourhoods, so we can find an open U_1 containing x (hence non-empty) and such that $\overline{U_1}$ is compact and contained in $U_0 - C_1$. Then we proceed recursively: if we have U_n, the argument above allows to find a non-empty open set U_{n+1} such that $\overline{U_{n+1}}$ is compact and contained in $U_n - C_{n+1}$. If $A = \cap_{n>0} \overline{U_n} \subset U_0$, by construction $A \subset X - C_n$ for all n, while Proposition 4.46 implies $A \neq \emptyset$. Therefore U_0 doesn't lie in the union of the nowhere-dense closed sets C_n.

Suppose now $\{C_n \mid n \in \mathbb{N}\}$ is a countable family of nowhere-dense closed sets in a complete metric space (X, d). By contradiction, suppose there's an open ball $B(x_0, r) \subset \cup_n C_n$. By assumption C_n has empty interior, so for any ball $B(x, s)$, $s > 0$, and any n we have $B(x, s) - C_n \neq \emptyset$. Pick $x_1 \in B(x_0, r/3) - C_1$ and a real number $0 < r_1 \leq r/3$ such that $B(x_1, r_1) \cap C_1 = \emptyset$. 'Extend' x_1 and r_1 to sequences $\{x_n\}, \{r_n\}$ defined as follows: choose $x_2 \in B(x_1, r_1/3) - C_2$ and $0 < r_2 \leq r_1/3$ with $B(x_2, r_2) \cap C_2 = \emptyset$. Now continue recursively, by taking the subsequent index. By the triangle inequality, if $n < m$

$$d(x_n, x_m) \leq d(x_n, x_{n+1}) + \cdots + d(x_{m-1}, x_m) \leq \frac{1}{3}(r_n + \cdots + r_{m-1})$$

$$\leq \frac{1}{3}\left(r_n + \frac{r_n}{3} \cdots + \frac{r_n}{3^{m-n-1}}\right) \leq \frac{1}{2}r_n \leq \frac{r}{3^n}.$$

Therefore $\{x_n\}$ is a Cauchy sequence, hence it converges to, say, h. Taking the limit as $m \to \infty$, the previous inequalities give $d(x_n, h) \leq r_n/2$, and so $h \notin C_n$ for every n. On the other hand $d(x_0, h) < r$, whence $h \in B(x_0, r) \subset \cup C_n$. $\qquad\square$

Remark 6.41 The two statements in Baire's theorem are completely independent. There are compact Hausdorff space that can't be metrised by any distance, and complete metric spaces that aren't locally compact (like infinite-dimensional Banach spaces).

The name Baire is French (so it should be spoken as the verb *faire*). René Baire (1874–1932) is also known for introducing, in 1899, the notion of semi-continuous functions and for proving that a lower semi-continuous function defined on a closed and bounded interval has a minimum.

Corollary 6.42 *Let* $\{F_m \mid m \in \mathbb{N}\}$ *be a countable family of subsets in* $\mathbb{R}^n$. *If* $\cup_m F_m = \mathbb{R}^n$, *there exists an integer m such that the closed set* $\overline{F_m}$ *has non-empty interior (in the Euclidean topology).*

Proof $\mathbb{R}^n$ is a Baire space. $\square$

Exercises

6.24 Prove that $\mathbb{R}^n$ cannot be written as a countable union of proper vector subspaces.

6.25 Prove Remark 3.32, i.e. that $\mathbb{Q}$ is not the countable intersection of open sets in $\mathbb{R}$.

6.26 Let $X = \cup U_i$ be an open cover of X. Show that the following are equivalent:

1. X is a Baire space;
2. any open set U_i is a Baire space;
3. any closed set $\overline{U_i}$ is a Baire space.

6.27 ($\heartsuit$) Is it possible to define a Hausdorff topology on $\mathbb{N}$ that makes it:

1. compact?
2. connected?
3. compact and connected?

6.28 ($\clubsuit$, $\heartsuit$) Let $A \subset \mathbb{R}$ be dense and countable (e.g. $\mathbb{Q}$). Given an arbitrary map $f \colon \mathbb{R} \to]0, +\infty[$, show that there exist $a \in A$ and $b \in \mathbb{R} - A$ such that $\min(f(a), f(b)) > |a - b|$.

6.6 Completions $\curvearrowright$

Definition 6.43 Let (X, d) and $(\widehat{X}, \widehat{d})$ be metric spaces. A map $\Phi \colon X \to \widehat{X}$ is called a **completion** of (X, d) if:

1. Φ is an isometry: $\widehat{d}(\Phi(x), \Phi(y)) = d(x, y)$ for all $x, y \in X$;
2. $(\widehat{X}, \widehat{d})$ is a complete metric space;
3. $\Phi(X)$ is dense in $\widehat{X}$.

Example 6.44 The inclusion maps $(\mathbb{Q}, d) \subset (\mathbb{R}, d)$ and $(]0, 1[, d) \subset ([0, 1], d)$ are completions (d is the Euclidean distance).

In this section we shall prove the existence, uniqueness and the main features of completions.

Lemma 6.45 *Let* $\{a_n\}$, $\{b_n\}$ *be Cauchy sequences in a metric space* (X, d). *The limit*

$$\lim_{n \to \infty} d(a_n, b_n) \in [0, +\infty[$$

exists and is finite.

Proof The quadrangle inequality (Exercise 3.32) implies

$$|d(a_n, b_n) - d(a_m, b_m)| \leq d(a_n, a_m) + d(b_n, b_m),$$

and so the real sequence $d(a_n, b_n)$ is Cauchy. $\square$

Given a metric space (X, d) we denote by $\mathfrak{c}(X, d)$ the set of all Cauchy sequences in it. Consider on $\mathfrak{c}(X, d)$ the equivalence relation

$$\{a_n\} \sim \{b_n\} \quad \text{if and only if} \quad \lim_{n \to \infty} d(a_n, b_n) = 0.$$

Write $\widehat{X} = \mathfrak{c}(X, d)/\sim$ for the corresponding quotient set and denote by $[a_n] \in \widehat{X}$ the coset of the Cauchy sequence $\{a_n\}$. For any $a \in X$ let $\Phi(a) \in \widehat{X}$ be the equivalence class of the constant sequence $a, a, a, \ldots$; note how $[a_n] = \Phi(a)$ if and only if $\lim_{n \to \infty} a_n = a$, so that the canonical inclusion

$$\Phi : X \to \widehat{X}$$

is onto if and only if (X, d) is complete. Lemma 6.45 ensures that

$$\hat{d} : \widehat{X} \times \widehat{X} \to \mathbb{R}, \qquad \hat{d}([a_n], [b_n]) = \lim_{n \to \infty} d(a_n, b_n)$$

is a well-defined map. Notice that $\widehat{X}$ depends on X but also on the chosen distance d.

Lemma 6.46 *Let A be a dense subset in a metric space (X, d). The natural inclusion $\widehat{A} \to \widehat{X}$ is one-to-one and onto.*

Proof The only non-obvious property to check is surjectivity, i.e. that every Cauchy sequence in X is equivalent to a Cauchy sequence in A. For this, let $\{x_n\}$ be a Cauchy sequence in X, and for any given n choose $a_n \in A$ so that $d(a_n, x_n) \leq 2^{-n}$. The quadrangle inequality implies

$$d(a_n, a_m) \leq d(x_n, x_m) + 2^{-N}, \qquad \text{for all} \quad n, m > N$$

and therefore $\{a_n\}$ is a Cauchy sequence. It is also clear that $[a_n] = [x_n]$. $\square$

Theorem 6.47 (Existence of completions) *Retaining the previous notation, the map $\hat{d}$ is a distance on $\widehat{X}$ and the inclusion Φ is a completion.*

Proof By definition $\hat{d}([a_n], [b_n]) \geq 0$, and $\hat{d}([a_n], [b_n]) = 0$ iff $[a_n] = [b_n]$. Given three Cauchy sequences $\{a_n\}, \{b_n\}, \{c_n\}$ we have:

$$\hat{d}([a_n], [b_n]) = \lim_{n \to \infty} d(a_n, b_n) = \lim_{n \to \infty} d(b_n, a_n) = \hat{d}([b_n], [a_n]).$$

$$\hat{d}([a_n], [b_n]) = \lim_{n \to \infty} d(a_n, b_n) \le \lim_{n \to \infty} (d(a_n, c_n) + d(c_n, b_n))$$

$$\le \lim_{n \to \infty} d(a_n, c_n) + \lim_{n \to \infty} d(c_n, b_n) = \hat{d}([a_n], [c_n]) + \hat{d}([c_n], [b_n]).$$

Hence $\hat{d}$ is a distance. Furthermore,

$$\hat{d}(\Phi(a), \Phi(b)) = \lim_{n \to \infty} d(a, b) = d(a, b)$$

for any $a, b \in X$, so Φ is an isometry.

Let's prove that $\Phi(X)$ is dense in $\widehat{X}$, i.e. for any $\varepsilon > 0$ and any $[a_n] \in \widehat{X}$ there exists $b \in X$ such that $\hat{d}([a_n], \Phi(b)) \le \varepsilon$. As $\{a_n\}$ is Cauchy, there exists N such that $d(a_N, a_n) \le \varepsilon$ for any $n \ge N$, and then we may as well just take $b = a_N$.

There remains to prove that $(\widehat{X}, \hat{d})$ is complete. The map $X \to \Phi(X)$ is a bijective isometry and $\Phi(X)$ is dense in $\widehat{X}$, so the natural maps

$$\widehat{X} \to \widehat{\Phi(X)} \to \widehat{\widehat{X}}$$

are bijections. In particular $\widehat{X} = \widehat{\widehat{X}}$, which is the same as saying $\widehat{X}$ is complete. $\square$

Theorem 6.48 (Uniqueness of completions) *If*

$$h \colon (X, d) \to (Y, \delta), \qquad k \colon (X, d) \to (Z, \rho)$$

are completions of the same metric space (X, d), *there exists a unique bijective isometry* $f \colon (Y, \delta) \to (Z, \rho)$ *such that* $k = fh$.

Proof The uniqueness of f descends from the fact that $h(X)$ is dense in Y. For the existence we can suppose that $Y = \widehat{X}$ is the completion built in Theorem 6.47, without loss of generality. The isometry $k \colon X \to Z$ has dense range, so it induces a bijective isometry

$$\hat{k} \colon \widehat{X} \to \widehat{Z}.$$

Moreover, since (Z, ρ) is complete, $\Phi \colon Z \to \widehat{Z}$ is a bijective isometry. Consequently $f = \Phi^{-1}\hat{k}$ has the properties we are seeking. $\square$

Corollary 6.49 (Universal property of completions) *Let* $\Phi \colon (X, d) \to (\widehat{X}, \hat{d})$ *be a completion. For any complete metric space* (Y, δ) *and any isometry* $h \colon (X, d) \to (Y, \delta)$ *there exists a unique isometry* $k \colon (\widehat{X}, \hat{d}) \to (Y, \delta)$ *such that* $k\Phi = h$.

Proof The map $h \colon X \to \overline{h(X)}$ is a completion, so it's enough to apply Theorem 6.48. $\square$

Exercises

6.29 Let $BC(X, \mathbb{R})$ the set of bounded, continuous maps $f \colon X \to \mathbb{R}$ defined on a space X. It has a natural metric structure given by the distance

$$\hat{d}(f, g) = \sup_{x \in X} |f(x) - g(x)|.$$

Suppose X is metric with distance d, and fix $a \in X$. Prove that the map

$$\Psi \colon X \to BC(X, \mathbb{R}), \qquad \Psi(x)(y) = d(x, y) - d(a, y),$$

is a well-defined isometry.

Remark 6.50 In analogy to the upcoming Theorem 6.51, it can be proved that $(BC(X, \mathbb{R}), \hat{d})$ is a complete space. Therefore the closure of $\Psi(X)$ is a completion of X.

6.30 Let $\mathbb{K}[x]$ be the ring of polynomials in x over the field $\mathbb{K}$. Prove that the function

$$d \colon \mathbb{K}[x] \times \mathbb{K}[x] \to \mathbb{R}, \qquad d(p, q) = \inf\{2^{-n} \mid x^n \text{ divides } p(x) - q(x)\},$$

is a distance, and that the completion of $(\mathbb{K}[x], d)$ is canonically isomorphic to the ring of formal power series $\mathbb{K}[[x]]$.

6.7 Function Spaces and Ascoli-Arzelà Theorem ↷

We indicate with $C(X, Y)$ the set of continuous mappings between the spaces X and Y; on it we can introduce many natural topological structures, some of which we shall encounter in the sequel (Example 7.4, Sect. 8.5).

This section is devoted to the special situation in which X is compact and (Y, d) is a metric space; given continuous maps $f, g \colon X \to Y$, the real function

$$X \to \mathbb{R}, \qquad x \mapsto d(f(x), g(x)),$$

is continuous, so it has a maximum and a minimum on X.

Theorem 6.51 *Let X be compact and (Y, d) a metric space. The function*

$$\rho \colon C(X, Y) \times C(X, Y) \to \mathbb{R}, \qquad \rho(f, g) = \max\{d(f(x), g(x)) \mid x \in X\},$$

is a distance on $C(X, Y)$. Moreover:

1. *the evaluation map*

$$e \colon C(X, Y) \times X \to Y, \qquad e(f, x) = f(x),$$

   is continuous;
2. *$(C(X, Y), \rho)$ is complete if and only if (Y, d) is complete.*

Proof The triangle inequality is the only non-evident fact if we are to prove that ρ is a distance. Take $f, g, h \in C(X, Y)$ and choose $x \in X$ such that $d(f(x), h(x)) = \rho(f, h)$; by the triangle inequality on Y

$$\rho(f, h) = d(f(x), h(x)) \leq d(f(x), g(x)) + d(g(x), h(x)) \leq \rho(f, g) + \rho(g, h).$$

Take $f \in C(X, Y)$, $x \in X$ and $\varepsilon > 0$. Pick a neighbourhood U of x such that $d(f(x), f(y)) < \varepsilon/2$ for every $y \in U$. If $\rho(f, g) < \varepsilon/2$,

$$d(f(x), g(y)) \leq d(f(x), f(y)) + d(f(y), g(y)) < \varepsilon$$

proving the evaluation mapping is continuous at the point (f, x).

Suppose now that (Y, d) is complete and let f_n be a Cauchy sequence in $C(X, Y)$. For any point $x \in X$

$$d(f_n(x), f_m(x)) \leq \rho(f_n, f_m),$$

so $\{f_n(x)\}$ is Cauchy in Y; letting $f(x)$ be the limit of $f_n(x)$, we have to show that $f : X \to Y$ is continuous at every $x_0 \in X$ and that $f_n \to f$ in $C(X, Y)$. Given $x_0 \in X$ and $\varepsilon > 0$, choose a positive integer n such that $\rho(f_n, f_m) \leq \varepsilon/3$ for any $m \geq n$ and a neighbourhood U of x_0 such that $d(f_n(x), f_n(x_0)) \leq \varepsilon/3$ for any $x \in U$. By the triangle inequality

$$d(f(x), f(x_0)) = \lim_m d(f_m(x), f_m(x_0))$$

$$\leq \lim_m d(f_n(x), f_m(x)) + \lim_m d(f_n(x_0), f_m(x_0)) + d(f_n(x), f_n(x_0)) \leq \varepsilon$$

for all $x \in U$, proving that f is continuous at x_0. To see that $f_n \to f$ in $C(X, Y)$, fix, for any $\varepsilon > 0$, an integer N such that $\rho(f_n, f_m) \leq \varepsilon$ for all $n, m \geq N$. Then

$$d(f_n(x), f(x)) = \lim_m d(f_n(x), f_m(x)) \leq \varepsilon$$

for any $x \in X$ and $n \geq N$, so $\rho(f_n, f) = \max\{d(f_n(x), f(x)) \mid x \in X\} \leq \varepsilon$.

Eventually, note that (Y, d) is isometric to the closed subspace of constant maps from X to Y. ☐

Corollary 6.52 *Let (Y, d) be complete. For every positive integer n the metric space (Y^n, ρ), where*

$$\rho((x_1, \ldots, x_n), (y_1, \ldots, y_n)) = \max_i d(x_i, y_i),$$

is complete.

Proof The space (Y^n, ρ) is canonically isometric to the complete space $C(\{1, \ldots, n\}, Y)$. ☐

The Ascoli-Arzelà theorem is the generalisation of Corollary 6.36 to the space $C(X, Y)$; first, though, a definition.

Definition 6.53 Let X be a space, (Y, d) a metric space. A family of continuous functions $\mathcal{F} \subset C(X, Y)$ is called **equicontinuous** if for any $x_0 \in X$ and $\varepsilon > 0$ there exists a neighbourhood U of x_0 such that

$$d(f(x), f(x_0)) < \varepsilon$$

for every $f \in \mathcal{F}$ and every $x \in U$. The family $\mathcal{F}$ is said to be **pointwise totally bounded** if the set
$$\{f(x) \mid f \in \mathcal{F}\}$$

is totally bounded in Y for every $x \in X$.

Observe that equicontinuity and pointwise total boundedness depend on the distance d.

Theorem 6.54 (Ascoli-Arzelà) *Let X be a compact topological space and (Y, d) a complete metric space. A family of continuous maps $\mathcal{F} \subset C(X, Y)$ is relatively compact in $C(X, Y)$ if and only if:*

1. *$\mathcal{F}$ is equicontinuous, and*
2. *$\mathcal{F}$ is pointwise totally bounded.*

Proof We begin by proving 1 and 2 are necessary, and suppose $\mathcal{F}$ is contained in a compact subspace $K \subset C(X, Y)$. Since the evaluation map at any $x \in X$

$$e_x \colon C(X, Y) \to Y, \quad e_x(f) = f(x),$$

is continuous, it follows that $\{f(x) \mid f \in \mathcal{F}\} = e_x(\mathcal{F})$ lies in the compact set $e_x(K)$.

Let $x_0 \in X$ be given, and consider the continuous function

$$\alpha \colon C(X, Y) \times X \to [0, +\infty[, \quad \alpha(f, x) = d(f(x_0), f(x)).$$

For any $\varepsilon > 0$ the compact set $K \times \{x_0\}$ is a subset in the open set $\alpha^{-1}([0, \varepsilon[)$, and Wallace's theorem gives us an open set $U \subset X$ such that $x_0 \in U$ and $K \times U \subset \alpha^{-1}([0, \varepsilon[)$; in particular $d(f(x_0), f(x)) < \varepsilon$ for every $f \in \mathcal{F}$ and any $x \in U$.

Now we show that if $\mathcal{F}$ fulfils 1 and 2, it is relatively compact in the complete space $C(X, Y)$. By Corollary 6.36 it suffices to prove that $\mathcal{F}$ is totally bounded. Fix a real $\varepsilon > 0$; the equicontinuity of $\mathcal{F}$ implies that for any $x \in X$ there's an open neighbourhood U_x with $d(f(x), f(y)) < \varepsilon$ for any $y \in U_x$ and any $f \in \mathcal{F}$. The space X is compact, so there are $x_1, \ldots, x_n \in X$ with

$$X = U_{x_1} \cup \cdots \cup U_{x_n}.$$

The range of
$$\mathcal{F} \to Y^n, \quad f \mapsto (f(x_1), \ldots, f(x_n))$$

is contained in the product $\prod e_{x_i}(\mathcal{F})$ of totally bounded sets, so it is totally bounded. We can find a finite $F \subset \mathcal{F}$ such that, for every $f \in \mathcal{F}$, there exists $g \in F$ with $d(f(x_i), g(x_i)) < \varepsilon$ for any i. We claim $\mathcal{F}$ lies in the union of open balls centred at $g \in F$ with radius 3ε: take $f \in \mathcal{F}$ and $g \in F$ such that $d(f(x_i), g(x_i)) < \varepsilon$ for every i. Then for any $x \in X$ there's an i such that $x \in U_{x_i}$, whence

$$d(f(x), g(x)) \leq d(f(x), f(x_i)) + d(f(x_i), g(x_i)) + d(g(x_i), g(x)) < 3\varepsilon. \qquad \square$$

Exercises

6.31 Say whether the family of continuous maps

$$\{f_n\} \subset C([0, 1], \mathbb{R}), \qquad f_n(x) = x^n,$$

is equicontinuous and if it has limit points in $C([0, 1], \mathbb{R})$.

6.32 (Dini's lemma) Let X be compact and $\{f_n\}$, $n \in \mathbb{N}$, a sequence in $C(X, \mathbb{R})$ such that:

1. $f_m(x) \geq f_{m+1}(x)$ for every $x \in X$ and every m;
2. $\lim_n f_n(x) = 0$ for every $x \in X$.

Prove that $\{f_n\}$ converges to 0 in $C(X, \mathbb{R})$. (Hint: consider the open family $U_n = \{x \in X \mid f_n(x) < \varepsilon\}$, $\varepsilon > 0$.)

6.33 ($\heartsuit$) Let X be a compact space and I a proper ideal in the ring $C(X, \mathbb{R})$. Prove that all maps $f \in I$ have a common zero $x \in X$.

6.8 Directed Sets and Nets (Generalised Sequences) ↶

Nets, also known as generalised sequences, extend the concept of a sequence in that they allow to apply most of the present chapter's results to spaces that may not be first countable.

Definition 6.55 An ordered set $(I, \leq)$ is called **directed** when, for every $i, j \in I$, there exists $h \in I$ such that $h \geq i$ and $h \geq j$.

Equivalently, a directed set is an ordered set in which every finite subset is upper bounded. Let us present a few examples:

1. the set $\mathbb{N}$ with the usual ordering and, more generally, $\mathbb{N}^n$ with the order relation $(a_1, \ldots, a_n) \leq (b_1, \ldots, b_n) \iff a_i \leq b_i$ for every i;
2. the set $\mathcal{P}_0(A)$ of finite subsets of A with the inclusion;
3. the set $\mathcal{I}(x)$ of neighbourhoods of a point x in a topological space, with the relation $U \leq V \iff V \subset U$.

Definition 6.56 A map $p\colon J \to I$ between directed sets is called a **cofinal morphism** if it preserves orderings ($p(j) \geq p(j')$ if $j \geq j'$) and for every $i \in I$ there's a $j \in J$ such that $p(j) \geq i$.

Example 6.57 Consider the family I of finite subsets inside a set A with even cardinality. The inclusion $I \hookrightarrow \mathcal{P}_0(A)$ is a cofinal morphism.

The composite of two cofinal morphisms is a cofinal morphism.

Definition 6.58 A **net** in a topological space X is a map $f\colon I \to X$, where I is directed.

For $I = \mathbb{N}$, a net $f\colon I \to X$ is nothing but a sequence in X. The name 'net' is inspired by the case $I = \mathbb{N}^n$.

Definition 6.59 Let X be topological space, $f\colon I \to X$ a net and x a point in X. One says that:

1. the net f **converges** to x if for any neighbourhood $U \in \mathcal{I}(x)$ there's an index $i \in I$ such that $f(j) \in U$ for every $j \geq i$;
2. x is a **limit point** of the net f if for any neighbourhood $U \in \mathcal{I}(x)$ and any $i \in I$ there's a $j \geq i$ such that $f(j) \in U$.

In a similar manner to what occurs with sequences, a converging net must converge to a limit point, whilst a limit point isn't necessarily a point to which the net converges, in general.

Lemma 6.60 *Let $f\colon I \to X$ be a net. Then $x \in X$ is a limit point for f if and only if there's a cofinal morphism $p\colon J \to I$ such that the net $fp\colon J \to X$ converges to x.*

Proof If fp converges to some x, then x is a limit point of f because p is cofinal.

As for the converse: suppose x is a limit point of f and call $\mathcal{I}(x)$ the family of neighbourhoods around x, as usual. Consider

$$J = \{(i, U) \in I \times \mathcal{I}(x) \mid f(i) \in U\},$$

with the relation

$$(i, U) \leq (j, V) \quad \text{if} \quad i \leq j \text{ and } V \subset U.$$

The set $(J, \leq)$ is directed: if $(i, U), (j, V) \in J$, there's an $h \in I$ such that $h \geq i$ and $h \geq j$. By definition of limit point, there exists $k \geq h$ such that $f(k) \in U \cap V$, so $(k, U \cap V) \geq (i, U)$, $(k, U \cap V) \geq (j, V)$. The projection onto the first factor $p\colon J \to I$ is surjective, hence a cofinal morphism (just notice that $p(i, X) = i$ for every $i \in I$). Let U be a neighbourhood of x and take $i \in I$ such that $f(i) \in U$. Then for every $(j, V) \geq (i, U)$ we have $fp(j, V) \in U$, and fp converges to x. $\square$

At this juncture we can generalise Proposition 6.18.

Proposition 6.61 *Let A be a subset in a space X. For any point $x \in X$ the following are equivalent requirements:*

1. *there is a net in A that converges to x;*
2. *there is a net in A with limit point x;*
3. *the point x belongs in the closure of A.*

Proof (1) $\Rightarrow$ (2) is trivial. If $x \in X$ is a limit point of some net $f: I \to A$, for any neighbourhood U of x there's an $i \in I$ such that $f(i) \in U$; hence $U \cap f(I) \neq \emptyset$, i.e. x belongs to the closure of $f(I)$. As $f(I) \subset A$ we have $x \in \overline{A}$, proving (2) $\Rightarrow$ (3).

Let us see that (3) $\Rightarrow$ (1). If $x \in \overline{A}$ and $\mathcal{I}(x)$ is the directed set of neighbourhoods around x, the axiom of choice guarantees that we may pick a net $f: \mathcal{I}(x) \to A$ such that $f(U) \in U \cap A$ for every neighbourhood U of x. Immediately, then, f converges to x. $\qquad\square$

Theorem 6.62 *A topological space X is compact if and only if every net has a limit point.*

Proof Suppose X is compact and there's a net $f: I \to X$ without limit points: this will lead to a contradiction. For any $x \in X$ (not a limit point) there are a neighbourhood $U(x) \in \mathcal{I}(x)$ and an index $i(x) \in I$ such that $f(j) \notin U(x)$ for any $j \geq i(x)$. By compactness we have $x_1, \ldots, x_n \in X$ such that $X = U(x_1) \cup \cdots \cup U(x_n)$. Let $h \in I$ be an upper bound for $i(x_1), \ldots, i(x_n)$. Then $f(h) \notin U(x_i)$ for any $i = 1, \ldots, n$, absurd.

Vice versa, assume X isn't compact and choose an open cover $X = \cup\{U_a \mid a \in A\}$ without finite subcovers. Consider the directed set $\mathcal{P}_0(A)$ of finite subsets of A. Using the axiom of choice we pick a net $f: \mathcal{P}_0(A) \to X$ such that $f(B) \notin U_a$ for any $B \in \mathcal{P}_0(A)$ and every $a \in B$. If $x \in U_a \subset X$ were a limit point, there would exist a finite subset $B \subset A$ containing a and such that $f(B) \in U_a$. But this cannot be, so the net f can't have limit points. $\qquad\square$

Example 6.63 To get an idea of but one possible application of nets, we outline the proof of the fact that *every finite-dimensional subspace of a topological vector space is closed*. The reader will find extremely worthwhile filling in all the gaps. For simplicity let's consider real vector spaces: for complex vectors spaces the argument is completely analogous.

A **topological vector space** V is a vector space furnished with a Hausdorff topology that makes the operation of taking linear combinations continuous; in particular, for any linearly independent vectors $v_1, \ldots, v_n$ the linear map $f: \mathbb{R}^n \to V$, $f(x_1, \ldots, x_n) = \sum x_i v_i$ is continuous and 1-1. Let us prove f is a closed immersion. Take $C \subset \mathbb{R}^n$ closed and $v \in \overline{f(C)}$; we have to show $v \in f(C)$. Let $s: I \to f(C)$ be a net converging to $v \in V$ and write $t = f^{-1} \circ s: I \to C$. Next, immerse $\mathbb{R}^n$ in its one-point compactification $\mathbb{R}^n \cup \{\infty\} \simeq S^n$; we may assume that $\lim_i t(i) \in \mathbb{R}^n \cup \{\infty\}$ exists, possibly composing with another cofinal morphism.

There are two cases to consider, in the first of which the limit of t is contained in $\mathbb{R}^n$; as C is closed in $\mathbb{R}^n$ we have $v = f(\lim_i t(i)) \in f(C)$.

In the second case $\lim_i t(i) = \infty$, so $\lim_i \|t(i)\| = +\infty$ and $\lim_i \dfrac{s(i)}{\|t(i)\|} = 0$.

On the other hand $\dfrac{s(i)}{\|t(i)\|} = f\left(\dfrac{t(i)}{\|t(i)\|}\right)$, and the net $i \mapsto \dfrac{t(i)}{\|t(i)\|}$ has at least one limit point $a \in S^{n-1}$. Therefore $f(a) = 0$, contradicting the injectivity of f.

Exercises

6.34 Prove that a space is Hausdorff if and only if every net converges to one point, at most.

6.35 Say if the following nets $a \colon \mathbb{N} \times \mathbb{N} \to \mathbb{R}$ converge, and if so compute the limits:

$$a(n, m) = \frac{1}{n + m}, \qquad a(n, m) = \frac{1}{(n - m)^2 + 1}, \qquad a(n, m) = ne^{-m}.$$

6.36 Let S be a non-empty set and $a \colon S \to \mathbb{R}$ any function. Explain the meaning of the expression 'the series $\sum_{s \in S} f(s)$ converges'. (Hint: consider the directed set of finite subsets of S.)

6.37 (✿) Let $\{a_n\}$ be a real sequence. Consider the directed set $\mathcal{P}_0(\mathbb{N})$ of finite subsets in $\mathbb{N}$ and the net

$$f \colon \mathcal{P}_0(\mathbb{N}) \to \mathbb{R}, \qquad f(A) = \sum_{n \in A} a_n.$$

Prove that f is convergent if and only if the series $\sum_{n=1}^{+\infty} a_n$ is absolutely convergent.

Chapter 7
Manifolds, Infinite Products and Paracompactness

7.1 Sub-bases and Alexander's Theorem

Definition 7.1 A **sub-basis** of a topological space is a family $\mathcal{P}$ of open sets such that finite intersections in $\mathcal{P}$ form a basis of the topology.

Every basis is also a sub-basis.

Example 7.2 A sub-basis for the Euclidean topology on $\mathbb{R}$ is given by the open sets $]-\infty, a[,\ ]b, +\infty[$ as $a, b \in \mathbb{R}$ vary, since open intervals $]a, b[$ form a basis, and we can write $]a, b[=]-\infty, b[\cap]a, +\infty[$.

Lemma 7.3 *Let X, Y be topological spaces and $\mathcal{P}$ a sub-basis of Y. A map $f : X \to Y$ is continuous if and only if $f^{-1}(U)$ is open for every $U \in \mathcal{P}$.*

Proof Just observe that f^{-1} commutes with union and intersection. $\qquad\square$

Let $\mathcal{P}$ be a cover of a set X, and consider the family $\mathcal{B}$ of finite intersections in $\mathcal{P}$. If $A, B \in \mathcal{B}$ then $A \cap B \in \mathcal{B}$, and $\mathcal{B}$ covers X. By Theorem 3.7 $\mathcal{B}$ is a basis of a topology that has $\mathcal{P}$ as sub-basis. It's easy to see that this topology is the coarsest one having $\mathcal{P}$ as open sets.

Example 7.4 Take a set S and a topological space X; in the set X^S of maps $f : S \to X$ consider the family $\mathcal{P}$ of subsets

$$P(s, U) = \{f : S \to X \mid f(s) \in U\}$$

for any $s \in S$ and any open set U. The coarsest topology on X^S containing $\mathcal{P}$ is called **pointwise-convergence topology**, and has $\mathcal{P}$ as sub-basis.

Theorem 7.5 (Alexander) *Let $\mathcal{P}$ be a sub-basis for X. If every cover of X made by elements of $\mathcal{P}$ has a finite subcover, then X is compact.*

© Springer International Publishing Switzerland 2015
M. Manetti, *Topology*, UNITEXT - La Matematica per il 3+2 91,
DOI 10.1007/978-3-319-16958-3_7

Proof Suppose X not compact and let's prove there's a cover consisting of elements of $\mathcal{P}$ that has no finite subcovers. Call $\mathcal{T}$ the family of open sets and $\mathbf{Z}$ the collection of subfamilies $\mathcal{A} \subset \mathcal{T}$ covering X but without finite subcovers. Non-compactness implies that $\mathbf{Z}$ is not empty. Consider on $\mathbf{Z}$ the inclusion order, i.e. $\mathcal{A} \leq \mathcal{A}'$ if $\mathcal{A} \subset \mathcal{A}'$. Let $\mathbf{C} \subset \mathbf{Z}$ be a chain. Then $\mathcal{C} = \cup\{\mathcal{A} \in \mathbf{C}\} \in \mathbf{Z}$ and $\mathcal{C}$ is an upper bound for $\mathbf{C}$: in fact if there were a finite subcover $\{A_1, \ldots, A_n\} \subset \mathcal{C}$, every A_i would belong to some $\mathcal{A}_i$ in the chain $\mathbf{C}$ and we'd have the contradiction $\{A_1, \ldots, A_n\} \subset \max(\mathcal{A}_1, \ldots, \mathcal{A}_n)$. By Zorn's lemma there is a maximal element $\mathcal{Z} \in \mathbf{Z}$.

Now we shall prove that $\mathcal{P} \cap \mathcal{Z}$ is an open cover, i.e. for any x there's a $P \in \mathcal{P} \cap \mathcal{Z}$ such that $x \in P$. Take $x \in X$ and choose $A \in \mathcal{Z}$ such that $x \in A$; by definition of sub-basis we have $P_1, \ldots, P_n \in \mathcal{P}$ such that $x \in P_1 \cap \cdots \cap P_n \subset A$; we have to show $P_i \in \mathcal{Z}$ for some $i = 1, \ldots, n$. So, suppose not: then for every $i = 1, \ldots, n$ the cover $\mathcal{Z}_i = \mathcal{Z} \cup \{P_i\}$ is strictly bigger than $\mathcal{Z}$, whence it cannot belong to $\mathbf{Z}$; therefore there's a finite subcover $X = P_i \cup A_{i,1} \cup \cdots \cup A_{i,s_i}$, where $A_{i,j} \in \mathcal{Z}$. We deduce that $X = (\cap_i P_i) \cup_{i,j} A_{i,j}$, making $X = A \cup \cup_{i,j} A_{i,j}$ a finite subcover of $\mathcal{Z}$.

Summarising, $\mathcal{P} \cap \mathcal{Z}$ is an open cover of X consisting of sub-basis sets; being contained in $\mathcal{Z}$, on the other hand, it cannot admit finite subcovers. But this is precisely what we had to prove. $\square$

It is important to remark, by what we saw in Example 2.8, that a space is second countable if and only if it admits a countable sub-basis.

Exercises

7.1 Let $\mathcal{P}(\mathbb{N})$ be the family of all subsets of $\mathbb{N}$. For any pair of *finite and disjoint* subsets $A, B \subset \mathbb{N}$ define

$$U(A, B) = \{S \in \mathcal{P}(\mathbb{N}) \mid A \subset S, \ S \cap B = \emptyset\}.$$

Prove that:

1. as A and B vary, the $U(A, B)$ form a basis for a topology $\mathcal{T}$ on $\mathcal{P}(\mathbb{N})$;
2. the subsets

$$P(n) = \{S \in \mathcal{P}(\mathbb{N}) \mid n \in S\}, \qquad Q(n) = \{S \in \mathcal{P}(\mathbb{N}) \mid n \notin S\},$$

   form, as $n \in \mathbb{N}$ varies, a sub-basis of $\mathcal{T}$;
3. the space $(\mathcal{P}(\mathbb{N}), \mathcal{T})$ is compact and Hausdorff; (Hint: consider a cover $\mathcal{P}(\mathbb{N}) = \cup_{n \in A} P(n) \cup_{m \in B} Q(m)$, it follows $B \in \cup_{n \in A} P(n)$, and therefore $A \cap B \neq \emptyset$; use Alexander's theorem.)
4. the function

$$f : \mathcal{P}(\mathbb{N}) \to \left[0, \frac{1}{r-1}\right], \qquad f(S) = \sum_{n \in S} \frac{1}{r^n},$$

   is continuous for any $r > 1$. When $r > 2$ it's also 1-1.

7.2 ($\heartsuit$) Consider the space X of maps $f : \mathbb{R} \to \mathbb{R}$ with the topology of pointwise convergence (Example 7.4). Prove that X is Hausdorff and not first countable.

7.2 Infinite Products

Given an arbitrary family $\{X_i \mid i \in I\}$ of sets one defines the Cartesian product $X = \prod_{i \in I} X_i$ as the set of all maps $x : I \to \cup_i X_i$ such that $x_i \in X_i$ for every index $i \in I$. That's to say that any element of the product is a collection $\{x_i\}_{i \in I}$ indexed by I such that $x_i \in X_i$ for every i. The axiom of choice ensures that X isn't empty provided each X_i is non-empty. The projections $p_i : X \to X_i$ are defined as $p_i(x) = x_i$.

When the X_i are all equal to some set X_0, the product $\prod_{i \in I} X_i = \prod_{i \in I} X_0$ coincides with X_0^I, the set of maps $I \to X_0$. For any set Y and any $f : Y \to \prod_{i \in I} X_i$ we write f_i to mean f followed by the projection p_i. Note that f is uniquely determined by the family $\{f_i : Y \to X_i \mid i \in I\}$.

If all X_i are topological spaces, we define product topology on X the coarsest one for which the projections are continuous. This amounts to say that $p_i^{-1}(U)$, for every $i \in I$ and U open in X_i, constitute a sub-basis that we shall call **canonical sub-basis**. A **canonical basis** is given by open sets that are finite intersections of the canonical sub-basis.

When the X_i are all equal the product topology coincides with the pointwise-convergence topology (Example 7.4), because the two have the same sub-basis.

Lemma 7.6 *In the previous notations, a map $f : Y \to \prod_{i \in I} X_i$ is continuous if and only if all its components $f_i : Y \to X_i$ are continuous.*

Proof The projections p_i are continuous, so f continuous implies that the components $f_i = p_i f$ are.

Conversely, suppose each component f_i is continuous. For any open set $p_i^{-1}(U)$ in the canonical sub-basis we have $f^{-1}(p_i^{-1}(U)) = f_i^{-1}(U)$ is open, implying f continuous. $\qquad\qquad\square$

Theorem 7.7 (Tychonov) *The product of an arbitrary family of compact spaces is compact.*

Proof Take $X = \prod_{i \in I} X_i$ with the product topology and suppose all X_i are compact. We'll prove compactness through Alexander's Theorem 7.5.

Consider a collection $\mathcal{A}$ of open sets from the canonical sub-basis: assigning $\mathcal{A}$ is the same as assigning, for every i, a collection $\mathcal{A}_i$ of open sets in X_i such that

$$\mathcal{A} = \{p_i^{-1}(U) \mid i \in I, \ U \in \mathcal{A}_i\}.$$

Suppose that $\mathcal{A}$ is a cover; then $\mathcal{A}_i$ covers X_i for some index i. In fact if $C_i = X_i - \cup\{U \mid U \in \mathcal{A}_i\}$ was non-empty for every i, by the axiom of choice there would exist $x \in X$ such that $x_i \in C_i$ for each i, so

$$x \notin \cup\{p_i^{-1}(U) \mid i \in I, \ U \in \mathcal{A}_i\} = \cup\{V \mid V \in \mathcal{A}\}.$$

Therefore take that $i \in I$ for which $\mathcal{A}_i$ is a cover of X_i. By compactness there's a finite subcover $X_i = U_1 \cup \cdots \cup U_n$ with $U_j \in \mathcal{A}_i$, hence $\{p_i^{-1}(U_j)\} \subset \mathcal{A}$ is a finite subcover. $\square$

Remark 7.8 There are proofs of Theorem 7.7 in the literature that don't require Alexander's result. Yet any other proof makes use of the axiom of choice or an equivalent statement: that's because *Tychonov's theorem is equivalent to the axiom of choice* (Exercise 7.6).

Theorem 7.9 *The product of any number of connected spaces is connected.*

Proof We already know that this is true for products of two spaces, and by induction the same carries through for a finite number. Consider now the product $X = \prod_{i \in I} X_i$ of an arbitrary family of connected spaces. If X is empty there's nothing to prove. Suppose $X \neq \emptyset$ and choose $x \in X$. Call $F(x) \subset X$ the subset of points y such that $y_i \neq x_i$ for finitely many indices $i \in I$ at most. Because the closure of a connected subset is connected, we can limit ourselves to proving that:

1. $F(x)$ is connected;
2. $F(x)$ is dense in X.

If $J \subset I$ is finite we denote by $p_J : X \to \prod_{j \in J} X_j$ the projection to the J-coordinates, and define

$$h_J : \prod_{j \in J} X_j \to X, \qquad h_J(z)_i = \begin{cases} z_i & \text{if } i \in J \\ x_i & \text{if } i \notin J \end{cases}.$$

Each h_J is continuous, so the range of h_J is connected and contains x. By definition $F(x)$ is the union of these ranges, and Lemma 4.23 implies $F(x)$ is connected.

By the definition of product topology, for every J finite in I and U open in $\prod_{j \in J} X_j$, the open sets $p_J^{-1}(U)$ contain the canonical basis and are therefore a basis. Moreover $h_J^{-1}(p_J^{-1}(U)) = U$, so if U isn't empty $F(x) \cap p_J^{-1}(U) \neq \emptyset$: in particular $F(x)$ intersects every non-empty open set of the canonical basis. $\square$

Exercises

7.3 Prove that the product of an arbitrary number of Hausdorff spaces is Hausdorff.

7.4 Prove that the product of countably many second-countable spaces is second countable.

7.5 Show that $\mathcal{P}(\mathbb{N})$, with the topology of Exercise 7.1, is homeomorphic to the product of $|\mathbb{N}|$ copies of $\{0, 1\}$.

7.6 Let $g\colon Y \to X$ be a surjective map between sets. Put on Y the topology whose closed sets are Y and all subsets $g^{-1}(A)$ with $A \subset X$ finite. Prove Y is compact.

Identify the space S of maps $f\colon X \to Y$ with the product of $|X|$ copies of Y; by Tychonov's theorem the latter topology turns S into a compact space.

If $A \subset X$ is a finite subset, there exist maps $f\colon X \to Y$ such that $g(f(x)) = x$ for every $x \in A$: to prove the existence of such an f one needs to make finitely many choices, so the axiom of choice doesn't come into play. As A varies among all finite subsets in X, prove that the subspaces

$$F(A) = \{f \in S \mid g(f(x)) = x \text{ for all } x \in A\}$$

are closed in the product and satisfy the finite intersection property (Exercise 4.25). Deduce that Tychonov's theorem implies the axiom of choice.

7.7 (✹, ♡) Let S be the set of maps $a\colon \mathbb{N} \to \{-1, 1\}$. Prove that the product of $|S|$ copies of $[-1, 1]$ is compact but not sequentially compact.

7.8 (✹) Let I be the interval $[0, 1]$ and endow $X = I^I = \{f\colon I \to I\}$ with the product topology. By Tychonov's theorem X is compact and Hausdorff.

Call $B \subset X$ the subspace of maps $f\colon I \to I$ such that $f(x) \neq 0$ for at most countably many $x \in I$. Show that B is dense in X, not compact but sequentially compact.

7.3 Refinements and Paracompactness ↻

The notion of being locally finite extends in a natural way to arbitrary families of subsets.

Definition 7.10 A family $\mathcal{A}$ of subsets in a space X is **locally finite** if every point $x \in X$ admits a neighbourhood $V \in \mathcal{I}(x)$ such that $V \cap A \neq \emptyset$ for at most finitely many $A \in \mathcal{A}$.

Since any neighbourhood contains an open set, and an open set intersects a subset A if and only if it intersects the closure, a family $\{A_i \mid i \in I\}$ is locally finite if and only if $\{\overline{A_i} \mid i \in I\}$ is locally finite.

Lemma 7.11 *For any locally finite family $\{A_i\}$ of subsets,*

$$\overline{\bigcup_i A_i} = \bigcup_i \overline{A_i}.$$

In particular the union of a locally finite family of closed sets is closed.

Proof The relation $\bigcup_i \overline{A_i} \subset \overline{\bigcup_i A_i}$ is always true because the closed set $\overline{\bigcup_i A_i}$ contains A_i, and so $\overline{A_i}$, for every i. There remains to prove that if $\{A_i\}$ is locally finite, then

$\cup_i \overline{A_i}$ is closed. We can find an open cover $X = \cup_j U_j$ such that U_j intersects finitely many sets A_i, whence $(\cup_i \overline{A_i}) \cap U_j = \cup_i (U_j \cap \overline{A_i})$ is closed in U_j. To conclude, recall that any open cover is an identification cover. $\square$

Definition 7.12 Let $\{U_i \mid i \in I\}$ and $\{V_j \mid j \in J\}$ be covers of some space. One says that $\{U_i \mid i \in I\}$ **refines** $\{V_j \mid j \in J\}$, or that $\{U_i \mid i \in I\}$ is a **refinement** of $\{V_j \mid j \in J\}$, if for every $i \in I$ there is a $j \in J$ such that $U_i \subset V_j$. In such a case we will call a map $f : I \to J$ such that $U_i \subset V_{f(i)}$ for any $i \in I$ a **refinement function**.

Example 7.13 Consider covers $\{U_i \mid i \in I\}$ and $\{V_j \mid j \in J\}$. Then the cover $\{U_i \cap V_j \mid (i, j) \in I \times J\}$ is a refinement of both. As refinement functions we may take the projections $I \times J \to I, I \times J \to J$.

Definition 7.14 A space is called **paracompact** if every open cover possesses a locally finite open refinement.

Any compact space is clearly paracompact. The reader might wonder why we insist on refinements and not just consider locally finite subcovers: the reason is that if every open cover in X has a locally finite subcover, X is compact (easy exercise). Similarly, if all open covers of X have finite refinements, again X is compact (super-easy exercise).

Theorem 7.15 *Let X be a Hausdorff space.*

1. *If X is exhausted by compact sets, then X is paracompact and locally compact.*
2. *If X is connected, paracompact and locally compact, it has an exhaustion by compacts sets.*

Proof Concerning (1), let $K_1 \subset K_2 \subset \cdots$ be a compact exhaustion of X and $\mathcal{B}$ an open basis. We claim that for any open cover $\mathcal{A}$ of X there's a subfamily $\mathcal{C} \subset \mathcal{B}$ that is a locally finite open refinement of $\mathcal{A}$.

We set, by convention, $K_n = \emptyset$ in case $n \le 0$. For every $n \in \mathbb{N}$ and any $x \in K_n - K_{n-1}^\circ$ let's choose an open set $A \in \mathcal{A}$ such that $x \in A$ and a basis element $B(n, x) \in \mathcal{B}$ such that $x \in B(n, x)$ and $B(n, x) \subset A \cap (K_{n+1}^\circ - K_{n-2})$. The $B(n, x)$ cover the compact set $K_n - K_{n-1}^\circ$, so there's a finite subcover $K_n - K_{n-1}^\circ \subset B(n, x_1) \cup \cdots \cup B(n, x_s)$. The union, as n varies, of all such covers is the required family $\mathcal{C}$.

Now we show (2). Suppose X connected, paracompact and locally compact. The latter property says we can find an open cover $\mathcal{A}$ of X such that $\overline{A}$ is compact for any $A \in \mathcal{A}$, and paracompactness allows to refine it with a locally finite open cover $\mathcal{B}$: clearly $\overline{B}$ is compact for every $B \in \mathcal{B}$. We also notice that by Lemma 7.11 the union

$$\cup \{\overline{B} \mid B \in \mathcal{C}\}$$

is closed in X, for any subfamily $\mathcal{C} \subset \mathcal{B}$. Let $\mathcal{B}_1$ be a finite subset in $\mathcal{B}$ such that the compact set $K_1 = \cup \{\overline{B} \mid B \in \mathcal{B}_1\}$ is non-empty. There exists a finite family $\mathcal{B}_2 \subset \mathcal{B}$ such that

$$K_1 \subset \cup \{B \mid B \in \mathcal{B}_2\}.$$

A recursive process yields a sequence of finite families $\mathcal{B}_n \subset \mathcal{B}$ such that

$$K_n = \cup \{\overline{B} \mid B \in \mathcal{B}_n\} \subset \cup \{B \mid B \in \mathcal{B}_{n+1}\}$$

for every n. To finish we have to show that $\cup_n K_n = X$. But

$$\cup_n K_n = \cup \{\overline{B} \mid B \in \cup_n \mathcal{B}_n\} = \cup \{B \mid B \in \cup_n \mathcal{B}_n\}$$

implies that $\cup_n K_n$ is open and closed in the connected space X. In conclusion, X is exhausted by the compact family $\{K_n\}$. □

Corollary 7.16 *Every locally compact, second-countable Hausdorff space is para-compact.*

Proof By Theorem 6.13 any locally compact Hausdorff space with a countable basis has an exhaustion by compact sets, and Theorem 7.15 guarantees its paracompactness. □

Lemma 7.17 *In paracompact Hausdorff spaces every point has a local basis of closed neighbourhoods.*

Proof Start with X paracompact and Hausdorff. We have to prove that if $U \subset X$ is open and contains x, there's an open $V \subset X$ such that $X - U \subset V$ and $x \notin \overline{V}$. It follows from this that the closed set $X - V$ is a neighbourhood of x and lies in U.

Write $C = X - U$; as X is Hausdorff, every $y \in C$ has an open neighbourhood $W_y \in \mathcal{I}(y)$ such that $x \notin \overline{W_y}$. The space is paracompact, so the cover $X = U \cup \{W_y \mid y \in C\}$ admits a locally finite refinement $X = \cup \{V_i \mid i \in I\}$. Consider the open set $V = \cup \{V_i \mid i \in I,\ V_i \cap C \neq \emptyset\}$: if $V_i \cap C \neq \emptyset$ then V_i lies in some W_y. Consequently $x \notin \overline{V_i}$ and $x \notin \overline{V} = \cup \{\overline{V_i} \mid V_i \cap C \neq \emptyset\}$. □

Theorem 7.18 (Refinement theorem) *Let $X = \cup \{U_i \mid i \in I\}$ be an open cover of a paracompact Hausdorff space. There exists a locally finite open cover $X = \cup \{V_i \mid i \in I\}$ such that $\overline{V_i} \subset U_i$ for every $i \in I$.*

Proof For any $x \in X$ choose $i(x) \in I$ with $x \in U_{i(x)}$, and also an open neighbourhood $W(x) \in \mathcal{I}(x)$ such that $\overline{W(x)} \subset U_{i(x)}$. If $X = \cup \{A_j \mid j \in J\}$ is a locally finite open refinement of the open cover $X = \cup \{W(x) \mid x \in X\}$, there exists a refinement function $f : J \to I$ satisfying $\overline{A_j} \subset U_{f(j)}$ for every $j \in J$. Consider, for any $i \in I$, the open set $V_i = \cup \{A_j \mid f(j) = i\}$: as $\{A_j\}$ is locally finite, Lemma 7.11 implies $\overline{V_i} = \cup \{\overline{A_j} \mid f(j) = i\} \subset U_i$. □

Eventually, we state a result due to A.H. Stone. The interested reader will find a proof in [Du66, Mu00].

Theorem 7.19 (Stone) *Every metrisable space is paracompact.*

Exercises

7.9 Prove that any open cover of a space X possesses refinements made by open sets from some basis $\mathcal{B}$.

7.10 Prove that every closed subspace in a paracompact space is paracompact.

7.11 (✋) Prove that the product of a compact space with a paracompact space is paracompact.

7.12 (✋) Prove that the inteval $[a, b[$ $(a < b)$ is paracompact in the lower-limit topology (Example 3.9). Conclude that the lower-limit line is paracompact. (Hint: it might be worth observing that every cover made by disjoint open sets is locally finite.)

7.4 Topological Manifolds

Definition 7.20 A space M is called an **n-dimensional topological manifold** if:

1. M is Hausdorff;
2. every point in M has an open neighbourhood homeomorphic to an open set of $\mathbb{R}^n$;
3. every connected component of M is second countable.

Example 7.21 Any open set in $\mathbb{R}^n$ is an n-dimensional topological manifold, and any open subset in $\mathbb{C}^n$ is a topological manifold of dimension $2n$.

Example 7.22 The sphere S^n is a topological manifold of dimension n: each point x lies in the open set $S^n - \{-x\}$, which is homeomorphic to $\mathbb{R}^n$ under stereographic projection.

Example 7.23 The real projective space $\mathbb{P}^n(\mathbb{R})$ is an n-dimensional topological manifold: every point lies in the complement of some hyperplane H, and $\mathbb{P}^n(\mathbb{R}) - H$ is an open set homeomorphic to $\mathbb{R}^n$.

Example 7.24 The complex projective space $\mathbb{P}^n(\mathbb{C})$ is a $2n$-dimensional topological manifold: any point lies in the complement of some hyperplane H and $\mathbb{P}^n(\mathbb{C}) - H$ is open and homeomorphic to $\mathbb{C}^n$.

Remark 7.25 The three conditions of Definition 7.20 are independent from one another, in that any two do not imply the third one. The space described in Exercise 5.8 is connected and second countable, any of its points has a neighbourhood homeomorphic to $\mathbb{R}$, but it is not Hausdorff. Exercise 6.6 provides an instance of a connected Hausdorff space that is locally homeomorphic to $\mathbb{R}^2$ but not second countable.

In many textbooks condition 3. is replaced by paracompactness, and in the rest of this section we set out to prove that the two definitions are equivalent.

Lemma 7.26 *Suppose every point in a space M has an open neighbourhood homeomorphic to an open set of $\mathbb{R}^n$. Then M is locally compact and locally connected.*

Proof Take x in M; by assumption there is an open neighbourhood U of 0 in $\mathbb{R}^n$ and an open immersion $f : U \to M$ such that $f(0) = x$. If $B(0, r) \subset U$ for some $r > 0$, the $f(\overline{B(0, t)})$ constitute, for every $0 < t < r$, a local basis of closed, compact and connected neighbourhoods of x. $\qquad\Box$

Proposition 7.27 *A topological manifold is locally compact and Hausdorff. Each connected component is open and exhausted by compact sets.*

Proof Local connectedness and Hausdorff-local compactness follow from Lemma 7.26. Lemma 4.28 guarantees that connected components are open, while Theorem 6.13 ensures they are exhausted by compact sets. $\qquad\Box$

Corollary 7.28 *Every topological manifold is paracompact.*

Proof A connected topological manifold is locally compact, Hausdorff and second countable, hence paracompact by Corollary 7.16. A topological manifold is also the disjoint union of its components, and it's straightforward from the definition that the disjoint union of paracompact spaces is still paracompact. $\qquad\Box$

Corollary 7.29 *Let M be paracompact, Hausdorff and such that any point has an open neighbourhood homeomorphic to an open set in $\mathbb{R}^n$. Then M is a topological manifold.*

Proof In Lemma 7.26 we proved that M is locally compact and all connected components are open. Let M_0 be a given connected component. By Theorem 7.15 M_0 is exhausted by compact sets K_n. We cover each K_n with a finite number of open sets homeomorphic to open sets of $\mathbb{R}^n$. Therefore M_0 is the countable union of open sets, each of which is second countable. But then also M_0 is second countable. $\qquad\Box$

Exercises

7.13 Prove that every 0-dimensional topological manifold is a discrete space, and conversely, every discrete topological space is a topological manifold of dimension zero.

7.14 Prove that a non-empty open set in a topological manifold is a topological manifold of the same dimension.

7.15 Prove that the product of two topological manifolds is a topological manifold.

7.16 Let M be a connected topological manifold of dimension bigger than 1 and $K \subset M$ a finite subset. Prove that $M - K$ is connected. (Hint: Example 4.5.)

7.17 (✿) Let M be a connected topological manifold and $p, q \in M$ two points. Prove that there exists a homeomorphism $f : M \to M$ such that $f(p) = q$.

7.18 (✿) Prove that the Grassmannian $G(k, \mathbb{R}^n)$ (Exercise 5.22) is a topological manifold of dimension $k(n - k)$.

7.5 Normal Spaces ↷

Definition 7.30 A space is said to be **normal** if it is Hausdorff and disjoint closed sets are separated by disjoint open sets.

Rephrasing, a Hausdorff space X is normal if any two disjoint closed sets $A, B \subset X$ are contained in disjoint open sets $U, V \subset X$: $A \subset U, B \subset V$ and $U \cap V = \emptyset$.

Proposition 7.31

1. *Every metrisable space is normal.*
2. *Every paracompact Hausdorff space is normal.*

Proof (1) We know that a metric space (X, d) is Hausdorff. If A, B denote disjoint closed sets, consider the map $f : X \to [0, 3]$,

$$f(x) = \frac{3d_A(x)}{d_A(x) + d_B(x)}, \quad \text{where} \quad d_Y(x) = \inf_{y \in Y} d(x, y).$$

According to Example 3.47 f is continuous, the open sets $U = f^{-1}([0, 1[)$, $V = f^{-1}(]2, 3])$ are disjoint and contain A and B respectively.

(2) Take X paracompact Hausdorff and A, B disjoint and closed. Consider the open cover $X = (X - A) \cup (X - B)$; by the refinement theorem 7.18 there exist open sets U, V such that

$$U \cup V = X, \quad \overline{U} \subset X - A \quad \text{and} \quad \overline{V} \subset X - B.$$

Hence $A \subset X - \overline{U}, B \subset X - \overline{V}$ and $(X - \overline{U}) \cap (X - \overline{V}) = \emptyset$.   □

Definition 7.32 Given two covers $\{U_i\}$ and $\{V_j\}$ of the same space X, we say that $\{V_j\}$ is a **star refinement** of $\{U_i\}$ if, for any $x \in X$, the open set

$$V(x) = \cup\{V_j \mid x \in V_j\}$$

lies in some U_i. A space is **fully normal** if it is Hausdorff and each open cover admits a star refinement.

Every fully normal space is normal: if C, D are closed and disjoint we consider a star refinement $\{V_j\}$ of $X = (X - C) \cup (X - D)$, and observe that the open sets

$$U = \cup\{V_j \mid V_j \cap C \neq \emptyset\}, \quad W = \cup\{V_j \mid V_j \cap D \neq \emptyset\}$$

don't intersect.

Theorem 7.33 *Every paracompact Hausdorff space is fully normal.*

Proof Let $\{U_i\}$ be an open cover of the paracompact Hausdorff space X, of which we seek a star refinement. Possibly refining further, we can assume $X = \cup\{U_i\}$ is locally finite. By the refinement theorem there's a closed cover $X = \cup\{C_i\}$ such that $C_i \subset U_i$ for every i; in particular $\{C_i\}$ is locally finite. Now let $W(x)$ be an open neighbourhood of $x \in X$ that intersects finitely many elements of $\{C_i\}$. Refining $W(x)$ again, if necessary, let's assume the following:

1. if $x \notin C_i$, then $W(x) \cap C_i = \emptyset$;
2. if $x \in C_i$, then $W(x) \subset U_i$.

Fix a point $x \in X$. There is an index i such that $x \in C_i$, so if $x \in W(y)$ then $W(y) \cap C_i \neq \emptyset$, $y \in C_i$ and $W(y) \subset U_i$. Thus we have proved that $\{W(y)\}$ is a star refinement of $\{U_i\}$. □

Exercises

7.19 Prove that every closed subspace in a normal space is normal.

7.20 Use Wallace's theorem to prove directly that compact plus Hausdorff implies normal.

7.21 Let X be Hausdorff and $\{A_n \mid n \in \mathbb{N}\}$ a countable family of connected, compact subspaces such that $A_{n+1} \subset A_n$, for all n. Prove that $\cap_n A_n$ is connected. (Hint: assume—without loss of generality—that $X = A_1$. If $\cap_n A_n$ were contained in the union of two open disjoint sets we could apply Exercise 4.26...)

7.22 Let X be compact, connected and Hausdorff. Prove that the family of closed and connected subspaces that contain a given $A \subset X$ has minimal elements for the inclusion. (Of course, when A is connected every minimal element as above must be equal to $\overline{A}$.)

7.23 (🐝, ♡) Let $\{U_i \mid i \in I\}$ be a locally finite open cover of a normal space X. Prove that there's an open cover $\{V_i \mid i \in I\}$ such that $\overline{V_i} \subset U_i$ for every i. (Hint: use Zorn's lemma on the set of pairs $(J, \{V_j \mid j \in J\})$, where $J \subset I$, $V_j \subset X$ is open with $\overline{V_j} \subset U_j$ for every $j \in J$, and $(\cup_{j \in J} V_j) \cup (\cup_{i \notin J} U_i) = X$.)

7.24 (🐝) Use Exercise 6.28 and the arguments in Exercise 3.62 to show that the lower-limit plane $\mathbb{R}_l \times \mathbb{R}_l$ is not normal, and hence neither paracompact.

7.6 Separation Axioms ∼

Being Hausdorff or normal are perhaps the most familiar instances of what are called in the literature **separation axioms**. The imagination of mathematicians has produced many more properties of this kind, and the next definition summarises only the most renowned, which Tietze introduced in 1921 under the name *Trennbarkeitsaxiome*.

Definition 7.34 A space is called:

- **T0** if distinct points have distinct closures;
- **T1** if every point is closed;
- **T2** (or Hausdorff[1]) if distinct points $c \neq d$ are contained in disjoint open sets: $c \in U, d \in V$ and $U \cap V = \emptyset$;
- **T3** if every closed set C and every point $d \notin C$ are contained in disjoint open sets: $C \subset U, d \in V$ and $U \cap V = \emptyset$;
- **T4** if every two closed disjoint sets C, D are contained in open disjoint sets: $C \subset U, D \subset V$ and $U \cap V = \emptyset$.

Definition 7.35 A space is called **regular** if it is T1 and T3.

Under axiom T1, i.e. if points are closed, T4 implies T3 and T3 implies T2. In particular, a space is normal if and only if it satisfies T1 and T4, so that

$$\text{metrisable} \Rightarrow \text{normal} \Rightarrow \text{regular} \Rightarrow \text{Hausdorff} \Rightarrow \text{T1} \Rightarrow \text{T0}.$$

Theorem 7.36 *Every second-countable regular space is normal. In particular, subspaces and finite products of second-countable normal spaces are normal.*

Proof Let X be regular with countable basis $\mathcal{B}$, and let C, D be disjoint closed sets. Regularity implies that for every $x \in C$ there's a $U \in \mathcal{B}$ with

$$x \in U \subset \overline{U} \subset X - D,$$

and similarly for every point $y \in D$. Consider the subfamilies

$$\mathcal{C} = \{U \in \mathcal{B} \mid \overline{U} \cap D = \emptyset\}, \qquad \mathcal{D} = \{V \in \mathcal{B} \mid \overline{V} \cap C = \emptyset\}.$$

Both $\mathcal{C}$ and $\mathcal{D}$ being countable allows to write $\mathcal{C} = \{U_n \mid n \in \mathbb{N}\}$, $\mathcal{D} = \{V_n \mid n \in \mathbb{N}\}$, and then $C \subset \cup\{U_n \mid n \in \mathbb{N}\}$, $D \subset \cup\{V_n \mid n \in \mathbb{N}\}$. For each n consider open sets

$$A_n = U_n - \bigcup_{j \leq n} \overline{V_j}, \qquad B_n = V_n - \bigcup_{j \leq n} \overline{U_j}.$$

Suppose $m \geq n$. Then $A_m \cap V_n = B_m \cap U_n = \emptyset$ and therefore $A_m \cap B_n = B_m \cap A_n = \emptyset$. But the last relation means that $A_n \cap B_m = \emptyset$ for any chosen $n, m \in \mathbb{N}$. Eventually, if $A = \cup A_n$ and $B = \cup B_n$ we have $C \subset A$, $D \subset B$ and $A \cap B = \emptyset$.

Now observe that subspaces and products of regular spaces are regular (Exercise 7.25). $\qquad\square$

Traditionally, Theorem 7.36 is served together with another result which we state without proof.

---

[1]Rather confusingly, the term 'separated' can sometimes be found to mean T2, whereas being 'separable' (as of Definition 6.5) is not a separation axiom. Most scholars agree, with hindsight, that using separability to indicate a countability axiom was not a particularly brilliant idea.

Theorem 7.37 (Urysohn) *Every second-countable normal space is metrisable.*

The original argument, published in [Ur25], was actually written up by Alexandrov based on the notes left by Urysohn, who drowned when he was 26 years old. Other proofs can be found in [Du66, Mu00].

Among the several preliminary results needed for Theorem 7.37, one is particularly famous and commonly referred to as *Urysohn's Lemma* (Lemma 8.29). It is proved in almost every textbook on point-set topology.

Exercises

7.25 (♡) Prove that subspaces and products of spaces satisfying axiom **Tx** are still **Tx** when x = 1, 2, 3.

7.26 Let X be regular but not normal. Prove the existence of a closed subspace $A \subset X$ for which the quotient X/A is Hausdorff but not regular.

7.27 Recall that the lower-limit line $\mathbb{R}_l$ is normal (Exercise 7.12), whilst the product $\mathbb{R}_l \times \mathbb{R}_l$ is not (Exercise 7.24). Deduce that the converse arrows in

$$\text{metrisable} \;\Rightarrow\; \text{normal} \;\Rightarrow\; \text{regular} \;\Rightarrow\; \text{Hausdorff}$$

are all false, in general.

7.28 (☛, ♡) Let $\{Y_n\}$ be a countable family of closed sets in a normal space. Prove that $Y = \cup_n Y_n$, with the subspace topology, is normal.

References

[Du66] Dugundji, J.: Topology. Allyn and Bacon, Inc., Boston (1966)
[Mu00] Munkres, J.R.: Topology, 2nd edn. Prenctice-Hall, Upper Saddle River (2000)
[Ur25] Urysohn, P.: Zum Metrisations problem. Math. Ann. **94**, 309–315 (1925)

Chapter 8
More Topics in General Topology ↷

8.1 Russell's Paradox

According to the unsophisticated and naïve definition, a set is determined by its elements (with no further conditions); therefore it might seem that the same would hold for the *set of all sets*, i.e. the set whose elements are all possible sets. Such a set, let us call it X for the moment, would have the curious property that $X \in X$. If we buy the fact that a set can belong to itself, we may as well take one step further and consider the subset

$$Y = \{A \in X \mid A \notin A\} \, .$$

Now wait: if $Y \notin Y$, then we have $Y \in Y$, and if $Y \in Y$, then $Y \notin Y$, in either case by the mere definition of Y.

What we have outlined above is a very popular antinomy called **Russell's paradox**. It teaches us that the set of all sets cannot exist, and invites us never to forget the divine rule: *the 'set of all sets' has no ontological meaning. When tempted to talk about it, we must by all means avoid to do so.*

And because any set possesses topological structures, neither a 'set of topological spaces' exists.

Exercises

8.1 Let X be a Hausdorff space. The *Stone-Čech compactification* of X is a continuous map $f \colon X \to c(X)$, where $c(X)$ is a compact and Hausdorff space, with the following universal property: *for any compact Hausdorff space Y and any continuous map $g \colon X \to Y$ there exists a unique continuous map $h \colon c(X) \to Y$ such that $g = hf$.*

We will present now an argument allegedly proving the existence of the Stone-Čech compactification. Alas, it contains a **deadly mistake** that the reader is asked to spot. A correct proof can be found in the books by Kelley [Ke55, p. 53] or Dugundji [Du66, Th. 8.2].

© Springer International Publishing Switzerland 2015
M. Manetti, *Topology*, UNITEXT - La Matematica per il 3+2 91,
DOI 10.1007/978-3-319-16958-3_8

Pseudo-proof. Call $\mathcal{A}$ the set of pairs (Z, ϕ), with Z compact Hausdorff and $\phi \colon X \to Z$ continuous. The product

$$W = \prod_{(Z,\phi) \in \mathcal{A}} Z$$

is a compact Hausdorff space by Tychonov's theorem. Now consider the projections $\pi_{(Z,\phi)} \colon W \to Z$ onto the various factors and the continuous map $f \colon X \to W$ of components $\pi_{(Z,\phi)} f = \phi$. At last, define $c(X) = \overline{f(X)}$. The subspace $c(X)$ is closed in W hence compact and Hausdorff. If $g \colon X \to Y$ is continuous and Y is compact Hausdorff, it is enough to consider $h = \pi_{(Y,g)}$ to have $g = hf$. The uniqueness of h descends from the facts that Y is Hausdorff and $f(X)$ dense in $c(X)$.

8.2 The Axiom of Choice Implies Zorn's Lemma

We observed earlier that Zorn's lemma implies the axiom of choice. Here we prove the converse, i.e. that Zorn's lemma also follows from the axiom of choice.

Definition 8.1 Let X be a set. A family $\mathcal{F} \subset \mathcal{P}(X)$ of subsets in X is called **strictly inductive** if the union of any non-empty chain $\mathcal{C} \subset \mathcal{F}$ belongs in the family:

$$\bigcup_{A \in \mathcal{C}} A \in \mathcal{F}.$$

Theorem 8.2 *Let $\mathcal{F}$ denote a strictly inductive family of subsets in X and $f \colon \mathcal{F} \to \mathcal{F}$ a map such that $A \subset f(A)$ for every $A \in \mathcal{F}$. Then for any $Q \in \mathcal{F}$ there exists $P \in \mathcal{F}$ such that $Q \subset P$ and $f(P) = P$.*

Proof It will suffice to find an element $P \in \mathcal{F}$ such that $Q \subset P$ and $f(P) \subset P$. We call *tower* a subfamily $\mathcal{A} \subset \mathcal{F}$ satisfying the following conditions:

S1: $Q \in \mathcal{A}$;
S2: $f(\mathcal{A}) \subset \mathcal{A}$;
S3: $\bigcup_{A \in \mathcal{C}} A \in \mathcal{A}$ for any non-empty chain $\mathcal{C} \subset \mathcal{A}$.

Let **T** indicate the collection of towers in $\mathcal{F}$, a non-empty family because it contains $\mathcal{F}$. Now observe that

$$\mathcal{M} = \bigcap_{\mathcal{A} \in \mathbf{T}} \mathcal{A}$$

satisfies S1, S2, S3, so we have $\mathcal{M} \in \mathbf{T}$. Moreover, the family $\{B \in \mathcal{F} \mid Q \subset B\}$ belongs to **T**, and hence $Q \subset B$ for every $B \in \mathcal{M}$.

For the time being let's call a subset T in X *good* if $T \in \mathcal{M}$ and $f(A) \subset T$ for every $A \in \mathcal{M}$, $A \subsetneq T$; we remind that $A \subsetneq B$ means that A is a proper subset of B. For example, Q is a good subset.

In order to continue we need two intermediate results.

Lemma 8.3 *For every good subset $T \in \mathcal{M}$ and every $A \in \mathcal{M}$, either $A \subset T$ or $f(T) \subset A$.*

Proof Let $T \in \mathcal{M}$ be good and consider

$$\mathcal{N} = \{A \in \mathcal{M} \mid A \subset T \text{ or } f(T) \subset A\}.$$

Since $\mathcal{N} \subset \mathcal{M}$, to prove $\mathcal{N} = \mathcal{M}$ we may just prove that $\mathcal{N}$ is a tower. Concerning S1 we've already seen that $Q \subset B$ for every $B \in \mathcal{M}$; in particular $Q \subset T$, so $Q \in \mathcal{N}$.

To prove that $f(A) \in \mathcal{N}$ for every $A \in \mathcal{N}$, i.e. S2, we separate cases:

1. if $A \subsetneq T$ then $f(A) \subset T$ (T is good), and $f(A) \in \mathcal{N}$ follows;
2. if $A = T$ then $f(T) = f(A)$, so trivially $f(T) \subset f(A)$ and then $f(A) \in \mathcal{N}$;
3. if $f(T) \subset A$, then $f(T) \subset A \subset f(A)$ and therefore $f(A) \in \mathcal{N}$.

Now S3. Let $\mathcal{C}$ be a non-empty chain in $\mathcal{N}$ and set $H = \bigcup_{A \in \mathcal{C}} A \in \mathcal{M}$. If $H \subset T$ we claim $H \in \mathcal{N}$: if not, there would exist an $A \in \mathcal{C}$ not contained in T, and then $f(T) \subset A$; *a fortiori* $f(T) \subset H$, so in the end $H \in \mathcal{N}$.   □

Lemma 8.4 *Every element in $\mathcal{M}$ is good.*

Proof Indicate with $\mathcal{T}$ the family of good subsets:

$$\mathcal{T} = \{T \in \mathcal{M} \mid f(A) \subset T \text{ for every } A \in \mathcal{M}, A \subsetneq T\}.$$

As in the previous lemma, to show $\mathcal{T} = \mathcal{M}$ we can prove $\mathcal{T}$ is a tower. We know already that Q is good, so S1 holds. As for S2, we need to show that T good implies $f(T)$ good, i.e. $f(A) \subset f(T)$ for every $A \in \mathcal{M}$ with $A \subsetneq f(T)$. Lemma 8.3 says that $A \in \mathcal{M}$ and $A \subsetneq f(T)$ imply $A \subset T$, so applying f we get $f(A) \subset f(T)$.

There remains S3. Let $\mathcal{C} \subset \mathcal{T}$ be a chain and define $H = \bigcup_{A \in \mathcal{C}} A \in \mathcal{M}$. We have to show that H is good, in other words that if $A \in \mathcal{M}$ and $A \subsetneq H$, then $f(A) \subset H$. As H is not contained in A there is a $T \in \mathcal{C}$ such that $T \not\subset A$; hence $f(T) \not\subset A$. By Lemma 8.3 $A \subset T$, which together with $T \not\subset A$ implies $A \subsetneq T$. But T is good, so $f(A) \subset T \subset H$.   □

Let's now resume the proof of Theorem 8.2. It suffices to prove that the tower $\mathcal{M}$ is also a chain in X. If so, namely, properties S3 and S2 give

$$P := \bigcup_{A \in \mathcal{M}} A \in \mathcal{M}, \qquad f(P) \in \mathcal{M},$$

and therefore $f(P) \subset P$.

So take $S, T \in \mathcal{M}$ and suppose $S \not\subset T$. Lemma 8.4 says T is good, and from Lemma 8.3 we obtain $f(T) \subset S$. As $T \subset f(T)$, we conclude $T \subset S$.   □

Remark 8.5 It's important to stress that the proof of Theorem 8.2 uses neither the axiom of choice, nor Zorn's lemma. As a matter of fact, invoking Zorn's lemma would've simplified the proof drastically, and reduced it to a few lines only, because any maximal element of $\mathcal{F}$ necessarily is a fixed point of f.

Corollary 8.6 *A non-empty, strictly inductive family $\mathcal{F} \subset \mathcal{P}(X)$ of subsets in a set X contains maximal elements.*

Proof Consider the family $\mathcal{R} \subset \mathcal{F} \times \mathcal{F}$ of pairs (A, B) with $A \subsetneq B$. Let's argue by contradiction and suppose there were no maximal elements in $\mathcal{F}$. Then the projection $\mathcal{R} \to \mathcal{F}$ on the first factor would be onto, and the axiom of choice would give a map $f : \mathcal{F} \to \mathcal{F}$ such that $A \subsetneq f(A)$ for every $A \in \mathcal{F}$. But this would contrast with Theorem 8.2.   □

Corollary 8.7 (Hausdorff's maximum principle) *The family of chains in an ordered set is strictly inductive. In particular, every chain in an ordered set is contained in a maximal chain.*

Proof Suppose $\mathcal{F} \subset \mathcal{P}(X)$ denotes the family of chains in an ordered set $(X, \leq)$. Given $\mathcal{C} \subset \mathcal{F}$ we need to show that the subset

$$E = \bigcup_{H \in \mathcal{C}} H$$

is a chain. Take $x, y \in E$ and two chains $H, K \in \mathcal{C}$ with $x \in H$ and $y \in K$. As $\mathcal{C}$ is a chain in $\mathcal{P}(X)$, either $H \subset K$ or $K \subset H$. If, for instance, $H \subset K$, then $x, y \in K$, and since K is a chain in X we also have $x \leq y$ or $y \leq x$. This proves that E is a chain in X. To finish we appeal to Corollary 8.6.   □

Corollary 8.8 (Zorn's lemma) *A non-empty ordered set, all of whose chains are bounded from above, contains maximal elements.*

Proof Let $(X, \leq)$ be ordered and not empty, and suppose every chain has an upper bound. By Corollary 8.7 there's a maximal chain $C \subset X$. Let m denote an upper bound of C. We claim m is maximal in X. If not, there would exist $x \in X$ with $x > m$, and $C \cup \{x\}$ would be strictly bigger than C.   □

Exercises

8.2 (♡) Let X denote an infinite topological space. Write A for the ring of continuous maps $f : X \to \mathbb{R}$ and $D \subset A$ for the subset of maps vanishing at infinitely many points. Prove that A has prime ideals that are contained in D.

8.3 (✒) Kilgore Trout finally managed to publish his latest sci-fi novel, about the adventures of Dr. Zorn, Minister of Transportation in a universe with infinitely many planets connected by an efficient shuttle service. Each spaceship travels back and forth between two planets and follows a precise route. This system allows to move from one planet to any other with a finite number of journeys and stopovers.

Budget cuts force Dr. Zorn to abolish certain routes, with the result that for any two given planets there's only one route to go from one to the other without taking the same spacecraft twice. 'If there were finitely many planets,' said Billy, 'my earthling brain would find the solution; but the more I think about it, the less I understand'.

Despite Billy's bafflement Dr. Zorn manages to get the job done. How did he do it?

8.3 Zermelo's Theorem

Consider an ordered set $(X, \leq)$ and a subset $S \subset X$: a point $s \in X$ is called the minimum of S, written $s = \min(S)$, if $s \in S$ and $s \leq x$ for every $x \in S$. Because of the anti-symmetric property of $\leq$ the minimum must be unique, if it exists.

Definition 8.9 An order relation $\leq$ on a set X is a **well-ordering** if every non-empty subset in X has a smallest element, i.e. every $\emptyset \neq S \subset X$ contains an $s \in S$ such that $s \leq x$ for any $x \in S$. In this case X is called **well ordered**.

For example, the well-ordering principle asserts that the set of natural numbers, with the usual ordering, is well ordered.

Write $\mathcal{P}(X)'$ for the family of non-empty subsets of some set X. To any well-ordering on X we associate the minimum function

$$\min \colon \mathcal{P}(X)' \to X,$$

which satisfies:

1. $\min(A) \in A$ for every $A \in \mathcal{P}(X)'$;
2. $\min(A \cup B) = \min(\min(A), \min(B))$ for every $A, B \in \mathcal{P}(X)'$.

From the minimum function we can reconstruct the order relation: given x, y we have $x \leq y$ iff $x = \min(x, y)$, while $y \leq x$ iff $y = \min(x, y)$. In particular any well-ordered set is totally ordered.

Lemma 8.10 *Let X be a non-empty set and $\lambda \colon \mathcal{P}(X)' \to X$ a map such that:*

1. *$\lambda(A) \in A$ for every $A \in \mathcal{P}(X)'$;*
2. *$\lambda(A \cup B) = \lambda(\lambda(A), \lambda(B))$ for every $A, B \in \mathcal{P}(X)'$.*

Then there exists a unique well-ordering on X for which λ is the minimum function.

Proof Define $x \le y$ if $x = \lambda(x, y)$, and let's show that $(X, \le)$ is well ordered and $\lambda = \min$. As $\lambda(x, x) = \lambda(x) = x$ for every $x \in X$, $\le$ is reflexive. If $x \le y$ and $y \le x$ then $x = \lambda(x, y)$ and $y = \lambda(x, y)$ by definition, so $x = y$. If $x \le y$ and $y \le z$ then

$$x = \lambda(x, y) = \lambda(\lambda(x), \lambda(y, z)) = \lambda(x, y, z) = \lambda(\lambda(x, y), \lambda(z)) = \lambda(x, z)$$

whence $x \le z$. At last, if $A \subset X$ is a non-empty subset and $a = \lambda(A) \in A$, then $A = \{x\} \cup A$ for every $x \in A$, and so

$$a = \lambda(A) = \lambda(A \cup \{x\}) = \lambda(\lambda(A), \lambda(x)) = \lambda(a, x),$$

implying $a \le x$. $\qquad\qquad\qquad\qquad\qquad\qquad\qquad\qquad\qquad\qquad\qquad\qquad\qquad\quad\square$

Theorem 8.11 *Any set X admits a function* $\lambda \colon \mathcal{P}(X)' \to X$ *such that:*

1. $\lambda(A) \in A$ *for every* $A \in \mathcal{P}(X)'$;
2. $\lambda(A \cup B) = \lambda(\lambda(A), \lambda(B))$ *for every* $A, B \in \mathcal{P}(X)'$.

Proof If $X = \emptyset$ there's nothing to prove. If $X \ne \emptyset$ we consider the family $\mathcal{A}$ of pairs (E, λ_E) where $E \subset X$ is non-empty and $\lambda_E \colon \mathcal{P}(E)' \to X$ satisfies properties 1 and 2. For $x \in X$ we always have $(\{x\}, \{x\} \mapsto x) \in \mathcal{A}$, showing that $\mathcal{A}$ is not empty. Let's introduce on $\mathcal{A}$ the order relation $(E, \lambda_E) \le (F, \lambda_F) \iff E \subset F$ and $\lambda_E(E \cap A) = \lambda_F(A)$ for every $A \subset F$ such that $A \cap E \ne \emptyset$.

With the aid of Zorn's lemma we shall prove that $\mathcal{A}$ has maximal elements. Take a chain $\mathcal{C}$ in $\mathcal{A}$ and define the pair (C, λ_C) as follows:

$$C = \bigcup \{E \mid (E, \lambda_E) \in \mathcal{C}\}$$

$$\lambda_C(A) = \lambda_E(A \cap E) \text{ for some } (E, \lambda_E) \in \mathcal{C} \text{ such that } A \cap E \ne \emptyset.$$

The reader can easily check that $(C, \lambda_C) \in \mathcal{A}$ is an upper bound for $\mathcal{C}$.

So take a maximal element (M, λ_M) and suppose by contradiction that there is an $m \in X - M$. Consider (N, λ_N), where $N = M \cup \{m\}$, $\lambda_N(\{m\}) = m$ and $\lambda_N(A) = \lambda_M(A \cap M)$ for every $A \subset N$ different from $\emptyset$ and $\{m\}$. Since $(M, \lambda_M) < (N, \lambda_N)$, we've contradicted the maximality of (M, λ_M). $\qquad\qquad\qquad\qquad\qquad\quad\square$

We point out that condition 1 in Theorem 8.11 is equivalent to the axiom of choice (see Exercise 2.13).

Corollary 8.12 (Zermelo's theorem) *Every non-empty set X admits a well-ordering* $\le$ *such that the cardinality of* $\{y \in X \mid y < x\}$, *for any* $x \in X$, *is strictly less than* $|X|$.

Proof By Theorem 8.11 and Lemma 8.10 X admits a well-ordering $\preceq$. Define $L(x) = \{z \in X \mid z \prec x\}$, for $x \in X$. If the cardinality of $L(x)$ is less than that of X, we can take $\preceq$ as our $\le$. Otherwise let $a \in X$ be the minimum of the non-empty set

$$\{x \in X \mid |L(x)| = |X|\}.$$

By construction $L(a)$ has the cardinality of X, and $\preceq$ induces on $L(a)$ a well-ordering with the additional property that $|L(x)| < |L(a)|$ for every $x \in L(a)$. Now it suffices to take an arbitrary bijection $f : X \to L(a)$ and set

$$x \le y \iff f(x) \preceq f(y).$$

$\square$

Exercises

8.4 Let $K \subset \mathbb{R}^2$ be a connected set. Prove that if K has at least two points then it has the cardinality of $\mathbb{R}$.

8.5 Let $A \subset \mathbb{R}^2$ be path disconnected. Show that $|\mathbb{R}^2 - A| = |\mathbb{R}|$.

8.6 (✻) Let $\mathcal{C}$ be the family of closed sets in $\mathbb{R}^2$ with the cardinality of $\mathbb{R}$.

1. If $C, D \in \mathcal{C}$ and $C \cup D = \mathbb{R}^2$, prove that $C \cap D \in \mathcal{C}$. (Hint: if neither is contained in the other, use Exercise 8.5.)
2. Prove that $\mathcal{C}$ has the cardinality of $\mathbb{R}$. (Hint: $\mathbb{R}^2$ is second countable.)
3. Fix an order $\le$ on $\mathcal{C}$ that satisfies the hypotheses of Zermelo's theorem. One says that $\mathcal{U} \subset \mathcal{C}$ is a lower set if, for any given $x \in \mathcal{U}$, $y \le x \Longrightarrow y \in \mathcal{U}$. Denote by **A** the set of triples $(\mathcal{U}, f, g)$ where $\mathcal{U}$ is a lower set in $\mathcal{C}$ and $f, g : \mathcal{U} \to \mathbb{R}^2$ satisfy:

    (a) $f(\mathcal{U}) \cap g(\mathcal{U}) = \emptyset$;
    (b) $f(C), g(C) \in C$ for every closed set $C \in \mathcal{U}$.

    Show that **A**, ordered by extension, contains maximal elements. Use statement 2. to conclude that there exist disjoint subsets $A, B \subset \mathbb{R}^2$ such that $C \cap A \ne \emptyset$, $C \cap B \ne \emptyset$ for every $C \in \mathcal{C}$.
4. Using the previous points and Exercise 8.4 prove that there exists a dense connected subset $A \subset \mathbb{R}^2$ such that any path $\alpha : [0, 1] \to A$ is constant.

Remark 8.13 The reader familiar with measure theory may also prove that the subsets A, B in Exercise 8.6 are not Lebesgue measurable.

8.7 (✻, ♡) Let X be an infinite set. Prove that there is a family $\mathcal{A}$ of subsets of X such that:

1. $|A| = |X|$ for every $A \in \mathcal{A}$;
2. $|A \cap B| < |X|$ for every $A, B \in \mathcal{A}$, $A \ne B$;
3. $|X| < |\mathcal{A}|$.

(Hint: consider a maximal family among those satisfying conditions 1 and 2 and with cardinality larger than or equal to $|X|$.)

8.4 Ultrafilters

This section can be skipped at first (even second) reading: the topics dealt with should be considered merely as 'topological folklore' and a bizarre exercise.

The notion of an ultrafilter was introduced by Henri Cartan in 1937 and has applications in topology, logic and measure theory. It was, by the way, used by Gödel for the 'proof' of the existence of God!

The definition we will give is not the standard one, albeit equivalent.

Definition 8.14 A non-empty family $\mathcal{U}$ of subsets in a set X is called an **ultrafilter** if these conditions hold:

1. $A \neq \emptyset$ for every $A \in \mathcal{U}$;
2. if $A \subset B \subset X$ and $A \in \mathcal{U}$, then $B \in \mathcal{U}$;
3. if $A, B \in \mathcal{U}$ then $A \cap B \in \mathcal{U}$;
4. if $U \subset X$ and $A \cap U \neq \emptyset$ for every $A \in \mathcal{U}$, then $U \in \mathcal{U}$.

Example 8.15 Let p be a point in X. The family $\mathcal{U} = \{A \subset X \mid p \in A\}$ is an ultrafilter.

Lemma 8.16 *Let $\mathcal{U}$ be an ultrafilter in X. Then for every subset $A \subset X$ either $A \in \mathcal{U}$ or $X - A \in \mathcal{U}$.*

Proof Argue by contradiction and suppose $A \notin \mathcal{U}$ and $X - A \notin \mathcal{U}$. Condition 4 in Definition 8.14 gives $C, D \in \mathcal{U}$ such that $A \cap C = \emptyset$, $D \cap (X - A) = \emptyset$. Then $C \subset X - A$, $D \subset A$,

$$C \cap D \subset A \cap (X - A) = \emptyset,$$

violating 1 and 3.   □

Theorem 8.17 (Ultrafilter lemma) *Let $\mathcal{A}$ be a family of non-empty subsets in X. There exists an ultrafilter containing $\mathcal{A}$ if and only if every finite subfamily of $\mathcal{A}$ has non-empty intersection.*

Proof One direction is trivial. If $A_1 \cap \cdots \cap A_n \neq \emptyset$ for every $A_1, \ldots, A_n \in \mathcal{A}$, the family $\mathcal{F}$ of subsets containing a finite intersection of $\mathcal{A}$ satisfies 1, 2 and 3 of Definition 8.14.

Let **U** be the collection of subfamilies $\mathcal{G} \subset \mathcal{P}(X)$ containing $\mathcal{F}$ and that satisfy 1, 2 and 3. Then **U** is ordered by the inclusion, and has a maximal element $\mathcal{U}$ by Zorn's lemma. We claim **U** satisfies condition 4 as well. Take a subset A in X. If $A \cap U \neq \emptyset$ for every $U \in \mathcal{U}$, the family of subsets containing $A \cap U$ for some $U \in \mathcal{U}$ belongs in **U**, and by maximality $A \in \mathcal{U}$.   □

Before passing to the exercises, which we recommend to solve in the given order, we need the notion of convergence of an ultrafilter.

Definition 8.18 An ultrafilter $\mathcal{U}$ in a topological space X converges to a point $x \in X$ if every neighbourhood of x belongs in $\mathcal{U}$.

It is immediate from definition that in a Hausdorff space an ultrafilter converges at most to one point.

Exercises

8.8 Show that any ultrafilter inside a finite set must be of the type described in Example 8.15.

8.9 Let $\mathcal{A}$ be an open cover of a space X without finite subcovers. Prove that there exists an ultrafilter $\mathcal{U}$ in X such that $\mathcal{A} \cap \mathcal{U} = \emptyset$. Conclude that not every ultrafilter has the form of Example 8.15.

8.10 Let $\mathcal{U}$ be an ultrafilter that converges to no point. Prove that the family of open sets not belonging in $\mathcal{U}$ is a cover with no finite subcovers. Conclude that a space is compact if and only if each of its ultrafilters converges.

8.11 Let $\mathcal{P}$ be a sub-basis in X and $\mathcal{U}$ an ultrafilter. Prove that $\mathcal{U}$ converges to x if and only if $\{P \in \mathcal{P} \mid x \in P\} \subset \mathcal{U}$.

8.12 Let $\mathcal{P}$ be a sub-basis of X and $\mathcal{U}$ an ultrafilter. Suppose $\mathcal{U}$ doesn't converge in X. Prove that $\mathcal{P} - \mathcal{U}$ is an open cover of X without finite subcovers.

8.13 Use the previous exercises to give an alternative proof of Alexander's Theorem 7.5.

8.14 Let $\mathcal{U}$ be an ultrafilter in a set X, $f : X \to Y$ a surjective map and define $f(\mathcal{U}) = \{f(A) \mid A \in \mathcal{U}\}$. Prove that $f(\mathcal{U})$ is an ultrafilter in Y, that $f(\mathcal{U}) = \{A \subset Y \mid f^{-1}(A) \in \mathcal{U}\}$, and that for every ultrafilter $\mathcal{V}$ in Y there's an ultrafilter $\mathcal{U}$ in X such that $f(\mathcal{U}) = \mathcal{V}$.

8.15 Let $X = \prod_i X_i$ be a product space and $p_i : X \to X_i$ the projections. Prove that an ultrafilter $\mathcal{U}$ in X is convergent if and only if $p_i(\mathcal{U})$ is convergent for every i.

8.16 Use Exercises 8.10 and 8.15 to prove Tychonov's Theorem 7.7.

Warning: the ultrafilter lemma is not equivalent to the axiom of choice. In the proof of Tychonov's theorem given in Exercise 8.16 the ultrafilter lemma does not fully replace the axiom of choice: the latter is still necessary, for instance, to solve Exercise 8.15 since the spaces X_i are not assumed Hausdorff.

8.5 The Compact-Open Topology

We retrieve the notations of Sect. 6.7 and recall that $C(X, Y)$ denoted the space of continuous maps between topological spaces X, Y. The subsets

$$W(K, U) = \{f \in C(X, Y) \mid f(K) \subset U\},$$

for any K compact in X and U open in Y, form the sub-basis of the so-called **compact-open topology** of $C(X, Y)$.

In this section we will show that the compact-open topology on $C(X, Y)$ has nice properties when X is locally compact and Hausdorff.

Lemma 8.19 *Let X, Y be spaces with X locally compact Hausdorff, and $\mathcal{P}$ a sub-basis on Y. The family $\{W(K, U)\}$, with $U \in \mathcal{P}$ and K compact in X, is a sub-basis of the compact-open topology on $C(X, Y)$.*

Proof We'll show that for every $f: X \to Y$ continuous, $K \subset X$ compact and $U \subset Y$ open with $f(K) \subset U$, there exist compact sets $K_1, \ldots, K_n$ and open sets $U_1, \ldots, U_n \in \mathcal{P}$ such that

$$f \in W(K_1, U_1) \cap \cdots \cap W(K_n, U_n) \subset W(K, U).$$

Let $\mathcal{B}$ be the family of finite intersections of open sets from $\mathcal{P}$. Then $\mathcal{B}$ is a basis of Y, and since $f(K)$ is compact, we may find $V_1, \ldots, V_m \in \mathcal{B}$ such that

$$f(K) \subset V_1 \cup \cdots \cup V_m \subset U.$$

Any $x \in K$ has a compact neighbourhood K_x such that $f(K_x) \subset V_i$ for some i. But the family of interiors of the K_x is an open cover of K, so there's a finite subset $S \subset K$ such that $K \subset \cup\{K_x \mid x \in S\}$. For every $i = 1, \ldots, m$ set

$$K_i = \cup\{K_x \mid x \in S, \ f(K_x) \subset V_i\}.$$

Each K_i is compact, and

$$f \in W(K_1, V_1) \cap \cdots \cap W(K_m, V_m) \subset W(K, U).$$

For every index i there are $U_{i1}, \ldots, U_{is} \in \mathcal{P}$ such that $V_i = U_{i1} \cap \cdots \cap U_{is}$, and now we can use the trivial relation

$$W(K_i, U_{i1}) \cap \cdots \cap W(K_i, U_{is}) = W(K_i, V_i)$$

to conclude.   □

Theorem 8.20 (Exponential law) *Let X, Y, Z be spaces, and assume all function spaces have the compact-open topology. There exists a natural one-to-one map*

$$\widehat{\ } : C(X \times Y, Z) \to C(X, C(Y, Z)).$$

Furthermore:

1. *if Y is locally compact Hausdorff, $\widehat{\ }$ is bijective;*
2. *if X is locally compact Hausdorff, $\widehat{\ }$ is continuous;*
3. *if X, Y are locally compact Hausdorff, $\widehat{\ }$ is a homeomorphism.*

Proof Given $f: X \times Y \to Z$ continuous, the mapping

$$\widehat{f}: X \to C(Y, Z), \qquad \widehat{f}(x)(y) = f(x, y)$$

is well defined. Easily, $\widehat{f}$ is continuous: let $W(K, U)$ belong in the sub-basis of $C(Y, Z)$ and take $x \in X$ such that $\widehat{f}(x) \in W(K, U)$, i.e. $f(\{x\} \times K) \subset U$. According to Wallace's theorem there's an open set $A \subset X$ such that $x \in A$ and $A \times K \subset f^{-1}(U)$, i.e. $\widehat{f}(A) \subset W(K, U)$. This proves the continuity of $\widehat{f}$.

Thus we've defined a natural 1-1 map

$$\widehat{} : C(X \times Y, Z) \to C(X, C(Y, Z)).$$

If Y is locally compact, then $\widehat{}$ is bijective. We will prove that for every $g: X \to C(Y, Z)$ continuous,

$$f: X \times Y \to Z, \qquad f(x, y) = g(x)(y),$$

is continuous. Take $U \subset Z$ open and $(x, y) \in f^{-1}(U)$; then $g(x)$ is continuous and Y is locally compact Hausdorff. Hence y has a compact neighbourhood B such that $g(x)(B) \subset U$, so $g(x) \in W(B, U)$. The map g is continuous, whence x has a neighbourhood A in X such that $g(A) \subset W(B, U)$; this implies $f(A \times B) \subset U$.

By Lemma 8.19, if X is locally compact Hausdorff, we can take as sub-basis on $C(X, C(Y, Z))$ the open sets $W(H, W(K, U))$, where H, K are compact in X and Y respectively, and U is open in Z. Since $\widehat{}$ identifies $W(H \times K, U)$ with $W(H, W(K, U))$, it is necessarily continuous.

As for claim 3, it suffices to show that the $W(H \times K, U)$ constitute a sub-basis for the compact-open topology on $C(X \times Y, Z)$. Take $T \subset X \times Y$ compact and $f \in W(T, U)$. For every $t \in T$ there exist compact sets $K_t \subset X$ and $H_t \subset Y$ such that $f(K_t \times H_t) \subset U$ and t is in the interior of $K_t \times H_t$. By passing to a finite subcover we may find finitely many compact sets $K_1, \ldots, K_n \subset X$, $H_1, \ldots, H_n \subset Y$ with

$$T \subset \bigcup_i K_i \times H_i, \qquad f(K_i \times H_i) \subset U,$$

so that, eventually,

$$f \in W(K_1 \times H_1, U) \cap \cdots \cap W(K_n \times H_n, U) \subset W(T, U).$$

$\square$

Exercises

In the following exercises, unless otherwise stated, all function spaces will implicitly have the compact-open topology.

8.17 Consider topological spaces X, Y, Z with X locally compact Hausdorff. Prove that $C(X, Y \times Z)$ is homeomorphic to $C(X, Y) \times C(X, Z)$.

8.18 Let X, Y be spaces, X locally compact Hausdorff. Prove that

$$C(X, Y) \times X \to Y, \qquad (f, x) \mapsto f(x),$$

is continuous.

8.19 Let X be a finite set with n points equipped with the discrete topology. Show that $C(X, Y)$ is homeomorphic to the product Y^n.

8.20 Let X, Y, Z, W be spaces. Prove that

$$C(X, Y) \to C(Z, W), \qquad f \mapsto G \circ f \circ F,$$

is continuous, for any pair of continuous maps $F: Z \to X, G: Y \to W$,

8.21 ($\maltese$, $\heartsuit$) Let X be compact Hausdorff and (Y, d) a metric space. Prove that the distance (see Theorem 6.51)

$$\delta: C(X, Y) \times C(X, Y) \to \mathbb{R}, \qquad \delta(f, g) = \max\{d(f(x), g(x)) \mid x \in X\},$$

induces the compact-open topology.

8.22 ($\maltese$)

1. Let X, Y, Z be locally compact Hausdorff. Prove that the composition of maps

$$C(Y, Z) \times C(X, Y) \to C(X, Z), \qquad (f, g) \mapsto fg,$$

   is continuous for the compact-open topology.
2. Let X be compact Hausdorff. Show that the compact-open topology induces on Homeo(X) the structure of a topological group, see Sect. 5.3. (Hint: $f(A) \subset B$ iff $f^{-1}(X - B) \subset X - A$.)
3. Prove that the compact-open topology turns Homeo$(\mathbb{R})$ into a topological group.

8.6 Noetherian Spaces

The upcoming result is very much used in algebraic geometry and commutative algebra.

Proposition 8.21 *On an ordered set $(X, \leq)$ the following are equivalent:*

1. *every non-empty subset of X contains maximal elements;*
2. *every countable ascending chain $\{x_1 \leq x_2 \leq \cdots\} \subset X$ stabilises (i.e. there's an index $m \in \mathbb{N}$ such that $x_n = x_m$ for every $n \geq m$).*

Proof For (1)$\Rightarrow$(2) it's enough to notice that any countable ascending chain $\{x_1 \leq x_2 \leq \cdots\}$ contains a maximal element, say x_m, and therefore $x_n = x_m$ for every $n \geq m$.

Let us prove (2)$\Rightarrow$(1). By contradiction, suppose we have a non-empty $S \subset X$ with no maximal elements, i.e. $\{y \in S \mid y > x\} \neq \emptyset$ for every $x \in S$. By the axiom of choice there's a function $f: S \to S$ such that $f(x) > x$ for every $x \in S$. Take $x_0 \in S$: the ascending chain $\{x_n = f^n(x_0) \mid n \in \mathbb{N}\}$ does not stabilise. $\square$

Definition 8.22 A space is called **Noetherian** if every non-empty family of open sets has a maximal element for the inclusion.

By Proposition 8.21 a space is Noetherian if and only if every countable ascending chain stabilises.

Example 8.23 Let $\mathbb{K}$ be any field. The affine space $\mathbb{K}^n$ equipped with the Zariski topology (Example 3.11) is Noetherian. The proof is a simple consequence of Hilbert's basis theorem,[1] and as such we leave it to lecture courses on algebraic geometry and commutative algebra.

Lemma 8.24 *If X is a Noetherian space:*

1. *X is compact;*
2. *any continuous image of X is Noetherian;*
3. *any subspace of X is Noetherian.*

Proof To prove 1 we need to see that every open cover $\mathcal{U}$ has a finite subcover: for this we can pick a maximal element in the family of finite unions in $\mathcal{U}$.

Property 2 is trivial, and for 3 take $Y \subset X$. Call $\mathcal{T}(X)$ and $\mathcal{T}(Y)$ the open sets in X and Y, and let

$$r: \mathcal{T}(X) \to \mathcal{T}(Y), \qquad r(U) = U \cap Y,$$

denote the restriction map. If $\mathcal{F} \subset \mathcal{T}(Y)$ is a non-empty collection of open sets and $U \in \mathcal{T}(X)$ a maximal element in $r^{-1}(\mathcal{F})$, we claim $r(U) = U \cap Y$ is maximal in $\mathcal{F}$. Let V be open in X such that $r(U) \subset r(V)$ and $r(V) \in \mathcal{F}$; then $U \cup V \in r^{-1}(\mathcal{F})$, by maximality $V \subset U$ and therefore $r(U) = r(V)$. □

Definition 8.25 A topological space is said to be **irreducible** if any pair of open, non-empty sets has non-empty intersection. Equivalently, a space is irreducible if it cannot be written as finite union of proper closed subsets. A subspace is called irreducible if it is an irreducible space for the induced topology.

Examples include the empty set, points, and more generally, any space with the trivial topology.

Lemma 8.26 *Let X be a space and $Y \subset X$ an irreducible subspace. Then:*

1. *the closure $\overline{Y}$ is irreducible;*
2. *if $U \subset X$ is open, $Y \cap U$ is irreducible;*
3. *if $f: X \to Z$ is continuous, $f(Y)$ is irreducible.*

Proof We will prove only 1, and leave the rest to the reader. Take open, non-empty sets U, V in $\overline{Y}$; as any subspace is dense in its closure, $U \cap Y$ and $V \cap Y$ aren't empty. If Y is irreducible then $(U \cap Y) \cap (V \cap Y) \neq \emptyset$, and so $U \cap V \neq \emptyset$. □

---

[1] **Hilbert's basis theorem:** *Every non-empty family of ideals in the ring $\mathbb{K}[x_1, \ldots, x_n]$ contains maximal elements for the inclusion.*

Definition 8.27 The **irreducible components** of a topological space are the maximal elements in the family of irreducible closed subsets ordered by inclusion.

The family of irreducible closed subsets is never empty, for it contains the empty set. Exercise 8.30 shows that any space can be viewed as the union of its irreducible components. The following theorem studies this feature when the ambient is Noetherian.

Theorem 8.28 *A Noetherian space X has a finite number of irreducible components $X_1, \ldots, X_n$. Moreover, $X = X_1 \cup \cdots \cup X_n$, and any component X_i is not contained in the union of the other components X_j, $j \neq i$.*

Proof We start by showing that every closed set in X can be written as finite union of irreducible closed subsets; for that consider the family $\mathcal{C}$ of closed sets in X and the subfamily $\mathcal{F} \subset \mathcal{C}$ of finite unions of irreducible closed subsets. By contradiction, if $\mathcal{F} \neq \mathcal{C}$ there exists a minimal element $Z \in \mathcal{C} - \mathcal{F}$. But $Z \notin \mathcal{F}$, so the closed set Z is not irreducible, and then there are proper closed sets Z_1, Z_2 such that $Z = Z_1 \cup Z_2$. By minimality $Z_1, Z_2 \in \mathcal{F}$, so $Z \in \mathcal{F}$.

So we have $X = X_1 \cup \cdots \cup X_n$, where X_i is closed and irreducible; with the proviso of eliminating superfluous closed sets we can assume X_i doesn't lie in the union of the remaining X_j. The claim is that $X_1, \ldots, X_n$ are the irreducible components of X.

Take $Z \subset X$ irreducible and closed. Since $Z = (Z \cap X_1) \cup \cdots \cup (Z \cap X_n)$, we must have $Z = Z \cap X_i$ for some index i: in other terms there's an i such that $Z \subset X_i$. The same holds if Z is an irreducible component. By maximality, then, any irreducible component of X must be one of the X_i.

Conversely, if an X_i is not an irreducible component, there's a proper inclusion $X_i \subset Z$ with Z irreducible; the previous argument tells that Z is contained in some X_j, contradicting the assumption. □

Exercises

8.23 Prove that the cofinite topology, on any set, is Noetherian.

8.24 Prove that the finite union of Noetherian spaces is Noetherian.

8.25 Prove that every non-empty irreducible subspace in a Hausdorff space contains only one point.

8.26 Show that if $\mathbb{K}$ is an infinite field, the space $\mathbb{K}^n$ with the Zariski topology is irreducible, for every $n > 0$.

8.27 ($\heartsuit$) Take irreducible spaces X, Y and a topology on the Cartesian product $X \times Y$ such that, for every $(x_0, y_0) \in X \times Y$, the inclusions

$$X \to X \times Y, \qquad x \mapsto (x, y_0),$$

and

$$Y \to X \times Y, \qquad y \mapsto (x_0, y),$$

are continuous. Prove that $X \times Y$ is irreducible.

8.28 Let $f \colon X \to Y$ be continuous, open and with irreducible fibres. If $Z \subset Y$ is irreducible, prove that $f^{-1}(Z)$ is irreducible as well.

8.29 Give an example of closed identification $f \colon X \to Y$ where Y is irreducible, $f^{-1}(y)$ is irreducible for every $y \in Y$ and X is not irreducible.

8.30 ($\heartsuit$) Let $\mathcal{C}$ be the family of closed, irreducible subsets in X ordered by inclusion. Show that any point of X is contained in one maximal element of $\mathcal{C}$ at least.

8.7 A Long Exercise: Tietze's Extension Theorem

Recall that a space is called normal when it is Hausdorff and disjoint closed sets have disjoint neighbourhoods. The exercises at the end of the section, solved in the given order, provide a proof of the following two results.

Lemma 8.29 (Urysohn's lemma) *Let A, C be disjoint closed sets in a normal space X. There exists a continuous map $f \colon X \to [0, 1]$ such that $f(x) = 0$ when $x \in A$ and $f(x) = 1$ when $x \in C$.*

Theorem 8.30 (Tietze extension) *Let B be closed in a normal space X, $J \subset \mathbb{R}$ a convex subspace and $f \colon B \to J$ a continuous map. Then there exists a continuous map $g \colon X \to J$ such that $g(x) = f(x)$ for every $x \in B$.*

For metric spaces Urysohn's lemma is easy to prove: it suffices to recycle the argument of Proposition 7.31.

Let us remark that Lemma 8.29 is a special case of Theorem 8.30 when $J = [0, 1]$ and $B = A \cup C$. The classical proof of Theorem 8.30, for which we suggest consulting [Mu00, Du66], uses Urysohn's lemma and the completeness of the space $BC(X, \mathbb{R})$ of continuous and bounded real functions on X.

Exercises

8.31 Let $f \colon B \to [0, 1]$ be a continuous mapping, $S \subset [0, 1]$ a dense subset and define $B(s) = \{x \in B \mid f(x) < s\}$, for $s \in S$. Prove that

$$\begin{cases} f(x) = \inf\{s \in S \mid x \in B(s)\} & \text{if } x \in \cup\{B(s) \mid s \in S\} \\ f(x) = 1 & \text{otherwise} \end{cases}$$

for every $x \in B$.

8.32 ($\heartsuit$) Let X be a space and $S \subset [0, 1]$ a dense subset. Suppose to have, for every $s \in S$, open sets $X(s) \subset X$ so that $\overline{X(s)} \subset X(t)$ if $s < t$ in S. Now define $g : X \to [0, 1]$ by

$$\begin{cases} g(x) = \inf\{s \in S \mid x \in X(s)\} & \text{if } x \in \bigcup\{X(s) \mid s \in S\} \\ g(x) = 1 & \text{otherwise.} \end{cases}$$

Prove g is continuous.

8.33 ($\heartsuit$) Consider closed sets A, B, C in a normal space X and a continuous map $f : B \to \mathbb{R}$. Assume $A \cap C = \emptyset$, and that there exist real numbers $a < c$ such that:

$$A \cap B \subset \{x \in B \mid f(x) \le a\}, \qquad C \cap B \subset \{x \in B \mid f(x) \ge c\}.$$

Prove that for any $b \in {]a, c[}$ there are open *disjoint* sets $U, V \subset X$ with

$$A \subset U, \quad U \cap B = \{x \in B \mid f(x) < b\}, \quad \overline{U} \cap B \subset \{x \in B \mid f(x) \le b\},$$

$$C \subset V, \quad V \cap B = \{x \in B \mid f(x) > b\}, \quad \overline{V} \cap B \subset \{x \in B \mid f(x) \ge b\}.$$

(Hint: write $A \cup \{x \in B \mid f(x) < b\}$ and $C \cup \{x \in B \mid f(x) > b\}$ as countable unions of closed sets, then argue as for the proof of Theorem 7.36.)

8.34 Take B closed in a normal space X, and let $f : B \to [0, 1]$ be continuous. Indicate with $\Delta \subset [0, 1]$ the set of rational numbers of the form $\frac{r}{2^n}$, for $n \ge 0$ and $0 \le r \le 2^n$. For every $s \in \Delta$ set $B(s) = \{x \in B \mid f(x) < s\}$. Choose an open set $X(1) \subset X$ such that $X(1) \cap B = B(1)$ and define $X(0) = \emptyset$.

Prove that one can extend the pair $X(0), X(1)$ to an open family $\{X(s) \mid s \in \Delta\}$ such that

$$X(s) \cap B = B(s), \qquad \overline{X(s)} \cap B \subset \{x \in B \mid f(x) \le s\}$$

for every $s \in \Delta$, and also that $\overline{X(s)} \subset X(t)$ if $s < t$.

Hint: any element of $\Delta - \{0, 1\}$ can be written in a unique way as $r/2^n$, with $n > 0$ and r odd. Construct the open sets $X(r/2^n)$ by induction on n by employing Exercise 8.33 with

$$A = \overline{X\left(\frac{r-1}{2^n}\right)}, \qquad C = X - X\left(\frac{r+1}{2^n}\right),$$

$$a = \frac{r-1}{2^n}, \qquad b = \frac{r}{2^n}, \qquad c = \frac{r+1}{2^n}.$$

8.35 Let B be closed in a normal space X and $f : B \to [0, 1]$ a continuous mapping. Combine the results of Exercises 8.34, 8.32 and 8.31 to show that f extends to a continuous map $g : X \to [0, 1]$.

8.36 Prove Urysohn's lemma.

8.37 Let B be closed in X normal, J an open convex subset of $[0, 1]$ and $f : B \to J$ a continuous map. Call $k : X \to [0, 1]$ a continuous extension of f and write $A = k^{-1}([0, 1] - J)$. Now take $h : X \to [0, 1]$ continuous and such that $h(A) = 0$, $h(B) = 1$. Given a point $j \in J$ consider the convex combination

$$g(x) = h(x)k(x) + (1 - h(x))j.$$

Prove that $g : X \to J$ is a continuous extension of f.

8.38 Prove Theorem 8.30.

8.39 If B is closed in a normal space X, show that any continuous mapping $f : B \to S^n$ extends over an open neighbourhood of B.

8.40 ($\heartsuit$) Prove that a normal space X admits an exhaustion by compact sets if and only if there exists a proper map $f : X \to \mathbb{R}$.

8.41 (☛) Let B be closed in a normal space X, and

$$f : X \to S^n, \qquad F : B \times [0, 1] \to S^n,$$

two continuous maps such that $f(x) = F(x, 0)$ for every $x \in B$. Prove that

$$g : B \to S^n, \qquad g(x) = F(x, 1),$$

extends to a continuous map $X \to S^n$. (Hint: extend F to a suitable open set in $X \times [0, 1]$ and then restrict to the graph of another suitable map $X \to [0, 1]$.)

References

[Du66] Dugundji, J.: Topology. Allyn and Bacon, Inc., Boston (1966)
[Ke55] Kelley, J.L.: General Topology. D. Van Nostrand Company Inc, Toronto-New York-London (1955)
[Mu00] Munkres, J.R.: Topology, 2nd edn. Prenctice-Hall, Upper Saddle River (2000)

Chapter 9
Intermezzo ↷

In this shorter chapter, and only here, we will allows ourself to be less rigorous in the arguments and a little more vague in defining notions. The purpose is to explain in an intuitive manner, sometimes even slightly punningly, a few foundational ideas of the branch of mathematics called 'algebraic topology'.

9.1 Trees

We met in Chap. 1 the intuitive ideas of graph, connected graph and walk inside a graph. Let's call length of a walk the number of edges in it. A **simple path** is a walk without repeated nodes. In a connected graph any two distinct nodes are the endpoints of a simple path: just take a shortest walk among all walks joining the two nodes. A walk is said to be closed if the last and first nodes coincide. We shall call **circuit** a closed walk without repeated edges. A **tree** is then a connected graph containing no circuits.

Definition 9.1 Let Γ be a graph with V nodes and S edges. The number $e(\Gamma) = V - S$ is called the **Euler-Poincaré characteristic** of Γ.

Theorem 9.2 *If Γ is a connected graph, $e(\Gamma) \leq 1$. Equality holds if and only if Γ is a tree.*

Proof If a graph Γ contains a circuit, the operation of removing one edge of the circuit from Γ will produce a new graph Γ', still connected, with $e(\Gamma) = e(\Gamma') - 1$. This shows, in particular, that if a graph Γ contains circuits, we can remove finitely many edges and eventually obtain a tree Λ. Then $e(\Gamma) < e(\Lambda)$, and to finish the proof it suffices to show that $e = 1$ for trees.

We proceed by induction on the number of nodes; trees with one node have no edges, so the claim is trivial. Any tree Γ contains simple paths (e.g. the single nodes) whose length is bounded by the edge number S. Let H be the set of nodes of a simple path of maximum length in Γ, and $v_0 \in H$ an end node; v_0 is joined to $v_1 \in H$ by

© Springer International Publishing Switzerland 2015
M. Manetti, *Topology*, UNITEXT - La Matematica per il 3+2 91,
DOI 10.1007/978-3-319-16958-3_9

an edge l_1. We claim that l_1 is the unique edge in Γ passing through v_0. If v_0 were connected to a node u by some edge $l \neq l_1$ then either $u \in H$, in which case Γ would contain a circuit, or $u \notin H$ and then we could take a longer trail by adding the edge l and the node u. In any case we would have a contradiction. Hence the graph Γ' obtained by removing l_1 and v_0 from Γ is still connected, it has no circuits and by inductive hypothesis $e(\Gamma) = e(\Gamma') = 1$. $\qquad\qquad\qquad\square$

Exercises

9.1 Prove that also the converse to Theorem 1.1 holds: if a connected graph contains at most two nodes of odd degree, then it's an Eulerian graph. (Hint: the proof is similar to that of Theorem 9.2).

9.2 Polybricks and Betti Numbers

This section discusses concepts that are typical of advanced lectures on topology, and so might prove a bit challenging to less-experienced readers. At any rate the material presented here doesn't prepare for the remaining chapters, so skipping it wouldn't do any harm.

Billy has a magnificent collection of LEGO bricks, all shaped as parallelopipeds, wonderfully coloured and numbered from 1 to n. He spends lots of time playing with them and expresses his creativity by assembling various objects $X \subset \mathbb{R}^3$, which we'll call polybricks. He's so pleased with his creations that he takes note of the relative positions of the single pieces to be able to replicate his best works at international optometry conferences.

After long conversations with Mr. Čech, Billy has started to write down the set $N_0 \subset \{1, \ldots, n\}$ of bricks used to build every X, and the set $N_1 \subset N_0 \times N_0$ of pairs of bricks (a, b), $a < b$, that touch at least at one point. Generalising, for every $i \geq 0$ we call N_i the family of $i + 1$-tuples $(a_0, a_1, \ldots, a_i) \in N_0^{i+1}$ such that $a_0 < a_1 < \cdots < a_i$ and the $a_0, \ldots, a_n$ all touch at one point at least.

Mr. Čech has a recipe requiring that we indicate, for each index i, with C^i the vector space of maps $f : N_i \to \mathbb{R}$ and define linear maps $d_i : C^i \to C^{i+1}$ as follows:

$$d_0 f(a, b) = f(b) - f(a), \qquad d_1 f(a, b, c) = f(b, c) - f(a, c) + f(a, b),$$

and more generally

$$d_i f(a_0, \ldots, a_{i+1}) = \sum_{j=0}^{i+1} (-1)^j f(a_0, \ldots, \widehat{a_j}, \ldots, a_{i+1}),$$

where the hat $\widehat{}$ reminds we have to omit that variable. With the help of his father-in-law Billy managed to prove that the composition of d_i and d_{i+1} is always 0 (exercise for the reader), so it is possible to define vector spaces

$$H^i(X) = \frac{\text{kernel of } d_i : C^i \to C^{i+1}}{\text{image of } d_{i-1} : C^{i-1} \to C^i}, \quad i \geq 0.$$

It's rather surprising that the dimensions of the real vector spaces $H^i(X)$ only depend on the homeomorphism class of the polybrick X. The numbers $b_i(X) = \dim H^i(X)$, $i = 0, 1, \ldots$, are called **Betti numbers** of X, in honour of Enrico Betti (Pistoia 1823 - Pisa 1892), the forefather, with Riemann, of modern topology. Readers may prove as exercise that b_0 coincides with the number of connected components of X; the unabridged and rigorous proof that the Betti numbers are invariant under homeomorphisms is far from easy, and was achieved around 1940.

The Euler-Poincaré characteristic of a polybrick is defined as the alternating sum of Betti numbers

$$e(X) = b_0(X) - b_1(X) + b_2(X) + \cdots + (-1)^i b_i(X) + \cdots$$

and is a topological invariant. Its core property is that (exercise in linear algebra for the reader) it equals the alternating sum of the cardinalities of the sets N_i

$$e(X) = |N_0| - |N_1| + |N_2| + \cdots + (-1)^i |N_i| + \cdots,$$

and so it's quite easy to compute.

Suppose we have a polybrick in which each point belongs to not more than two bricks, that is, $N_2 = \emptyset$. Such an object can be represented as a graph with N_0 as set of vertices and N_1 as set of edges (meaning that two vertices $a, b \in N_0$ are joined by an edge iff one of the ordered pairs (a, b), (b, a) belongs to N_1.) Then the Euler-Poincaré characteristics of polybrick and graph coincide.

One can generalise Betti numbers to an arbitrary topological space as the dimensions of suitable vector spaces, and in such a way that they remain topological invariants. Regrettably, these vector spaces can have infinite dimension, so that $b_i(X) \in \mathbb{N}_0 \cup \{+\infty\}$ in general.

9.3 What Algebraic Topology Is

Imagine having a family $\mathcal{F}$ of topological spaces and wanting to understand when elements of $\mathcal{F}$ are homeomorphic; for the sake of simplicity $\mathcal{F}$ will consist of connected subspaces in $\mathbb{R}^3$.

We are seeking functions $f : \mathcal{F} \to I$ such that if $X, Y \in \mathcal{F}$ are homeomorphic, in symbols $X \sim Y$, then $f(X) = f(Y)$. A function of this sort is called a **topological invariant** of $\mathcal{F}$, and its usefulness depends on two factors: how complete the invariant is, and how concrete it is. A constant map $f : \mathcal{F} \to I$ is an invariant with no use whatsoever; at the other end of the spectrum are so-called **complete** invariants, characterised by the fact that $X \sim Y$ precisely when $f(X) = f(Y)$.

Complete invariants always exist: take $I = \mathcal{F}/\sim$ and the canonical quotient map f. Yet the whole point of finding invariants is to understand the deeper structure of $\mathcal{F}/\sim$, which is usually complicated and mysterious. That's why one requires that invariants be *concrete*, i.e. that the codomain I be a simpler set than $\mathcal{F}/\sim$, and that f correspond to an explicit procedure or geometrical criterion. In the previous chapters we witnessed many invariants like that, where $I = \{\text{true, false}\}$ and the maps f were representing compactness, connectedness, metrisability and so on.

Another example is the invariant $b_0 \colon \mathcal{F} \to I = \mathbb{N}_0 \cup \{+\infty\}$ that maps a space to the number of its connected components. More generally, every Betti number b_n is a topological invariant with values in $I = \mathbb{N}_0 \cup \{+\infty\}$.

Ever since Poincaré's work mathematicians have discovered a wealth of topological invariants $f \colon \mathcal{F} \to I$ where the elements of I are algebraic entities, like groups, vector spaces, polynomials et c. An instance are the **Čech cohomology** spaces $H^i(X)$ we met in the previous section about polybricks, but whose definition can be broadened to hold for any paracompact Hausdorff space. An excellent introduction to Čech cohomology can be found in Kodaira's book [Ko86, Sect. 3.1].

Algebraic topology deals with this kind of invariants and other related issues. The ensuing chapters will discuss the fundamental group, an invariant introduced by Poincaré and denoted by $\pi_1 \colon \mathcal{F} \to I$, where $\mathcal{F}$ are path-connected spaces and I isomorphism classes of groups.

Reference

[Ko86] Kodaira, K.: Complex Manifolds and Deformation of Complex Structures. Springer, Berlin (1986)

Chapter 10
Homotopy

There's an intuitive notion of 'equivalence of shapes' in topology that is broader than the concept of homeomorphism. Two connected regular subsets in $\mathbb{R}^2$ — where regular is meant naively, as the opposite of complicated, strange, pathological, unreasonable etc.—have equivalent shape if they have the same number of holes. To be concrete, the letters in the word HOMOTOPY, for instance, each thought as a connected union of curves in $\mathbb{R}^2$, can be grouped in two equivalence classes of shapes: O, P have the same shape because they have one hole, while H, M, T, Y have no holes. Here's another example: the letter B, a circle with its diameter, two touching circles (Fig. 10.1).

In this chapter we'll define in precise terms the notion of **homotopy**: having the same homotopy type will correspond to the intuitive idea of having equivalent shapes.

10.1 Locally Connected Spaces and the Functor π_0

Definition 10.1 A space is **locally connected** if every point has local basis of connected neighbourhoods.

From Lemma 4.28 the connected components of a locally connected space are open. While general connected spaces may not be locally connected (Exercise 10.1), open sets in $\mathbb{R}^n$ are locally connected, and the product of two locally connected spaces is locally connected.

Definition 10.2 Let X be a topological space. Denote by $\pi_0(X) = X/\sim$ the quotient space under the relation $\sim$ that identifies points connected by a path in X.

To be more precise, for any two points $x, y \in X$ one defines the set of paths from x to y:

$$\Omega(X, x, y) = \{\alpha \colon [0, 1] \to X \mid \alpha \text{ continuous}, \alpha(0) = x, \alpha(1) = y\}$$

© Springer International Publishing Switzerland 2015
M. Manetti, *Topology*, UNITEXT - La Matematica per il 3+2 91,
DOI 10.1007/978-3-319-16958-3_10

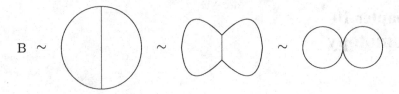

B ~

Fig. 10.1 Subsets in the plane with equivalent shapes

and then

$$\pi_0(X) = X/\sim, \quad \text{where} \quad x \sim y \iff \Omega(X, x, y) \neq \emptyset.$$

We have to make sure $\sim$ is an equivalence relation.

Reflexivity. To prove $x \sim x$ we consider the constant path

$$1_x \colon [0, 1] \to X, \quad 1_x(t) = x \text{ for every } t \in [0, 1].$$

Symmetry. For every $x, y \in X$ we have the path-reverting operator

$$i \colon \Omega(X, x, y) \to \Omega(X, y, x), \quad i(\alpha)(t) = \alpha(1 - t),$$

that is clearly invertible. In particular $\Omega(X, x, y)$ is empty precisely when $\Omega(X, y, x)$ is empty.

Transitivity. We just consider the product of paths (or composite)

$$* \colon \Omega(X, x, y) \times \Omega(X, y, z) \to \Omega(X, x, z), \quad (\alpha, \beta) \mapsto \alpha * \beta,$$

where

$$\alpha * \beta(t) = \begin{cases} \alpha(2t) & \text{if } 0 \leq t \leq 1/2, \\ \beta(2t - 1) & \text{if } 1/2 \leq t \leq 1. \end{cases}$$

Definition 10.3 Let X be a space and $\alpha \colon [a, b] \to X$ a path, a, b real numbers. We shall call **standard parametrisation** of α the path

$$\tilde{\alpha} \in \Omega(X, \alpha(a), \alpha(b)), \quad \tilde{\alpha}(t) = \alpha((1 - t)a + tb).$$

For example, if $\alpha \in \Omega(X, x, y)$ and $\beta \in \Omega(X, y, z)$, then α can be recovered as the standard parametrisation of $\alpha * \beta$ restricted to the interval $[0, 1/2]$.

The choice of $[0, 1]$ in the definition of $\sim$ is clearly only conventional, for nothing essential would change if we took paths defined on an arbitrary compact interval $[a, b]$, $a < b$. The only drawback would be worse formulas for inverting and composing paths.

Equivalence classes for the relation $\sim$ are called **path components** of X, which are usually neither open nor closed. Therefore $\pi_0(X)$ is the set of path components of X.

Definition 10.4 A space X is **locally path connected** if any point has a local basis of path connected neighbourhoods.

An open set in $\mathbb{R}^n$ is locally path connected: if $X \subset \mathbb{R}^n$ is open and $x \in X$, there exists $r > 0$ such that $B(x, r) \subset X$. The balls $B(x, t), 0 < t < r$, constitute a local path connected basis.

Proposition 10.5 *Let X be locally path connected. Its path components in X are open, and coincide with the connected components.*

Proof Take $x \in X$ and let $A = \{y \mid \Omega(X, x, y) \neq \emptyset\}$ be the path component containing it. We want to show that A is open, or equivalently that it is a neighbourhoods of any of its points. Take $y \in A$ and choose a path $\alpha \in \Omega(X, x, y)$; by assumption there's a path connected neighbourhood U of y, so for every $z \in U$ we have a $\beta \in \Omega(U, y, z)$. The path $\alpha * \beta$ belongs to $\Omega(X, x, z)$, so $z \in A$, and then $U \subset A$. The complement of A is the union of all (open) path components other than A, which makes A closed. Let $C(x)$ be the connected component containing x. As $A \cap C(x)$ is open and closed in $C(x)$, $C(x) \subset A$. On the other hand A is path connected, so A is connected and $A \subset C(x)$. $\qquad\square$

Consider a continuous map $f : X \to Y$ and a path $\alpha : [0, 1] \to X$; their composition is the path $f\alpha : [0, 1] \to Y$, so if $x_1, x_2 \in X$ are joined by a path then $f(x_1), f(x_2)$ belong to the same path component. Therefore f induces a mapping

$$\pi_0(f) : \pi_0(X) \to \pi_0(Y), \qquad \pi_0(f)([x]) = [f(x)]$$

between the quotient spaces. The next properties are immediate:

F1. let $Id : X \to X$ be the identity map. Then $\pi_0(Id) : \pi_0(X) \to \pi_0(X)$ is the identity map;
F2. if $f : X \to Y$ and $g : Y \to Z$ are continuous, then

$$\pi_0(gf) = \pi_0(g)\pi_0(f) : \pi_0(X) \to \pi_0(Z).$$

Properties **F1** and **F2** entitle one to say that the assignment

$$\pi_0 : \{ \text{topological spaces} \} \to \{ \text{sets} \}, \qquad X \mapsto \pi_0(X),$$

is a **functor**. We shouldn't call π_0 a map because its domain and codomain are not sets but **categories** (precise definitions are postponed to Sect. 10.4).

Example 10.6 Let $K_1 \subset K_2 \subset \cdots$ be an exhaustion by compact sets of a space X such that the inclusion $i: X - K_{n+1} \to X - K_n$ induces a bijection

$$\pi_0(i): \pi_0(X - K_{n+1}) \xrightarrow{\simeq} \pi_0(X - K_n)$$

for every n. Then the cardinality of $\pi_0(X - K_{n+1})$ is a topological invariant of X, i.e. it doesn't depend on the chosen exhaustion. Suppose we have another such exhaustion $H_1 \subset H_2 \subset \cdots$. There exist integers $n < m$ and $h < k$ such that $H_n \subset K_h \subset H_m \subset K_k$. Hence we have inclusions

$$X - K_k \xrightarrow{\alpha} X - H_m \xrightarrow{\beta} X - K_h \xrightarrow{\gamma} X - H_n.$$

Applying the functor π_0 gives

$$\pi_0(X - K_k) \xrightarrow{\pi_0(\alpha)} \pi_0(X - H_m) \xrightarrow{\pi_0(\beta)} \pi_0(X - K_h) \xrightarrow{\pi_0(\gamma)} \pi_0(X - H_n).$$

By assumption $\pi_0(\beta)\pi_0(\alpha)$ and $\pi_0(\gamma)\pi_0(\beta)$ are bijections, and it's not hard to conclude from this that $\pi_0(\alpha)$ is bijective (see Lemma 11.21 as well).

Example 10.7 As promised earlier, we prove that $\mathbb{R}^m - \{s \text{ points}\}$ is not homeomorphic to $\mathbb{R}^m - \{t \text{ points}\}$ if $m > 1$ and $s \neq t$.

Take distinct points $p_1, \ldots, p_t$ in $\mathbb{R}^m$. We claim that the number $t + 1$ is a topological invariant of $\mathbb{R}^m - \{p_1, \ldots, p_t\}$. For any $n > 0$ consider the compact set

$$K_n = \left\{ x \in \mathbb{R}^m \mid \|x\| \leq n, \ \|x - p_i\| \geq \frac{1}{n} \right\}.$$

When $n >> 0$, $\mathbb{R}^m - K_n$ has $t + 1$ connected components, and the inclusion $X - K_{n+1} \to X - K_n$ induces a bijection between the corresponding π_0's. The topological invariance of $t + 1$ is now a consequence of Example 10.6.

Exercises

10.1 Prove that the union $X \subset \mathbb{R}^2$ of all straight lines $ax = by$, $a, b \in \mathbb{Z}$ not both zero, is path connected but not locally path connected.

10.2 ($\heartsuit$) Let X be a space with the following property: for every point $x \in X$ and every neighbourhood U of x there exists a neighbourhood V of x such that any two points of V can be joined by a path in U. Prove X is locally path connected.

10.3 Let $A \subset \mathbb{R}^n$ be open and connected and take two points $p, q \in A$. Prove that there is a zig-zag from p to q (a continuous, piecewise-linear path $\gamma: [0, 1] \to A$ with $\gamma(0) = p$ and $\gamma(1) = q$).

Fig. 10.2 Every
homeomorphism fixes $(0, 0)$

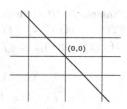

10.4 (**Local π_0,** 👆) Let $x \in X$ be a point in some space and $\mathcal{A}$ a local basis of neighbourhoods at x. Define the set

$$\pi_0(X - \{x\}, x) = \{s \in \prod_{U \in \mathcal{A}} \pi_0(U - \{x\}) \mid s \text{ is coherent}\},$$

where 'coherent' means that for every $U, V \in \mathcal{A}$ such that $U \subset V$ we have $\pi_0(i)(s_U) = s_V$, where $i : U - \{x\} \to V - \{x\}$ is the inclusion morphism.

Prove that $\pi_0(X - \{x\}, x)$ doesn't depend on the chosen neighbourhood basis.

10.5 Let $X \subset \mathbb{R}^2$ be defined by the equation $xy(x + y)(x^2 - 4)(y^2 - 1) = 0$ (Fig. 10.2). Use the result of Exercise 10.4 to show that any self-homeomorphism of X has $(0, 0)$ as fixed point.

10.2 Homotopy

Definition 10.8 Two continuous maps $f_0, f_1 : X \to Y$ are said to be **homotopic** if there is a continuous function

$$F : X \times [0, 1] \to Y$$

such that $F(x, 0) = f_0(x)$ and $F(x, 1) = f_1(x)$ for every $x \in X$. Such an F is called a **homotopy** between f_0 and f_1.

To help one 'visualise' the meaning of the above definition let's write $f_t(x) = F(x, t)$ for every $(x, t) \in X \times [0, 1]$. Then for any $t \in [0, 1]$ the map

$$f_t : X \to Y$$

is continuous. When $t = 0$ we recover f_0, which deforms in a continuous way, as t varies, until it becomes f_1 for $t = 1$.

Example 10.9 Let $Y \subset \mathbb{R}^n$ be a convex subspace. For any topological space X, two continuous maps $f_0, f_1 : X \to Y$ are homotopic: it suffices to define the homotopy as

$$F : X \times [0, 1] \to Y, \qquad F(x, t) = (1 - t) f_0(x) + t f_1(x).$$

Example 10.10 Consider continuous mappings $f, g\colon X \to \mathbb{R}^n - \{0\}$ such that $\|f(x) - g(x)\| < \|f(x)\|$ for every $x \in X$. Then f and g are homotopic, for $\|t(f(x) - g(x))\| < \|f(x)\|$ for every $t \in [0, 1]$ and we can take

$$F\colon X \times [0, 1] \to \mathbb{R}^n - \{0\},$$
$$F(x, t) = (1 - t)f(x) + tg(x) = f(x) - t(f(x) - g(x)),$$

as homotopy between them.

Notation From now on, and unless stated differently, the capital letter I will always indicate the closed interval $[0, 1]$ with the Euclidean topology.

Lemma 10.11 *Let X, Y be topological spaces. The homotopy relation is an equivalence relation on the set $C(X, Y)$ of continuous maps from X to Y.*

Proof Take $f\colon X \to Y$ continuous. Then

$$F\colon X \times I \to Y, \qquad F(x, t) = f(x),$$

is a homotopy between f and f. If $F(x, t)$ is a homotopy between f and g, $F(x, 1-t)$ is a homotopy between g and f. At last, if $F(x, t)$ is a homotopy between f and g, and $G(x, t)$ a homotopy between g and h,

$$H\colon X \times I \to Y, \qquad H(x, t) = \begin{cases} F(x, 2t) & \text{if } 0 \leq t \leq 1/2, \\ G(x, 2t - 1) & \text{if } 1/2 \leq t \leq 1, \end{cases}$$

is a homotopy between f and h. $\qquad\qquad\qquad\qquad\qquad\qquad\qquad\qquad\square$

Example 10.12 Consider the antipodal map $f\colon S^n \to S^n$, $f(x) = -x$. When $n = 2k - 1$ is odd f is homotopic to the identity map, for we can write $S^n = \{z \in \mathbb{C}^k \mid \|z\| = 1\}$ and $F(z, t) = ze^{\pi i t}$ is a homotopy. When n is even, instead, f is not homotopic to the identity, but this is—as one can well imagine—much harder to show.

The homotopy relation is stable under composition of maps, in the sense explained below.

Lemma 10.13 *Given continuous maps as in the diagram*

$$X \underset{f_1}{\overset{f_0}{\rightrightarrows}} Y \underset{g_1}{\overset{g_0}{\rightrightarrows}} Z \,,$$

if f_0 is homotopic to f_1 and g_0 is homotopic to g_1, then $g_0 f_0$ is homotopic to $g_1 f_1$.

Proof Let $F\colon X \times I \to Y$ and $G\colon Y \times I \to Z$ be the involved homotopies: $F(x, 0) = f_0(x)$, $F(x, 1) = f_1(x)$, $G(y, 0) = g_0(y)$ and $G(y, 1) = g_1(y)$. It's easy to see that

$$H: X \times I \to Z, \qquad H(x,t) = G(F(x,t),t),$$

is a homotopy between $g_0 f_0$ and $g_1 f_1$. □

Definition 10.14 A continuous map $f: X \to Y$ is a **homotopy equivalence** if there exists a continuous map $g: Y \to X$ such that fg is homotopic to the identity map of Y and gf is homotopic to the identity of X. Two topological spaces are called **homotopy equivalent** if they admit a homotopy equivalence between them.

It's quite common to say that two spaces have the same **homotopy type** if they are homotopy equivalent.

By definition, then, homeomorphic spaces are homotopy equivalent. The next example shows that the opposite is in general false.

Example 10.15 All non-empty convex subsets in $\mathbb{R}^n$ are homotopy equivalent. Take $X \subset \mathbb{R}^n, Y \subset \mathbb{R}^m$ convex and not empty, then pick two continuous maps $f: X \to Y$ and $g: Y \to X$ arbitrarily (e.g., constant ones). As X is convex, from Example 10.9 we know that $gf: X \to X$ is homotopic to the identity, and similarly fg.

Lemma 10.16 *If two continuous maps $f, g: X \to Y$ are homotopic, then*

$$\pi_0(f) = \pi_0(g): \pi_0(X) \to \pi_0(Y).$$

In particular, $\pi_0(f)$ is a bijection if $f: X \to Y$ is a homotopy equivalence.

Proof Saying $\pi_0(f) = \pi_0(g)$ means that the points $f(x)$ and $g(x)$ belong in the same path component of Y for every $x \in X$. Let $F: X \times I \to Y$ be a homotopy between f and g. Then $f(x)$ and $g(x)$ are joined by the path

$$\alpha: I \to Y, \qquad \alpha(t) = F(x,t).$$

Suppose now that $f: X \to Y$ is a homotopy equivalence, $g: Y \to X$ is continuous and both gf, fg are homotopic to the identity. The lemma's first statement ensures that

$$\pi_0(fg) = \pi_0(f)\pi_0(g): \pi_0(Y) \to \pi_0(Y),$$

$$\pi_0(gf) = \pi_0(g)\pi_0(f): \pi_0(X) \to \pi_0(X)$$

are identity maps, whence $\pi_0(f)$ is invertible with inverse $\pi_0(g)$. □

Definition 10.17 A space is **contractible** if it is homotopy equivalent to a point. This is the same as demanding that the identity map is homotopic to a constant map.

Example 10.15 shows that convex subsets of $\mathbb{R}^n$ are contractible. Any contractible space is, by Lemma 10.16, path connected, and it is natural to ask whether there exist path connected spaces that are not contractible. The answer is yes, they do exist: we'll have to suffer a bit to prove that the unit circle S^1 is not contractible. Other examples include the spheres S^n, $n \geq 2$; the proof that these aren't contractible goes beyond the scopes of this textbook.

Exercises

10.6 Prove, as claimed in Definition 10.17, that for X non-empty the following statements are equivalent:

1. X has the homotopy type of a point;
2. for any $p \in X$ the map $f: X \to X$, $f(x) = p$, is homotopic to the identity;
3. there is a point $p \in X$ such that $f: X \to X$, $f(x) = p$, is homotopic to the identity.

10.7 Let $A \subset C(X, Y)$ be a family of continuous maps from X to Y. One says that $f_0, f_1 \in A$ are *homotopic relatively to* A if there's a homotopy $F: X \times I \to Y$ between f_0 and f_1 such that $f_t \in A$ for every $t \in I$, where $f_t(x) = F(x, t)$. Show that this is an equivalence relation.

10.8 In the notation of Definition 10.14, the map g is called **homotopic inverse** of f. Let $f: X \to Y$ be a homotopy equivalence. Prove that the homotopic inverse of f is unique up to homotopy.

10.9 Prove that homotopy equivalence is transitive, i.e. if X has the homotopy type of Y and Y the homotopy type of Z, then X has the homotopy type of Z.

10.10 Given spaces X, Y one writes $[X, Y]$ to denote the set of homotopy classes of maps from X to Y. Show that if X is contractible there exists a natural bijection $[X, Y] = \pi_0(Y)$.

10.11 Prove that the product of two contractible spaces is contractible.

10.12 Let X be a space and $f, g: X \to S^n$ continuous maps. Using the algebraic expression

$$\frac{tf(x) + (1 - t)g(x)}{\|tf(x) + (1 - t)g(x)\|}, \qquad t \in [0, 1],$$

show that if $f(x) \neq -g(x)$ for every $x \in X$, then f is homotopic to g.

10.13 (✿) Let $S^\infty \subset \ell^2(\mathbb{R})$ (see Example 6.8) be the space of square-integrable sequences of norm one:

$$S^\infty = \{\{a_n\} \in \ell^2(\mathbb{R}) \mid \sum a_n^2 = 1\}.$$

Generalise the arguments used for Exercise 10.12 in order to show that the identity map on S^∞ and the constant map $(a_1, a_2, \ldots) \mapsto (1, 0, \ldots)$ are both homotopic to $(a_1, a_2, a_3, \ldots) \mapsto (0, a_1, a_2, a_3, \ldots)$. Conclude that S^∞ is contractible.

10.3 Retractions and Deformations

Definition 10.18 A subspace $Y \subset X$ is a **retract** of X if there is a continuous map $r \colon X \to Y$, called **retraction**, such that $r(y) = y$ for every $y \in Y$.

Example 10.19 Let $A, B \subset \mathbb{R}^2$ be two circles touching at a unique point p. Then

$$r \colon A \cup B \to A, \qquad r(x) = \begin{cases} x & \text{if } x \in A, \\ p & \text{if } x \in B, \end{cases}$$

is a retraction.

Definition 10.20 A subspace $Y \subset X$ is called a **deformation retract** of X if there is a continuous map $R \colon X \times I \to X$, called **deformation** of X into Y, such that:

1. $R(x, 0) \in Y$ and $R(x, 1) = x$ for every $x \in X$;
2. $R(y, t) = y$ for every $y \in Y, t \in I$.

Example 10.21 Let $A \subset \mathbb{R}^n$ be a star-shaped subset with respect to the point $p \in A$ (Exercise 4.8). Then p is a deformation retract of A: one possible deformation is

$$R \colon A \times I \to A, \qquad R(x, t) = tx + (1 - t)p.$$

Proposition 10.22 *A deformation retract $Y \subset X$ is a retract of X, and the inclusion $i \colon Y \hookrightarrow X$ is a homotopy equivalence.*

Proof If $R \colon X \times I \to X$ denotes the deformation of X to Y, the map $r \colon X \to Y$ defined by $R(x, 0) = i(r(x))$ is a retraction, and R a homotopy between ir and the identity on X. As $ri = Id_Y$, i and r are homotopy equivalences. □

Example 10.23 The sphere S^n is a deformation retract of $\mathbb{R}^{n+1} - \{0\}$, via the deformation

$$R \colon (\mathbb{R}^{n+1} - \{0\}) \times I \to \mathbb{R}^{n+1} - \{0\}, \qquad R(x, t) = tx + (1 - t)\frac{x}{\|x\|}.$$

Example 10.24 Suppose Y is the union of two sides in a triangle X. Then Y is a deformation retract of X, see Fig. 10.3.

We can assume X in $\mathbb{R}^2$ is given by the intersection of the half-planes $x \geq 0$, $y \geq 0, x + y \leq 1$, and Y is X intersected with the two axes $x = 0, y = 0$. A possible deformation is

$$R(x, y, t) = t(x, y) + (1 - t)(x - \min(x, y), y - \min(x, y)).$$

In a similar way one proves that any non-degenerate simplex $X \subset \mathbb{R}^n$ can be deformed to m of its faces $X_1, \ldots, X_m$, for every $1 \leq m \leq n$. In fact, up to affine transformations and permutation of indices, we can write

Fig. 10.3 Deformation of a
triangle to two of its sides

$$X = \{(x_1, \ldots, x_n) \in \mathbb{R}^n \mid x_i \geq 0, \ x_1 + \cdots + x_n \leq 1\},$$

$X_i = X \cap \{x_i = 0\}, i = 1, \ldots, m$, and take the deformation

$$R(x, t) = x - (1 - t)(\min(x_1, \ldots, x_m), \ldots, \min(x_1, \ldots, x_m), 0, \ldots, 0).$$

Exercises

10.14 ($\heartsuit$) Prove that in a Hausdorff space every retract is closed.

10.15 ($\heartsuit$) Prove that an empty glass is a deformation retract of the full glass. More precisely, show that $Y = D^2 \times \{0\} \cup S^1 \times [0, 1] \subset \mathbb{R}^3$ is a deformation retract of $X = D^2 \times [0, 1] \subset \mathbb{R}^3$.

10.16 ($\heartsuit$) Set $X = \{(tx, t) \in \mathbb{R}^2 \mid t \in [0, 1], \ x \in \mathbb{Q}\}$. Prove that:

1. $(0, 0)$ is a deformation retract of X, so X is, in particular, contractible;
2. $(0, 1)$ is not a deformation retract of X.

10.17 Prove that $X = \{(p, q) \in S^n \times S^n \mid p \neq q\}$ has the homotopy type of S^n. (Hint: the graph of the antipodal map is a deformation retract.)

10.18 Prove that $SL(n, \mathbb{R})$ is a deformation retract of $GL^+(n, \mathbb{R})$.

10.19 One denotes by $S^1 \vee S^1$ the space obtained by glueing together two circles at one point (the figure eight).

Convince yourself that the four spaces $\mathbb{R}^2 - \{2 \text{ points}\}$, $S^2 - \{3 \text{ points}\}$, $S^1 \times S^1 - \{1 \text{ point}\}$ and $S^1 \vee S^1$ have the same homotopy type. We don't require a formal proof. (Hint: show that the first three have a deformation retract homeomorphic to $S^1 \vee S^1$; interpret $S^1 \times S^1 - \{1 \text{ point}\}$ as quotient of $I^2 - \{\text{one interior point}\}$ by the equivalence relation discussed in the section on topological sewing.)

10.20 ($\clubsuit, \heartsuit$) Prove that $X = \{(a, b, c) \in \mathbb{R}^3 \mid b^2 > 4ac\}$ has the homotopy type of S^1.

10.21 ($\clubsuit, \heartsuit$) Let X be the space of real $n \times (n - 1)$-matrices of maximal rank. Prove X is homotopy equivalent to $GL^+(n, \mathbb{R})$.

10.22 ($\clubsuit, \heartsuit$) Let $X \subset \text{Homeo}(D^n)$ be the set of self-homeomorphisms of the closed unit ball that equal the identity on S^{n-1}. Put on X the subspace topology induced by $C(D^n, D^n)$ (Theorem 6.51). Prove that the identity $Id \in X$ is a deformation retract of X.

10.4 Categories and Functors

The terms 'category' and 'functor' began to be employed in mathematics around 1940 as part of a new language capable of simplifying the theory underlying certain phenomena occurring in algebraic topology. Only later, owing to the genius ideas of A. Grothendieck, category theory allowed to make such tremendous advancements in the most abstract areas (to be clear: logic, topology, algebra and algebraic geometry) that it became an essential part of these. Nowadays we are witnesses of a slow but unstoppable expansion of category theory into every field of mathematical knowledge.

Assigning a **category A** means assigning:

1. a collection $Ob(\mathbf{A})$, whose members are called **objects** of the category $\mathbf{A}$;
2. for any pair of objects X, Y in $\mathbf{A}$, a set $Mor_{\mathbf{A}}(X, Y)$ whose elements are called **morphisms** between X and Y in the category $\mathbf{A}$;
3. for any triple $X, Y, Z \in Ob(\mathbf{A})$, a map, called composition,

$$Mor_{\mathbf{A}}(X, Y) \times Mor_{\mathbf{A}}(Y, Z) \to Mor_{\mathbf{A}}(X, Z), \quad (f, g) \mapsto gf.$$

All these data must obey the following axioms:

1. $Mor_{\mathbf{A}}(X, Y) \cap Mor_{\mathbf{A}}(Z, W) = \emptyset$ unless $X = Z$ and $Y = W$;
2. for any object X there is an identity morphism $1_X \in Mor_{\mathbf{A}}(X, X)$ with the property that $1_X f = f$ and $g1_X = g$ for every $f \in Mor_{\mathbf{A}}(Y, X)$ and $g \in Mor_{\mathbf{A}}(X, Z)$;
3. the composition of morphisms is associative: $h(gf) = (hg)f$ for every $f \in Mor_{\mathbf{A}}(X, Y)$, $g \in Mor_{\mathbf{A}}(Y, Z)$ and $h \in Mor_{\mathbf{A}}(Z, W)$.

Just as one proves that the neutral element in a group is unique, so the identity morphism is unique. A morphism $f \in Mor_{\mathbf{A}}(X, Y)$ is called an **isomorphism** if there exists $f^{-1} \in Mor_{\mathbf{A}}(Y, X)$ such that $ff^{-1} = 1_Y$ and $f^{-1}f = 1_X$. In this case f^{-1} is unique, and is called the **inverse** of f.

To indicate that X is an object of the category $\mathbf{A}$ one almost always writes, abusing notations, $X \in \mathbf{A}$.

Example 10.25 The category of open sets in a topological space X is the category with open subsets in X as objects and inclusions as morphisms.

Example 10.26 The category **Set** has sets as objects and maps as morphisms.

Example 10.27 The category **Grp** has groups as objects and group homomorphisms as morphisms.

Example 10.28 The category **Top** has topological spaces as objects and continuous mas as morphisms.

Example 10.29 The category **KTop** has the same objects as **Top** and homotopy classes of continuous maps as morphisms. Lemma 10.13 ensures that the composition law is well defined.

Example 10.30 To any set S we can associate a category having S as collection of objects and identity maps as morphisms.

Example 10.31 Given a group G, **G** denotes the category with one object $*$ and $\mathrm{Mor}_G(*, *) = G$ as morphisms. The composition law is the product in G.

Remark 10.32 In the definition of category and the previous examples we have implicitly assumed the existence of a hierarchy between sets and collections—mathematical and hence also linguistic: every set is a collection, but there are collections that aren't sets, like the collection of all sets. Sometimes it's useful to push this farther, and introduce classes, conglomerates and so on, along the naive picture

$$\{\text{ sets }\} \subset \{\text{ classes }\} \subset \{\text{ conglomerates }\} \subset \cdots.$$

A collection of sets is a class, a collection of classes is a conglomerate. The advantage of this additional distinction is that it allows, in some sense, to treat classes and sets alike: for example, we can have an axiom of choice for classes. Needless to say, all this can be formalised and packaged into solid axiomatic set theories [En77].

Many authors require, in defining categories, that the collection of objects be a class.

The observation that category's axioms do not change if morphisms are 'reversed' leads to the notion of opposite category.

Definition 10.33 Let **C** be a category. The **opposite category** $\mathbf{C}^o$ has the same objects $\mathrm{Ob}(\mathbf{C}) = \mathrm{Ob}(\mathbf{C}^o)$ and

$$\mathrm{Mor}_{\mathbf{C}^o}(X, Y) = \mathrm{Mor}_{\mathbf{C}}(Y, X) \quad \text{for every} \quad X, Y \in \mathrm{Ob}(\mathbf{C}) = \mathrm{Ob}(\mathbf{C}^o)$$

as morphisms.

Note that $(\mathbf{C}^o)^o = \mathbf{C}$.

Definition 10.34 Let **A**, **B** be categories. A **functor** from **A** to **B**, written $F : \mathbf{A} \to \mathbf{B}$, consists of two functions (both denoted with F):

1. the *function on objects*, a law mapping an object X in **A** to one, and one only, object $F(X)$ of **B**;
2. the *function on morphisms*, consisting of a map

$$F : \mathrm{Mor}_{\mathbf{A}}(X, Y) \to \mathrm{Mor}_{\mathbf{B}}(F(X), F(Y)), \qquad f \mapsto F(f)$$

for every pair of objects X, Y in **A**.

The function on morphisms must preserve identity maps and composition laws: $F(1_X) = 1_{F(X)}$ and $F(fg) = F(f)F(g)$.

Example 10.35 π_0 is a functor from the category of topological spaces to the category of sets.

Example 10.36 Let G, H be groups. In the notations of Example 10.31, there is a natural bijection between the set of homomorphisms $F: G \to H$ and the set of functors $F: \mathbf{G} \to \mathbf{H}$.

Remark 10.37 Until recently a functor as of Definition 10.34 was called a *covariant functor*. A *contravariant functor* was a functor $F: \mathbf{A} \to \mathbf{B}$ with the following properties replacing the respective covariant ones: if $f \in \mathrm{Mor}_{\mathbf{A}}(X, Y)$ then $F(f) \in \mathrm{Mor}_{\mathbf{B}}(F(Y), F(X))$, and action on composites $F(fg) = F(g)F(f)$.

To any contravariant functor $F: \mathbf{A} \to \mathbf{B}$ corresponds tautologically a covariant functor $F: \mathbf{A}^\circ \to \mathbf{B}$, and the other way around. Contravariant functors are not very popular nowadays and one prefers to use opposite categories instead.

Example 10.38 Let $\mathbf{Vect}_{\mathbb{K}}$ denote the category of $\mathbb{K}$-vector spaces: objects are vector spaces over $\mathbb{K}$ and morphisms the linear maps. The function $\mathrm{Hom}_{\mathbb{K}}(-, \mathbb{K})$ that maps a vector space V to its dual $\mathrm{Hom}_{\mathbb{K}}(V, \mathbb{K})$ defines a functor

$$\mathrm{Hom}_{\mathbb{K}}(-, \mathbb{K}): \mathbf{Vect}_{\mathbb{K}}^\circ \to \mathbf{Vect}_{\mathbb{K}}.$$

Exercises

10.23 For any integer $n \geq 0$ let $[\mathbf{n}]$ be the category having as objects the numbers $0, 1, \ldots, n$ and as morphisms

$$\mathrm{Mor}_{[\mathbf{n}]}(a, b) = \begin{cases} \emptyset & \text{if } a > b \\ \text{one morphism} & \text{if } a \leq b \end{cases}.$$

Prove that there exists a natural bijection between functors $[\mathbf{n}] \to [\mathbf{m}]$ and non-decreasing maps $\{0, 1, \ldots, n\} \to \{0, 1, \ldots, m\}$.

10.24 Prove that two topological spaces are homotopy equivalent if and only if they are isomorphic in **KTop**.

10.25 Let Y be a given topological space. Prove that the recipe that assigns to a space X the set $C(Y, X)$ of continuous maps $Y \to X$ is a functor from the category of topological spaces to the category of sets.

10.26 For any integer $n \geq 0$ write $[n] = \{0, 1, \ldots, n\}$, equipped with the usual ordering. We indicate with Δ the category of **finite ordinals**, i.e. that with objects $\mathrm{Ob}(\Delta) = \{[n] \mid n \geq 0\}$ and morphisms

$$\mathrm{Mor}_{\Delta}([n], [m]) = \{f: [n] \to [m] \text{ non-decreasing}\}.$$

For a given $n \geq 0$ write

$$\Delta^n = \{(t_0, \ldots, t_n) \in \mathbb{R}^{n+1} \mid t_i \geq 0, \ t_0 + \cdots + t_n = 1\}$$

for the **standard simplex** of dimension n.

Check that $[n] \mapsto \Delta^n$ defines a functor from Δ to the category of topological spaces, where a morphism $f : [n] \to [m]$ is associated to the continuous mapping

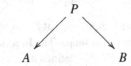

$$f_* : \Delta^n \to \Delta^m, \qquad f_*(t_0, \ldots, t_n) = \left(\sum_{f(i)=0} t_i, \sum_{f(i)=1} t_i, \ldots, \sum_{f(i)=m} t_i \right),$$

and the sum $\sum_{f(i)=j} t_i$ is zero if $f(i) \neq j$, for every $i \in [n]$.

10.27 (Products) A diagram of three objects and two morphisms

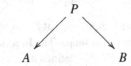

in a category $\mathbf{C}$ is called a **product** if, given any other diagram $A \leftarrow X \to B$, there's a unique morphism $X \to P$ rendering

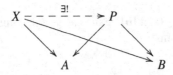

a commutative diagram. The object P is called the product of A and B in the category $\mathbf{C}$, written $P = A \times B$. The product of two objects might not exist: but if it does, then it is unique up to isomorphisms.

Check that products always exist in the categories **Set**, **Grp** and **Top**, and they correspond to Cartesian products of sets, products of groups and topological products respectively.

Prove that the intersection of two open sets coincides with their product in the category of Example 10.25.

Lastly, tell whether the product $[1] \times [1]$ exists in the category of finite ordinals (Exercise 10.26).

10.28 Consider the category **CGHaus** of compactly generated Hausdorff spaces and continuous maps (Exercise 4.36). It can be shown that the topological product of two compactly generated spaces may not be compactly generated.[1] That being said, use Kelleyfication (Exercise 4.40) to show that **CGHaus** has products.

10.29 (Coproducts) A diagram of three objects and two morphisms

$$A \to Q \leftarrow B$$

in a category **A** is called a **coproduct** if it is a product in the opposite category $\mathbf{A}^o$. State the universal property of coproducts. Show that in the categories **Set** and **Top** coproducts always exist, and correspond to disjoint unions.

10.5 A Detour $\curvearrowright$

Sometimes the greatest ideas in mathematics stem from simple, when not trivial, observations. Consider the singleton $\{*\}$ as topological space. Any space Y is homeomorphic in a natural way to $C(\{*\}, Y)$, the space of continuous maps from $\{*\}$ to Y endowed with the compact-open topology. Therefore

$$\pi_0(Y) = \pi_0(C(\{*\}, Y)).$$

Who says we have to stop at $\{*\}$? We may as well fix a locally compact Hausdorff space X, consider continuous maps $C(X, Y)$ with the compact-open topology and define the set

$$[X, Y] = \pi_0(C(X, Y)).$$

Each homeomorphism $Y \cong Z$ induces a homeomorphism $C(X, Y) \cong C(X, Z)$, hence a bijection $[X, Y] \cong [X, Z]$. So if we wanted to prove that the sphere isn't homeomorphic to the torus, we could show that $[S^2, S^1 \times S^1]$ contains one point only, whereas $[S^2, S^2]$ is a countably infinite space. The idea is certainly fascinating, and Theorem 8.20 implies that two continuous maps $f_0, f_1 \colon X \to Y$ belong in the same path component in $C(X, Y)$ if and only if they are homotopic.

For technical reasons to be clarified later, it's more convenient to work with pointed topological spaces, i.e. pairs (X, x_0) where $x_0 \in X$. We define $[(X, x_0), (Y, y_0)] = \pi_0(C((X, x_0), (Y, y_0)))$, where

$$C((X, x_0), (Y, y_0)) = \{f \in C(X, Y) \mid f(x_0) = y_0\}.$$

The case where $X = S^n$ is a sphere and $x_0 = N$ the North pole will play a particularly important role. For any $n \geq 0$ one writes

---

[1]The examples are highly non-trivial. Results analogous to the exponential law (Theorem 8.20) and to Exercises 8.17, 8.18 hold in the category of compactly generated spaces, without further hypotheses such as local compactness; this is perhaps the main reason for introducing **CGHaus**, see [Ste67].

$$\pi_n(Y, y_0) = [(S^n, N), (Y, y_0)].$$

As a matter of fact it isn't hard to prove that for $n \geq 1$ the space π_n is a group, called **nth homotopy group** of Y.

While the study of π_n, for $n \geq 2$, lies beyond the reach of the present text, we will discuss the group π_1 to great extent in the ensuing chapters.

References

[En77] Enderton, H.B.: Elements of Set Theory. Academic Press, New York (1977)
[Ste67] Steenrod, N.E.: A convenient category of topological spaces. Mich. Math. J. **14**, 133–152 (1967)

Chapter 11
The Fundamental Group

We keep on writing I to denote the closed unit interval $[0, 1]$. Given a path $\alpha \colon I \to X$, the points $\alpha(0)$, $\alpha(1)$ are called endpoints, in particular $\alpha(0)$ is the initial point and $\alpha(1)$ the end, or final, point. In case $\alpha(0) = \alpha(1) = a$ the path α is called a **loop** with **base point** $a \in X$.

11.1 Path Homotopy

We have already introduced the **space of paths** in X between points $a, b \in X$

$$\Omega(X, a, b) = \{\alpha \colon I \to X \mid \alpha \text{ continuous}, \alpha(0) = a, \alpha(1) = b\}.$$

We also have defined the product and the inversion

$$*\colon \Omega(X, a, b) \times \Omega(X, b, c) \to \Omega(X, a, c), \quad \alpha * \beta(t) = \begin{cases} \alpha(2t) & \text{if } 0 \le t \le \dfrac{1}{2}, \\ \beta(2t - 1) & \text{if } \dfrac{1}{2} \le t \le 1. \end{cases}$$

$$i\colon \Omega(X, a, b) \to \Omega(X, b, a), \quad i(\alpha)(t) = \alpha(1 - t).$$

Note that $i(i(\alpha)) = \alpha$ and $i(\alpha * \beta) = i(\beta) * i(\alpha)$.

Definition 11.1 Two paths $\alpha, \beta \in \Omega(X, a, b)$ are **path homotopic** if there is a continuous map $F \colon I \times I \to X$ such that:

1. $F(t, 0) = \alpha(t)$, $F(t, 1) = \beta(t)$ for every $t \in I$;
2. $F(0, s) = a$, $F(1, s) = b$ for every $s \in I$.

Such an F is called a **path homotopy** between α and β.

© Springer International Publishing Switzerland 2015
M. Manetti, *Topology*, UNITEXT - La Matematica per il 3+2 91,
DOI 10.1007/978-3-319-16958-3_11

Fig. 11.1 Path homotopy

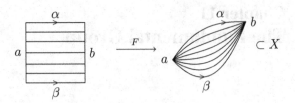

We remark that the notion of path homotopy is more restrictive than that of homotopy of continuous maps. Here we additionally demand that intermediate paths

$$F_s : I \to X, \qquad F_s(t) = F(t, s),$$

have the same initial and end points, for every $s \in I$. We'll write $\alpha \sim \beta$ to mean that α and β are path homotopic (Fig. 11.1).

We know from earlier that homotopic maps define an equivalence relation; the same proof, with minimal changes, also shows that *path homotopy is an equivalence relation*.

Example 11.2 Let $X \subset \mathbb{R}^n$ be a convex subset and $\alpha, \beta \in \Omega(X, a, b)$ paths with the same endpoints. Then α and β are path homotopic:

$$F : I \times I \to X, \qquad F(t, s) = s\beta(t) + (1 - s)\alpha(t),$$

is a path homotopy.

Composition and inversion of paths commute with path equivalence, meaning the following:

1. given paths $\alpha, \alpha' \in \Omega(X, a, b)$, $\beta, \beta' \in \Omega(X, b, c)$, if $\alpha \sim \alpha'$ and $\beta \sim \beta'$, then $\alpha * \beta \sim \alpha' * \beta'$;
2. if $\alpha, \alpha' \in \Omega(X, a, b)$ and $\alpha \sim \alpha'$, then $i(\alpha) \sim i(\alpha')$.

In fact, if $F(t, s)$ is a path homotopy from α to α', and $G(t, s)$ a path homotopy from β to β', then $F(1 - t, s)$ is a path homotopy from $i(\alpha)$ to $i(\alpha')$, while the 'product of homotopies'

$$F * G(t, s) = \begin{cases} F(2t, s) & \text{if } 0 \leq t \leq 1/2, \\ G(2t - 1, s) & \text{if } 1/2 \leq t \leq 1, \end{cases}$$

is a path homotopy between $\alpha * \beta$ and $\alpha' * \beta'$.

Lemma 11.3 *Let $\alpha : I \to X$ be a path, $\phi : I \to I$ any continuous map such that $\phi(0) = 0$ and $\phi(1) = 1$. Then $\alpha(t)$ is path equivalent to $\beta(t) = \alpha(\phi(t))$.*

Proof It suffices to consider the path homotopy

$$F : I \times I \to X, \qquad F(t, s) = \alpha(s\phi(t) + (1 - s)t).$$

□

Proposition 11.4 *The product* $*$ *is associative up to path homotopy:*

$$\alpha * (\beta * \gamma) \sim (\alpha * \beta) * \gamma$$

for any $\alpha \in \Omega(X, a, b)$, $\beta \in \Omega(X, b, c)$ *and* $\gamma \in \Omega(X, c, d)$.
In particular, the homotopy class of the path $\alpha * \beta * \gamma$ *is well defined.*

Proof The path $(\alpha * \beta) * \gamma$ is a reparametrisation of $\alpha * (\beta * \gamma)$. More precisely,

$$((\alpha * \beta) * \gamma)(t) = (\alpha * (\beta * \gamma))(\phi(t)), \quad \text{where } \phi(t) = \begin{cases} 2t & \text{for } 0 \leq t \leq \dfrac{1}{4} \\[2mm] t + \dfrac{1}{4} & \text{for } \dfrac{1}{4} \leq t \leq \dfrac{1}{2} \\[2mm] \dfrac{t+1}{2} & \text{for } \dfrac{1}{2} \leq t \leq 1. \end{cases}$$

Now we can apply Lemma 11.3 to conclude. □

Lemma 11.5 *Take* $p_1, p_2, p_3 \in \mathbb{R}^n$ *and call* $T \subset \mathbb{R}^n$ *the (possibly degenerate) triangle of vertices* p_1, p_2, p_3, *i.e.*

$$T = \{t_1 p_1 + t_2 p_2 + t_3 p_3 \mid t_1, t_2, t_3 \geq 0, \ t_1 + t_2 + t_3 = 1\}.$$

Given a continuous map $f : T \to X$, *we write* f_{ij} *for the standard parametrisation of f restricted to the edge joining* p_i, p_j:

$$f_{ij} : [0, 1] \to X, \qquad f_{ij}(t) = f((1 - t)p_i + tp_j).$$

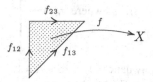

Then $f_{13} \sim f_{12} * f_{23}$.

Proof Consider

$$F : I \times I \to X, \qquad F(t, s) = f(q(t, s)),$$

where $q \colon I \times I \to T$ is defined piece-wise:

$$q(t,s) = \begin{cases} (1 - t - ts)p_1 + 2tsp_2 + (t - ts)p_3 & \text{for } t \le \dfrac{1}{2} \\[2mm] (1 - t - s + ts)p_1 + 2(1 - t)sp_2 + (t - s + ts)p_3 & \text{for } t \ge \dfrac{1}{2}. \end{cases}$$

Then

$$F(t,0) = \begin{cases} f((1-t)p_1 + tp_3) = f_{13}(t) & t \le \tfrac{1}{2} \\ f((1-t)p_1 + tp_3) = f_{13}(t) & t \ge \tfrac{1}{2} \end{cases}$$

$$F(t,1) = \begin{cases} f((1-2t)p_1 + 2tp_2) = f_{12}(2t) & t \le \tfrac{1}{2} \\ f((1-(2t-1))p_2 + (2t-1)p_3) = f_{23}(2t-1) & t \ge \tfrac{1}{2} \end{cases}$$

$$F(0,s) = f(p_1), \qquad F(1,s) = f(p_3),$$

and so F is a path homotopy from f_{13} to $f_{12} * f_{23}$. $\qquad\qquad\qquad\square$

Proposition 11.6 *For any path $\alpha \in \Omega(X, a, b)$:*

1. $1_a * \alpha \sim \alpha * 1_b \sim \alpha$;
2. $\alpha * i(\alpha) \sim 1_a$,

where 1_a, 1_b are the constant paths at a, b respectively.

Proof The paths $1_a * \alpha$ and $\alpha * 1_b$ arise from α after reparametrisation:

$$(1_a * \alpha)(t) = \alpha(\psi(t)), \qquad \text{where } \psi(t) = \begin{cases} 0 & \text{for } 0 \le t \le \dfrac{1}{2} \\[2mm] 2t - 1 & \text{for } \dfrac{1}{2} \le t \le 1. \end{cases}$$

$$(\alpha * 1_b)(t) = \alpha(\eta(t)), \qquad \text{where } \eta(t) = \begin{cases} 2t & \text{for } 0 \le t \le \dfrac{1}{2} \\[2mm] 1 & \text{for } \dfrac{1}{2} \le t \le 1. \end{cases}$$

By Lemma 11.3 we have equivalences $1_a * \alpha \sim \alpha * 1_b \sim \alpha$.

Consider the path homotopy $F \colon I \times I \to X$,

$$F(t,s) = \begin{cases} \alpha(2t) & \text{for } 0 \le t \le s/2 \\ \alpha(s) & \text{for } s/2 \le t \le 1 - s/2 \\ \alpha(2 - 2t) & \text{for } 1 - s/2 \le t \le 1. \end{cases}$$

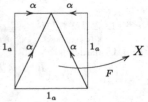

Table 11.1 Lemma 11.5 for a triangle with two vertices coinciding

$p_1 = p_2 = 0, p_3 = 1$	$\alpha \sim 1_a * \alpha$
$p_1 = 0, p_2 = p_3 = 1$	$\alpha \sim \alpha * 1_b$
$p_1 = p_3 = 0, p_2 = 1$	$1_a \sim \alpha * i(\alpha)$
$p_2 = 0, p_1 = p_3 = 1$	$1_b \sim i(\alpha) * \alpha$
$p_3 = 0, p_1 = p_2 = 1$	$i(\alpha) \sim 1_b * i(\alpha)$
$p_2 = p_3 = 0, p_1 = 1$	$i(\alpha) \sim i(\alpha) * 1_a$

As $F(t, 0) = 1_a(t)$ and $F(t, 1) = \alpha * i(\alpha)(t)$ for every t in $[0, 1]$, it follows that $\alpha * i(\alpha) \sim 1_a$.

Here's an alternative argument: imagine the interval I as a degenerate triangle of vertices $p_1, p_2, p_3 \in \{0, 1\}$ in $\mathbb{R}$, with the p_i not all equal, and apply Lemma 11.5 to $\alpha \colon I \to X$. This produces the six path homotopies of Table 11.1. $\qquad\square$

Corollary 11.7 *Let $\alpha \in \Omega(X, a, b)$ be a path and $p \in [0, 1]$ an arbitrary point. Write $d = \alpha(p) \in X$ and call α_0, α_1 the standard parametrisations of α restricted to $[0, p]$ and $[p, 1]$ respectively:*

$$\alpha_0 \in \Omega(X, a, d), \qquad \alpha_0(t) = \alpha(tp),$$

$$\alpha_1 \in \Omega(X, d, b), \qquad \alpha_1(t) = \alpha((1 - t)p + t).$$

Then

1. $\alpha_0 * \alpha_1 \sim \alpha$;
2. for any $c \in X$ and every path $\beta \in \Omega(X, d, c)$,

$$\alpha_0 * \beta * i(\beta) * \alpha_1 \sim \alpha.$$

Proof By definition $\alpha_0 * \alpha_1(t) = \alpha(\phi(t))$, where

$$\phi(t) = \begin{cases} 2tp & \text{if } 0 \leq t \leq 1/2 \\ p + (2t - 1)(1 - p) & \text{if } 1/2 \leq t \leq 1, \end{cases}$$

so $\alpha_0 * \alpha_1 \sim \alpha$ follows from Lemma 11.3.

The product $*$ is associative up to path homotopy, whence

$$\alpha_0 * \beta_1 * i(\beta_1) * \alpha_1 \sim (\alpha_0 * (\beta_1 * i(\beta_1))) * \alpha_1 \sim (\alpha_0 * 1_d) * \alpha_1 \sim \alpha_0 * \alpha_1 \sim \alpha.$$

$\qquad\square$

Corollary 11.8 *Let $\alpha \colon I \to X$ be a path, and $p_1, \ldots, p_n \in [0, 1]$ given points. Set $p_0 = 0$, $p_{n+1} = 1$ and for every $i = 0, \ldots, n$ let α_i be the standard parametrisation* *of the restriction of α to $[p_i, p_{i+1}]$, i.e. $\alpha_i(t) = \alpha((1-t)p_i + tp_{i+1})$. Then α is path homotopic to the product $\alpha_0 * \cdots * \alpha_n$.*

Proof Apply Lemma 11.5 to the triangle $T \subset \mathbb{R}$ with vertices $p_{n-1}, p_n, 1$ and the map $\alpha: T \to X$. The product $\alpha_{n-1} * \alpha_n$ is thus path homotopic to the standard parametrisation of α restricted to $[p_{n-1}, 1]$. The rest follows by induction on n. $\square$

At last we observe that the path homotopy equivalence relation and the operators $*$ and i behave well with respect to the composition of maps.

Proposition 11.9 *Let $f: X \to Y$ be a continuous mapping.*

1. *Given $\alpha, \beta \in \Omega(X, a, b)$, if $\alpha \sim \beta$ then $f\alpha \sim f\beta$.*
2. *Consider $\alpha \in \Omega(X, a, b)$ and $\beta \in \Omega(X, b, c)$. Then $f(\alpha * \beta) = f\alpha * f\beta$ and $i(f\alpha) = f(i(\alpha))$.*

Proof If $F: I \times I \to X$ is a path homotopy, so is $fF: I \times I \to Y$. The path operators of taking products and inverting involve exclusively operations on the domain I, so they commute with the composition of maps. $\square$

Exercises

11.1 Prove that $\alpha, \beta: I \to \mathbb{R}^2 - \{(0, 0)\}$,

$$\alpha(t) = (1+t)(\sin(8t), \cos(8t)), \qquad \beta(t) = (1+t^2)(\sin(8t), \cos(8t)),$$

are path homotopic.

11.2 Prove that the paths $\alpha_c: I \to S^2$,

$$\alpha_c(t) = (\sin(c)\sin(\pi t), \cos(c)\sin(\pi t), \cos(\pi t)),$$

are, for $c \in \mathbb{R}$, all path homotopic to one another.

11.3 Given paths $\alpha, \beta \in \Omega(X, a, b)$, prove that $\alpha \sim \beta$ iff $\alpha * i(\beta) \sim 1_a$.

11.4 ($\heartsuit$) Let $\alpha, \beta, \gamma \in \Omega(X, a, a)$ be such that $\alpha * (\beta * \gamma) = (\alpha * \beta) * \gamma$. Show that if X is Hausdorff, then α, β and γ are constant.

11.2 The Fundamental Group

Given a space X and a point $a \in X$, one defines $\pi_1(X, a)$ as the quotient of $\Omega(X, a, a)$ by path homotopy equivalence. For any loop $\alpha \in \Omega(X, a, a)$ we write $[\alpha] \in \pi_1(X, a)$ for the corresponding homotopy class.

Theorem 11.10 *The set $\pi_1(X, a)$ has a group structure with neutral element $[1_a]$ and operations*

$$[\alpha][\beta] = [\alpha * \beta], \qquad [\alpha]^{-1} = [i(\alpha)].$$

Proof Straightforward consequence of Propositions 11.4 and 11.6. □

Definition 11.11 The group $\pi_1(X, a)$ is called the **fundamental group**, or **first homotopy group** (more rarely **Poincaré group**) of X with base point a.

Note that $\pi_1(X, a)$ only depends on the path component of a in X.

Example 11.12 Let $X \subset \mathbb{R}^n$ be a convex subspace. For every $a \in X$ we have $\pi_1(X, a) = 0$: if $\alpha \in \Omega(X, a, a)$, in fact,

$$F: I \times I \to X, \qquad F(t, s) = sa + (1 - s)\alpha(t),$$

is a path homotopy from α to the constant path 1_a. Hence every loop is homotopic in X to a constant path.

What happens to the fundamental group if we change the base point? Clearly if a, b live in different path components there is no relationship between $\pi_1(X, a)$ and $\pi_1(X, b)$. Conversely,

Lemma 11.13 *Take $\gamma \in \Omega(X, a, b)$ and define*

$$\gamma_\sharp: \pi_1(X, a) \to \pi_1(X, b), \qquad \gamma_\sharp[\alpha] = [i(\gamma) * \alpha * \gamma].$$

The map $\gamma_\sharp$ is well defined and a group isomorphism.

Proof It was observed earlier that the product $*$ commutes with path equivalences, so $\gamma_\sharp: \pi_1(X, a) \to \pi_1(X, b)$ is well defined. Moreover $\gamma_\sharp$ is a group homomorphism since

$$\gamma_\sharp[\alpha]\gamma_\sharp[\beta] = [i(\gamma) * \alpha * \gamma][i(\gamma) * \beta * \gamma] = [i(\gamma) * \alpha * \gamma * i(\gamma) * \beta * \gamma]$$
$$= [i(\gamma) * \alpha * 1_a * \beta * \gamma] = [i(\gamma) * \alpha * \beta * \gamma]$$
$$= \gamma_\sharp[\alpha][\beta].$$

Eventually,

$$i(\gamma)_\sharp(\gamma_\sharp[\alpha]) = [\gamma * i(\gamma) * \alpha * \gamma * i(\gamma)] = [1_a * \alpha * 1_a] = [\alpha],$$

hence $\gamma_\sharp$ is an isomorphism with inverse $i(\gamma)_\sharp$. □

Thanks to Lemma 11.13 we are able to decide whether the fundamental group of a path connected space is isomorphic to a given group G, without specifying the base point. If X is path connected $\pi_1(X)$ denotes the *isomorphism class* of its fundamental group with respect to any base point.

Definition 11.14 A space is called **simply connected** if it is path connected and its fundamental group is trivial.

Equivalently, X is simply connected if path connected and $\pi_1(X, a) = 0$, with a in X arbitrary.

Example 11.15 The hypercube I^n is convex in $\mathbb{R}^n$, so Example 11.12 allows to say it is simply connected.

From now onwards we will, for simplicity, view the circle S^1 as the set of complex numbers of norm one, thus identifying it with the topological group $U(1, \mathbb{C})$.

Example 11.16 Given an integer $n \in \mathbb{Z}$ let $\alpha_n : I \to S^1$ be the path $\alpha_n(t) = \exp(2i\pi nt) = e^{2i\pi nt}$. Then

$$\mathbb{Z} \to \pi_1(S^1, 1), \qquad n \mapsto [\alpha_n],$$

is a group homomorphism (actually an isomorphism, see later). In fact α_0 is the constant path, and Corollary 11.7 implies that α_{n+m} is path homotopic to $\alpha_n * \alpha_m$, for any $n, m \geq 0$. Also observe that $\alpha_{-n} = i(\alpha_n)$ for every $n \in \mathbb{Z}$.

Proposition 11.17 *The fundamental group of the product of two spaces is isomorphic to the product of the fundamental groups:*

$$\pi_1(X \times Y, (a, b)) = \pi_1(X, a) \times \pi_1(Y, b).$$

Proof Any path $\alpha : I \to X \times Y$ is uniquely determined by its components $\alpha_1 : I \to X$, $\alpha_2 : I \to Y$, giving rise to the natural bijection

$$\Omega(X \times Y, (a, b), (a, b)) = \Omega(X, a, a) \times \Omega(Y, b, b).$$

Similarly, every homotopy $F : I^2 \to X \times Y$ is uniquely determined by the components $F_1 : I^2 \to X$, $F_2 : I^2 \to Y$, and the bijection above induces a group isomorphism

$$\pi_1(X \times Y, (a, b)) = \pi_1(X, a) \times \pi_1(Y, b)$$

on the quotients. $\square$

Exercises

11.5 Prove that $A = \{(x, y) \in \mathbb{R}^2 \mid y \geq x^2\}$ and $B = \{(x, y) \in \mathbb{R}^2 \mid y \leq x^2\}$ are simply connected.

11.6 Take a path connected space X and $a, b \in X$. Prove that $\pi_1(X, a)$ is an Abelian group if and only if the isomorphism $\gamma_\sharp : \pi_1(X, a) \to \pi_1(X, b)$ doesn't depend on the choice of $\gamma \in \Omega(X, a, b)$.

11.7 Let G be a connected topological group with neutral element e and $f: \mathbb{R} \to G$ a continuous homomorphism. For every $n \in \mathbb{Z}$ let $\alpha_n: I \to G$ be the path $\alpha_n(t) = f(nt)$. Prove that if $\mathbb{Z} \subset \ker(f) = f^{-1}(e)$, the map

$$\mathbb{Z} \to \pi_1(G, e), \qquad n \mapsto [\alpha_n],$$

is a group homomorphism.

11.8 ($\heartsuit$) Consider a space X and a base point $a \in X$. Show that there is a natural 1-1 and onto map between the space of loops $\Omega(X, a, a)$ and the set of continuous functions $f: S^1 \to X$ mapping 1 to a.

11.9 ($\maltese$) Let X be path connected, and denote by $[S^1, X]$ the set of homotopy classes of continuous maps $S^1 \to X$. Prove that there's a natural surjective map $\pi_1(X, a) \to [S^1, X]$, for every $a \in X$. Show further that $[S^1, X]$ is in 1-1 correspondence with the conjugacy classes of the group $\pi_1(X, a)$.

11.3 The Functor π_1

Amongst other things Proposition 11.9 implies that for any continuous map $f: X \to Y$ and every base point $a \in X$ the mapping

$$\pi_1(f): \pi_1(X, a) \to \pi_1(Y, f(a)), \qquad \pi_1(f)([\alpha]) = [f\alpha],$$

is a well-defined group homomorphism. To simplify the notation f_* is often used instead of $\pi_1(f)$, in absence of ambiguity.

Example 11.18 Consider a space X, the inclusion $i: A \to X$ and a base point $a \in A$. In general the homomorphism $i_*: \pi_1(A, a) \to \pi_1(X, a)$ is not 1-1 because it may happen there are homotopically non-trivial paths in A that are homotopically trivial in X. However:

1. if A is a retract of X, then $i_*: \pi_1(A, a) \to \pi_1(X, a)$ is injective;
2. if A is a deformation retract of X, $i_*: \pi_1(A, a) \to \pi_1(X, a)$ is an isomorphism.

To prove these assertions let α be a loop in A with base point a and such that $i_*[\alpha] = 0$. There exists a path homotopy $F: I^2 \to X$ such that $F(t, 0) = \alpha(t)$ and $F(t, 1) = a$. If $r: X \to A$ is a retraction, $rF: I^2 \to A$ is a path homotopy, so $[\alpha] = 0$ in $\pi_1(A, a)$, proving 1.

Suppose now $R: X \times I \to X$ is a deformation of X into A, and take a loop $\beta \in \Omega(X, a, a)$. The continuous map

$$F: I^2 \to X, \qquad F(t, s) = R(\beta(t), s)$$

is a path homotopy between β and $r\beta \in \Omega(A, a, a)$, where $r = R(-, 0)$. This proves $i_*([r\beta]) = [\beta]$, whence i_* is onto.

The functorial properties of $f \mapsto f_*$ are self-evident:

1. if $Id: X \to X$ is the identity, then $Id_*: \pi_1(X, a) \to \pi_1(X, a)$ is the identity, for every $a \in X$;
2. let $f: X \to Y, g: Y \to Z$ be continuous, and $a \in X$ a base point. Then

$$g_* f_* = (gf)_*: \pi_1(X, a) \to \pi_1(Z, gf(a)).$$

It's rather instructive to use the previous properties towards another proof that if $r: X \to A, A \subset X$, is a retraction, then $r_*: \pi_1(X, a) \to \pi_1(A, a)$ is onto, for any $a \in A$. Suppose $i: A \to X$ is the inclusion; then $r_* i_* = (ri)_* = (Id_{|A})_* = Id$, so r_* must be onto and i_* one-to-one.

Concerning homotopy invariance the situation now is slightly more involved than for π_0.

Proposition 11.19 *Let* $F: X \times I \to Y$ *be a homotopy between continuous maps* $f = F(-, 0)$ *and* $g = F(-, 1)$, *and* $a \in X$ *a base point. We denote by* $\gamma \in \Omega(Y, f(a), g(a))$ *the path* $\gamma(s) = F(a, s)$. *Then the diagram*

$$
\begin{array}{ccc}
 & \pi_1(X, a) & \\
 {\scriptstyle f_*} \swarrow & & \searrow {\scriptstyle g_*} \\
\pi_1(Y, f(a)) & \xrightarrow{\quad \gamma_\sharp \quad} & \pi_1(Y, g(a))
\end{array}
$$

is commutative.

Proof The claim is that the $\gamma * g\alpha$ and $f\alpha * \gamma$ are path homotopic, for any $\alpha \in \Omega(X, a, a)$. Let's look at the continuous map

$$I^2 \to Y, \qquad (t, s) \mapsto F(\alpha(t), s),$$

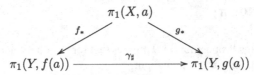

restricted to the edges of the square I^2. Lemma 11.5, applied to the triangles depicted, gives path equivalences $\gamma * g\alpha \sim \delta$ and $\delta \sim f\alpha * \gamma$, where δ is the path $\delta(t) = F(\alpha(t), t)$. Now it's enough to recall that homotopy equivalence is transitive. $\qquad \square$

Corollary 11.20 *Let* $g: X \to X$ *be a continuous map homotopic to the identity. For any* $a \in X$ *the map* $g_*: \pi_1(X, a) \to \pi_1(X, g(a))$ *is a group isomorphism.*

Proof Call $F: X \times I \to X$ a homotopy between $F(x, 0) = x$ and $F(x, 1) = g(x)$ and γ the path $\gamma(t) = F(a, t)$. Proposition 11.19 implies

$$g_* = \gamma_\sharp: \pi_1(X, a) \to \pi_1(X, g(a)),$$

making g_* an isomorphism. □

We are now in the position of proving that the isomorphism class of the fundamental group is a homotopy invariant; before that, though, we'll need a simple lemma about sets.

Lemma 11.21 *Given maps*

$$A \xrightarrow{f} B \xrightarrow{g} C \xrightarrow{h} D$$

between sets, if gf is bijective and hg one-to-one, then f is bijective.

Proof Both gf and hg are 1-1, so also f and g are injective. As gf is onto, for every $b \in B$ there's an $a \in A$ with $gf(a) = g(b)$, and the injectivity of g implies $f(a) = b$; hence f is onto. □

Theorem 11.22 *Let $f : X \to Y$ be a homotopy equivalence between two spaces. Then $f_* : \pi_1(X, a) \to \pi_1(Y, f(a))$ is a group isomorphism, for any $a \in X$.*

Proof Let $g : Y \to X$ be a homotopy inverse to f, i.e. a continuous map such that fg and gf are both homotopic to the identity. Then

$$\pi_1(X, a) \xrightarrow{f_*} \pi_1(Y, f(a)) \xrightarrow{g_*} \pi_1(X, gf(a)) \xrightarrow{f_*} \pi_1(Y, fgf(a))$$

are group homomorphisms. By Corollary 11.20 $f_* g_*$ and $g_* f_*$ are isomorphisms, so Lemma 11.21 tells that f_* is invertible. □

Exercises

11.10 Suppose—hypothetically—that S^1 were simply connected (it's not). Under this assumption prove that any path connected space X would have $\pi_1(X) = 0$.

11.11 Prove that a continuous map homotopic to a constant induces the zero homomorphism between the fundamental groups.

11.12 (♡) Let X be path connected. Prove that X is simply connected if and only if every continuous map $f : S^1 \to X$ extends to a continuous map $D^2 \to X$.

11.13 Let $f : X \to Y$ be continuous. Prove that for every path $\gamma \in \Omega(X, a, b)$

$$
\begin{array}{ccc}
\pi_1(X, a) & \xrightarrow{\gamma_\sharp} & \pi_1(X, b) \\
\downarrow{f_*} & & \downarrow{f_*} \\
\pi_1(Y, f(a)) & \xrightarrow{(f\gamma)_\sharp} & \pi_1(Y, f(b))
\end{array}
$$

is a commutative diagram of group homomorphisms.

11.14 Let X, Y be path connected, $f: X \to Y$ continuous. Prove that the fact that the homomorphism $f_*: \pi_1(X, a) \to \pi_1(Y, f(a))$ may be 1-1, or onto, doesn't depend upon the choice of the base point a.

11.15 ($\heartsuit$) Let Y be a retract of X, $i: Y \to X$ the inclusion, and fix a point $y \in Y$. Prove that if $i_*\pi_1(Y, y)$ is a normal subgroup in $\pi_1(X, x)$, then for every retraction $r: X \to Y$ we have an isomorphism

$$\pi_1(X, y) \simeq \pi_1(Y, y) \times \ker(r_*).$$

11.4 The Sphere S^n Is Simply Connected ($n \geq 2$)

In this section we'll prove that for every $n \geq 2$ the sphere S^n is simply connected. The idea is rather simple: take a path $\alpha \in \Omega(S^n, a, a)$ and assume $\alpha: I \to S^n$ isn't onto, i.e. that there's a point $b \in S^n$ such that $\alpha(I) \subset S^n - \{b\}$. As $S^n - \{b\}$ is homeomorphic to the simply connected space $\mathbb{R}^n$, it follows that α is homotopically trivial in $S^n - \{b\}$, hence homotopically trivial in S^n. But I has 'dimension 1' whereas S^n has 'dimension n', so the hypothesis that α isn't surjective for $n \geq 2$ is quite reasonable.

For every $n \geq 2$, alas, there exist paths $I \to S^n$ that are onto: the so-called *Peano curves* (Theorem 15.6 and Exercise 15.9).

On the bright side, we do have ways to overcome the proplem, the most widely known being the following:

1. prove that α is path homotopic to a C^∞ path, then call on Sard's theorem to show that a smooth path $I \to S^n$ can't be surjective if $n \geq 2$;
2. prove that α is path homotopic to the product of finitely many loops that aren't surjective.

Here we will choose the latter, and go on to prove a more general result known as **van Kampen's theorem**. We refer to [Mi65, GP74], instead, for more information on the first method.

Theorem 11.23 (Lebesgue number) *Let (Y, d) be a compact metric space, $f: Y \to X$ a continuous map and $\mathcal{A}$ an open cover of X. There exists a positive real number δ such that $f(B(y, \delta))$ is entirely contained in an open set of $\mathcal{A}$, for any $y \in Y$.*

Proof For any positive integer n indicate with Y_n for the set of points $y \in Y$ such that $B(y, 2^{-n}) \subset f^{-1}(U)$ for some $U \in \mathcal{A}$. All we have to prove is that $Y_n = Y$ for some n.

Clearly $Y_n \subset Y_{n+1}$ for every n, and since $\mathcal{A}$ is a cover, $\cup_n Y_n = Y$. The subspace Y_n is contained in the interior of Y_{n+1}: take $y \in Y_n$ and pick $U \in \mathcal{A}$ such that $B(y, 2^{-n}) \subset f^{-1}(U)$. By the triangle inequality, if $d(y, z) < 2^{-n-1}$ then $B(z, 2^{-n-1}) \subset B(y, 2^{-n}) \subset f^{-1}(U)$, and then $z \in Y_{n+1}$. Therefore $Y_n \subset Y_{n+1}^o$ and $\cup_n Y_n^o = Y$. As Y is compact, $Y = Y_n^o$ for some n, and *a fortiori* $Y = Y_n$. $\square$

Corollary 11.24 *Let $\alpha: I \to X$ be a path and $\mathcal{A}$ an open cover of X. There exists a positive integer n such that, for each $i = 0, \ldots, n-1$, the range of $\alpha: \left[\dfrac{i}{n}, \dfrac{i+1}{n}\right] \to X$ is contained in some element of $\mathcal{A}$.*

Proof The interval I is a compact metric space and Theorem 11.23 can be applied. $\square$

Theorem 11.25 (van Kampen: part I) *Let A, B be open sets in a space X such that $X = A \cup B$, $x_0 \in A \cap B$ a given point and $f_*: \pi_1(A, x_0) \to \pi_1(X, x_0)$, $g_*: \pi_1(B, x_0) \to \pi_1(X, x_0)$ the homomorphisms induced by inclusions $A \subset X$ and $B \subset X$.*

If A, B and $A \cap B$ are path connected, the group $\pi_1(X, x_0)$ is generated by the images of f_* and g_*.

Recall that a group G is said to be generated or spanned by a subset $S \subset G$ if every element of G arises as (finite) product of elements of $S \cup S^{-1}$, where $S^{-1} = \{g^{-1} \mid g \in S\}$. Analogously, G is generated by subsets S_1 and S_2 if it is spanned by their union $S = S_1 \cup S_2$.

Proof Let $\alpha \in \Omega(X, x_0, x_0)$ be a path. The claim is that α is path homotopic to the product of a finite number of paths $\gamma_1, \ldots, \gamma_n \in \Omega(X, x_0, x_0)$, each entirely contained either in A or in B. By Corollary 11.24 there's an $n > 0$ such that the restriction of α to each interval $\left[\dfrac{i-1}{n}, \dfrac{i}{n}\right]$, $i = 1, \ldots, n$, is contained in A or B. Call α_i the standard parametrisation of α restricted to $\left[\dfrac{i-1}{n}, \dfrac{i}{n}\right]$, i.e. $\alpha_i(t) = \alpha\left(\dfrac{i-1+t}{n}\right)$.

For every $i = 1, \ldots, n - 1$ define $x_i = \alpha(i/n)$. Since $A, B, A \cap B$ are path connected, for any i we can find a path $\beta_i \in \Omega(X, x_i, x_0)$ so that:

1. if $x_i \in A \cap B$, then $\beta_i \in \Omega(A \cap B, x_i, x_0)$;
2. if $x_i \in A - B$, then $\beta_i \in \Omega(A, x_i, x_0)$;
3. if $x_i \in B - A$, then $\beta_i \in \Omega(B, x_i, x_0)$.

These conditions imply, in particular:

1. if $x_i \in A$, then $\beta_i \in \Omega(A, x_i, x_0)$;
2. if $x_i \in B$, then $\beta_i \in \Omega(B, x_i, x_0)$.

Therefore α is path equivalent to $\gamma_1 * \cdots * \gamma_n$, where

$$\gamma_1 = \alpha_1 * \beta_1, \quad \gamma_2 = i(\beta_1) * \alpha_2 * \beta_2,$$

$$\ldots, \gamma_j = i(\beta_{j-1}) * \alpha_j * \beta_j, \ldots$$

$$\gamma_{n-1} = i(\beta_{n-2}) * \alpha_{n-1} * \beta_{n-1}, \quad \gamma_n = i(\beta_{n-1}) * \alpha_n.$$

Every γ_i belongs in $\Omega(A, x_0, x_0) \cup \Omega(B, x_0, x_0)$, and the proof is finished. $\square$

Corollary 11.26 *Let A, B be open in X with $X = A \cup B$ and $A \cap B \neq \emptyset$. If A, B, $A \cap B$ are path connected and A, B simply connected, then X is simply connected.*

Proof That X is path connected is clear. Choose a base point $x_0 \in A \cap B$. As $\pi_1(A, x_0) = \pi_1(B, x_0) = 0$, by Theorem 11.25 the group $\pi_1(X, x_0)$ is generated by the neutral element, and so it is trivial. $\qquad\square$

Corollary 11.27 *For every $n \geq 2$ the sphere S^n is simply connected.*

Proof Write $S^n = A \cup B$, where $A = S^n - \{(1, 0, \ldots, 0)\}$ and $B = S^n - \{(-1, 0, \ldots, 0)\}$ are open. The latter are homeomorphic to $\mathbb{R}^n$, hence simply connected, while $A \cap B$ is homeomorphic to $\mathbb{R}^n - \{0\}$, a connected set for every $n \geq 2$. $\square$

Corollary 11.28 *The complement of a finite set in $\mathbb{R}^n$ is simply connected when $n \geq 3$.*

Proof Suppose $n \geq 3$ and let's prove that $X = \mathbb{R}^n - \{p_1, \ldots, p_m\}$ is simply connected, where $p_1, \ldots, p_m$ are distinct. The argument is by induction on m. When $m = 0$ X is convex so simply connected. When $m = 1$ the sphere $\{x \in \mathbb{R}^n \mid \|x - p_1\| = 1\} \simeq S^{m-1}$ is a deformation retract of X, and the claim follows from Corollary 11.27.

Now assume $m \geq 2$ and that the result holds for $\mathbb{R}^n - \{k \text{ points}\}$, $k < m$. Since $p_1 \neq p_2$ there's a linear map $f \colon \mathbb{R}^n \to \mathbb{R}$ such that $f(p_1) \neq f(p_2)$; possibly flipping the sign of f (or swapping p_1 and p_2) we may suppose $f(p_1) < f(p_2)$. Now write $X = A \cup B$, where

$$A = \{x \in X \mid f(x) < f(p_2)\}, \qquad B = \{x \in X \mid f(x) > f(p_1)\}.$$

Each open set A, B, $A \cap B$ is homeomorphic to $\mathbb{R}^n - \{k \text{ points}\}$ for some $k < m$. The inductive hypothesis and Corollary 11.26 end the proof. $\qquad\square$

Corollary 11.29 *The complex projective space $\mathbb{P}^n(\mathbb{C})$ is simply connected for every $n \geq 0$.*

Proof We use induction on n, noting that $\mathbb{P}^0(\mathbb{C})$ consists of one point and hence is simply connected.

Let $z_0, \ldots, z_n$ be homogeneous coordinates on $\mathbb{P}^n(\mathbb{C})$, and call $H \subset \mathbb{P}^n(\mathbb{C})$ the hyperplane defined by $z_0 = 0$. Since $H \cong \mathbb{P}^{n-1}(\mathbb{C})$, by inductive hypothesis H can be taken to be simply connected. Write $\mathbb{P}^n(\mathbb{C}) = A \cup B$, with

$$A = \mathbb{P}^n(\mathbb{C}) - H, \qquad B = \mathbb{P}^n(\mathbb{C}) - \{[1, 0, \ldots, 0]\}.$$

Observe the following:

1. $\mathbb{C}^n \to A$, $(z_1, \ldots, z_n) \mapsto [1, z_1, \ldots, z_n]$, is a homeomorphism, implying A is simply connected;
2. the intersection $A \cap B$, as subset in A, is homeomorphic to $\mathbb{C}^n - \{0\}$, and hence connected;
3. the map

$$R: B \times I \to B, \qquad R([z_0, z_1, \ldots, z_n], t) = [tz_0, z_1, \ldots, z_n],$$

is a deformation of B into H, making B simply connected.

Corollary 11.26 then says that $\mathbb{P}^n(\mathbb{C})$ is simply connected. $\qquad\square$

The proof above won't work for real projective spaces, because ($n = 1$) the space $A \cap B \cong \mathbb{R} - \{0\}$ is disconnected. As a matter of fact, we will prove that $\mathbb{P}^n(\mathbb{R})$ is not simply connected for every $n > 0$. Nevertheless, the construction of Corollary 11.29, applied to real projective spaces, shows that $\pi_1(\mathbb{P}^n)$, $n > 0$, is generated by the fundamental group of one of its lines (Exercise 11.20).

Exercises

11.16 Let A, B be open. Prove that if $A \cup B$ and $A \cap B$ are path connected, then A and B are path connected.

11.17 ($\heartsuit$) Compute the fundamental group of the union of S^2 and the three coordinate planes in $\mathbb{R}^3$.

11.18 Prove that for every pair of integers $0 \leq n \leq m - 3$, the complement in $\mathbb{R}^m$ of a vector subspace of dimension n is simply connected.

11.19

(1) Consider simply connected open sets $\{U_i\}_{i\in\mathbb{N}}$ such that $U_i \subset U_{i+1}$ for every i. Show that the union $\bigcup_i U_i$ is simply connected.
(2) Let $U = (-a, a)^n$ be an open hypercube in $\mathbb{R}^n$, $n \geq 3$, and A a finite subset in U. Show that $U - A$ is simply connected.
(3) Prove that the complement of a discrete closed set in $\mathbb{R}^n$, $n \geq 3$, is simply connected.

11.20

(1) Retain the assumptions of Theorem 11.25. Suppose, further, that the inclusion $A \cap B \to B$ induces a surjective homomorphism of the fundamental groups. Prove that $f_*: \pi_1(A, x_0) \to \pi_1(X, x_0)$ is onto.
(2) Take $a \in \mathbb{P}^n(\mathbb{R})$ and a hyperplane $H \subset \mathbb{P}^n(\mathbb{R})$ through a. Prove that for $n \geq 2$ the inclusion $H \hookrightarrow \mathbb{P}^n(\mathbb{R})$ induces a surjective map $\pi_1(H, a) \twoheadrightarrow \pi_1(\mathbb{P}^n(\mathbb{R}), a)$.

11.21 ($\heartsuit$) Let compact sets $K_1 \subset K_2 \subset \cdots$ exhaust X. Prove that if every K_n is simply connected, then also X is simply connected. More generally, show that if $x \in K_1$ and the inclusion $K_n \subset K_{n+1}$ induces an isomorphism $\pi_1(K_n, x) \simeq \pi_1(K_{n+1}, x)$ for every n, then every inclusion $K_n \subset X$ induces an isomorphism $\pi_1(K_n, x) \simeq \pi_1(X, x)$.

11.22 ($\clubsuit$, $\heartsuit$) Prove that the spaces (see Fig. 11.2):

$$S = \{(x, y, z) \in \mathbb{R}^3 \mid x^2 + y^2 = \sin^2(\pi z)\},$$

Fig. 11.2 Sausages and necklace

$$C = \{(x, y, z) \in \mathbb{R}^3 \mid \sqrt{x^2 + y^2} = \max(0, \sin(\pi z))\},$$

are simply connected.

11.5 Topological Monoids

Definition 11.30 An A_2 **space** is a triple (X, μ, e) consisting of a topological space X, a continuous map $\mu\colon X \times X \to X$ and a point $e \in X$, called **unit**, such that

$$\mu(x, e) = \mu(e, x) = x$$

for every $x \in X$. An A_2 space is called a **topological monoid** if μ is an associative product:

$$\mu(x, \mu(y, z)) = \mu(\mu(x, y), z)$$

for any $x, y, z \in X$.

The unit e is completely determined by the product μ: if $e_1, e_2 \in X$ are units, then $e_2 = \mu(e_2, e_1) = e_1$. Every topological group is a topological monoid; every retract of an A_2 space containing the unity is an A_2 space.

Theorem 11.31 *Let (X, μ, e) be an A_2 space. Then:*

1. *the group $\pi_1(X, e)$ is Abelian;*
2. *under the natural isomorphism $\pi_1(X \times X, (e, e)) = \pi_1(X, e) \times \pi_1(X, e)$ the homomorphism μ_* can be identified with the product of $\pi_1(X, e)$:*

$$\mu_*\colon \pi_1(X, e) \times \pi_1(X, e) \to \pi_1(X, e), \qquad \mu_*([\alpha], [\beta]) = [\alpha][\beta].$$

Proof For paths α, β with base point e, by definition we have $\mu_*([\alpha], [\beta]) = [\delta]$ where $\delta(t) = \mu(\alpha(t), \beta(t))$. If we consider the map

$$F\colon I^2 \to X, \qquad F(t, s) = \mu(\alpha(t), \beta(s)),$$

Lemma 11.5 yields path equivalences

$$\alpha * \beta \sim \delta \sim \beta * \alpha.$$ $\square$

Remark 11.32 Between a monoid and an A_2 space are several 'intermediate' notions. For example an A_3 space, which is a quadruple (X, μ, e, h) where (X, μ, e) is an A_2 space and $h \colon X^3 \times I \to X$ a homotopy between

$$(x, y, z) \mapsto \mu(\mu(x, y), z), \qquad (x, y, z) \mapsto \mu(x, \mu(y, z)).$$

In an A_3 space, in particular, the product is associative up to homotopy.

Another instance (far too advanced to be discusses thoroughly here) are A_∞ spaces, introduced in the 1960s by Stasheff [Sta63]. They have the remarkable property of being invariant under homotopy. This and other features turned A_∞ spaces and their relatives (A_∞ algebras, A_∞ categories) into a hot topic, with a profusion of applications in all areas of contemporary mathematics.

Exercises

11.23 Prove that the fundamental groups of $SL(n, \mathbb{C})$, $U(n, \mathbb{C})$, $SL(n, \mathbb{R})$ and $SO(n, \mathbb{R})$ are Abelian.

11.24 Let (X, μ, e) be a topological monoid, and indicate with $p_n \colon X \to X$, $n > 0$, the map raising to the nth power. Prove that the group homomorphism $(p_n)_* \colon \pi_1(X, e) \to \pi_1(X, e)$ coincides with taking nth powers in the Abelian group $\pi_1(X, e)$.

11.25 Let G be a path connected topological group with neutral element e. For every $g \in G$ define the continuous map

$$\mathrm{Inn}_g \colon G \to G, \qquad Inn_g(a) = gag^{-1}.$$

Prove that the homomorphisms $(Inn_g)_* \colon \pi_1(G, e) \to \pi_1(G, e)$ equal the identity map.

11.26 Let X be a locally compact Hausdorff space. Show that $C(X, X)$, with the compact-open topology and composition of maps as product, is a topological monoid.

References

[Mi65]   Milnor, J.W.: Topology from the Differentiable Viewpoint. The University Press of Virginia, Charlottesville (1965)

[GP74]   Guillemin, V., Pollack, A.: Differential Topology. Prentice-Hall Inc, Englewood Cliffs (1974)

[Sta63] Stasheff, J.D.: On the homotopy associativity of H-spaces I, II. Trans. AMS **108**(275–292), 293–312 (1963)

Chapter 12
Covering Spaces

At this stage we are only able to show that certain spaces are simply connected: normally it's quite difficult to prove directly the existence of loops that aren't homotopic to constant paths. There are indirect methods for this, and this chapter is dedicated to one such.

We will define **covering spaces** and prove that the covering spaces of a simply connected space are of a special kind which we name 'trivial'. Then it'll be straightforward that many familiar spaces have non-trivial coverings, e.g. circles and real projective spaces.

12.1 Local Homeomorphisms and Sections

Definition 12.1 A continuous map $f : X \to Y$ is a **local homeomorphism** if for every $x \in X$ there exist open sets $A \subset X$, $B \subset Y$ such that $x \in A$, $f(A) = B$ and the restriction $f : A \to B$ is a homeomorphism.

Example 12.2 Every continuous, 1-1 and open map is a local homeomorphism.

Lemma 12.3 *Every local homeomorphism $f : X \to Y$ is open, and the fibres $f^{-1}(y)$, $y \in Y$, are discrete.*

Proof We want to show that the image $f(V)$ of an open set $V \subset X$ is a neighbourhood of each of its points. That is to say, for every $y \in f(V)$ there exists $U \subset Y$ open such that $y \in U \subset f(V)$.

Let $x \in V$ be such that $f(x) = y$; by assumption there are open sets $A \subset X$, $B \subset Y$ such that $x \in A$, $f(A) = B$ and the restriction $f : A \to B$ is a homeomorphism. In particular $y \in f(V \cap A)$, and $U = f(V \cap A)$ is open in B so also open in Y.

For every $y \in Y$ and $x \in f^{-1}(y)$ there exists an open neighbourhood $x \in A$ for which the restriction $f : A \to Y$ is 1-1. Hence $f^{-1}(y) \cap A = \{x\}$, proving that the subspace topology on the fibres $f^{-1}(y)$ is discrete. □

© Springer International Publishing Switzerland 2015
M. Manetti, *Topology*, UNITEXT - La Matematica per il 3+2 91,
DOI 10.1007/978-3-319-16958-3_12

If $p: X \to Y$ is a map between sets, a function $s: Y \to X$ is called a **section** of p if $p(s(y)) = y$ for every $y \in Y$. A necessary condition for p to have a section is that p be onto; *vice versa*, the axiom of choice says exactly that any map admits sections.

In contrast—moving back to the topological world—continuous sections of continuous surjective maps do not exist, in general.

Proposition 12.4 *Let $p: X \to Y$ be a continuous map and $s: Y \to X$ a continuous section of p, i.e. a continuous mapping with $p(s(y)) = y$ for every $y \in Y$. Then:*

1. *p is an identification;*
2. *if X is Hausdorff, s is a closed immersion.*

Proof (1) Clearly p must be onto. To show that it is an identification we should prove that if $A \subset Y$ is such that $p^{-1}(A)$ is open, then A is open in Y. But $s^{-1}(p^{-1}(A)) = A$, and s is continuous, so A is open in Y.

(2) It's obvious that s is 1-1; we need to show that if X is Hausdorff, s is closed. As $s(C) = s(Y) \cap p^{-1}(C)$ for every $C \subset Y$, it is enough to see that $s(Y)$ is closed in X. Since $x \in s(Y)$ iff $x = sp(x)$, we can write $s(Y) = f^{-1}(\Delta)$ where $\Delta \subset X \times X$ is the diagonal and

$$f: X \to X \times X, \qquad f(x) = (x, sp(x)).$$

Now recall that in a Hausdorff space the diagonal is closed. $\square$

So, if a continuous map p does have a continuous section, it must be an identification. But this is **not** sufficient. For example, the closed surjective map

$$p: [0, 1] \to S^1, \qquad p(t) = (\cos(2\pi t), \sin(2\pi t)),$$

doesn't admit sections: if, by contradiction, there were a continuous section, for set-theoretical reasons its range should be either $[0, 1[$ or $]0, 1]$, while from the topological viewpoint any continuous image of S^1 is compact. Similarly, since there's no continuous, 1-1 map $s: S^1 \to \mathbb{R}$ (Lemma 4.16), the open, onto map

$$p: \mathbb{R} \to S^1, \qquad p(t) = (\cos(2\pi t), \sin(2\pi t)),$$

cannot have continuous sections.

Exercises

12.1 Explain whether the map f of Exercise 3.47 is a local homeomorphism.

12.2 ($\heartsuit$) Let $p: E \to X$ be a local homeomorphism, with E Hausdorff. Prove that:

1. the diagonal in $E \times E$ is open and closed in

$$E \times_X E = \{(e_1, e_2) \in E \times E \mid p(e_1) = p(e_2)\};$$

2. let Y be connected and $f, g \colon Y \to E$ continuous maps such that $pf = pg$. Then
   $f = g$ or $f(y) \neq g(y)$ for every $y \in Y$;
3. let $U \subset E$ be open, connected and such that the restriction $p \colon U \to X$ is 1-1.
   Then U is a connected component of $p^{-1}(p(U))$.

12.3 Let $f \colon X \to Y$ be a local homeomorphism. Prove that:

1. if Y is first countable, so is X;
2. if Y is separable and every fibre is at most countable, X is separable.

12.4 Let $f \colon X \to Y$ and $g \colon Y \to Z$ be such that g, gf are continuous and g is a local homeomorphism. Prove that the set of points where f is continuous is open in X.

12.5 ($\heartsuit$) Show that the natural projection $p \colon S^n \to \mathbb{P}^n(\mathbb{R})$ doesn't have continuous sections, for any positive integer n.

12.2 Covering Spaces

Definition 12.5 Let X be a connected space. A space E together with a continuous map $p \colon E \to X$ is a **covering space of** X if every point $x \in X$ is contained in an open set $V \subset X$ whose pre-image $p^{-1}(V)$ is the disjoint union of open sets U_i with the property that $p \colon U_i \to V$ is a homeomorphism for every i.

The space X is called the **base (space)** of the covering space, E is the **total space** and p is the **covering map**. The sets $p^{-1}(x)$, $x \in X$, are called **fibres** of the covering space.

An open set $V \subset X$ is an **admissible (open)** set of the covering p if it fulfils the condition of Definition 12.5. With other words $V \subset X$ is admissible if we can write $p^{-1}(V) = \cup_i U_i$, where:

1. every U_i is open in E and the restrictions $p \colon U_i \to V$ are homeomorphisms;
2. $U_i \cap U_j = \emptyset$ for every $i \neq j$.

Clearly, an open set contained in an admissible open set is still admissible. We will say that the covering space is **trivial** if the whole base X is an admissible set.

If $p \colon E \to X$ is a covering space, from the definition every point $e \in E$ has an open neighbourhood homeomorphic to an open neighbourhood of $p(e)$. For later use we note that this implies that if X is locally path connected, also E is locally path connected.

Definition 12.6 A covering space $p \colon E \to X$ is **connected** if the total space E is connected.

Example 12.7 Let X be connected and F a non-empty discrete space. The projection on the first factor $X \times F \to X$ is a trivial covering space, and it's connected if and only if F consists of one point.

Fig. 12.1 By identifying $\mathbb{R}$ with the helix $E = \{(z, t) \in \mathbb{C} \times \mathbb{R} \mid z = e(t)\}$ under the homeomorphism $t \mapsto (e(t), t)$, the covering map $e \colon \mathbb{R} \to S^1$ coincides with the projection $p \colon \mathbb{C} \times \mathbb{R} \to \mathbb{C}$ restricted to E

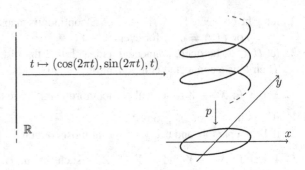

Example 12.8 Let's identify the circle S^1 with the set of unit complex numbers. Then

$$e \colon \mathbb{R} \to S^1, \qquad e(t) = e^{2\pi i t} = \cos(2\pi t) + i \sin(2\pi t)$$

defines a connected covering space. In fact, e is a surjective local homeomorphism, and for every open interval $]a, b[\subset \mathbb{R}$

$$e^{-1}(e(\,]a, b[\,)) = \cup\{\,]a + n, b + n[\mid n \in \mathbb{Z}\}.$$

If $|b - a| < 1$, for every $n \in \mathbb{Z}$ the interval $]a + n, b + n[$ is open in $\mathbb{R}$ and $e \colon]a + n, b + n[\to S^1$ is a homeomorphism on its range (Fig. 12.1).

Example 12.9 The natural projection $p \colon S^n \to \mathbb{P}^n(\mathbb{R})$ is a covering map. For every $x \in S^n$ the open set $U = \{y \in S^n \mid (x \cdot y) \neq 0\}$ is saturated by p and the open sets

$$\{y \in S^n \mid (x \cdot y) > 0\}, \qquad \{y \in S^n \mid (x \cdot y) < 0\}$$

are its connected components. The restriction of p to each one of the latter is a homeomorphism on the image.

Example 12.10 The map $p \colon \mathbb{C} - \{0\} \to \mathbb{C} - \{0\}$, $p(z) = z^2$, defines a covering space. An open set $U \subset \mathbb{C} - \{0\}$ is admissible if and only if U admits a continuous branch of the square-root function. Since this branch can be defined on the complement of any half-line starting at 0 in the complex plane, admissible open sets cover $\mathbb{C} - \{0\}$.

Example 12.11 The map

$$e \colon \mathbb{C} \to \mathbb{C} - \{0\}, \qquad e(z) = \exp(2\pi i z) = e^{2\pi i z},$$

is a covering map. The admissible sets are the open sets where the logarithm has a continuous branch.

Proposition 12.12 *Let* $p: E \to X$ *be a covering map. Then:*

1. *the map* p *is a local homeomorphism; in particular* p *is open and its fibres are discrete;*
2. *for every* $x, y \in X$, *the fibres* $p^{-1}(x)$ *and* $p^{-1}(y)$ *have the same cardinality;*
3. *for every connected* $Y \subset X$, *the restriction* $p: p^{-1}(Y) \to Y$ *is a covering map.*

Proof (1) If $V \subset X$ is an admissible open set of p, by definition of covering space the restriction $p: p^{-1}(V) \to V$ is a local homeomorphism. But the open sets $p^{-1}(V)$, with V admissible, cover E, so $p: E \to X$ is a local homeomorphism.

(2) Let $x_0 \in X$ be a given point and call $A \subset X$ the set of points $x \in X$ such that $p^{-1}(x)$ has the cardinality of $p^{-1}(x_0)$. As X is connected and A not empty, it suffices to show A is open and closed.

If $V \subset X$ is an admissible open set, the fibres $p^{-1}(x)$, $x \in V$, have the same cardinality. In fact if we write $p^{-1}(V) = \cup U_i$ with i in some index set J, the U_i open and disjoint and $p: U_i \to V$ a homeomorphism, then each fibre intersects every U_i in exactly one point. Hence we have a bijection between $p^{-1}(x)$ and J, and then $V \subset A$ or $V \cap A = \emptyset$. But admissible sets cover X, so A and $X - A$ are both open.

(3) For this it's enough to notice that the restriction of an admissible set to Y is admissible for the map $p: p^{-1}(Y) \to Y$. $\qquad\square$

Definition 12.13 Suppose every fibre of $p: E \to X$ is finite with cardinality d. Then one says that p is a **covering map of degree d**, or a **d-fold covering map**.

For instance: the projection $S^n \to \mathbb{P}^n(\mathbb{R})$ is a covering map of degree 2 (or a double covering), $p: S^1 \to S^1$, $p(z) = z^3$, is a three-fold covering map (degree 3), while $S^1 \times S^1 \to S^1 \times S^1$, $(z_1, z_2) \mapsto (z_1^2, z_2^2)$, has degree 4.

Lemma 12.14 *Let* $p: E \to X$ *be a covering space. For every admissible open set* $V \subset X$ *and every point* $e \in p^{-1}(V)$ *there exists a continuous section* $s_e: V \to p^{-1}(V)$ *of* $p: p^{-1}(V) \to V$ *such that* $s_e(p(e)) = e$. *If* V *is connected,* s_e *is unique.*

Proof As V is admissible we can write $p^{-1}(V) = U \cup W$ with U, W open disjoint, $e \in U$ and $p_{|U}: U \to V$ a homeomorphism. The inverse $s_e = p_{|U}^{-1}$ is a section with the required features.

If V is connected, then U is connected and also a connected component of $p^{-1}(V)$. Suppose $u: V \to E$ is another continuous section such that $u(p(e)) = e$. Since $u(V) \subset p^{-1}(V)$ and V is connected it follows that $u(V)$ is contained in the connected component of $p^{-1}(V)$ that contains e. Hence $u: V \to U$ is the inverse of $p_{|U}$. $\qquad\square$

By Lemma 12.14 we can define, for every $x \in X$ and any connected admissible set $V \subset X$ containing x, a continuous map

$$p^{-1}(x) \times V \xrightarrow{\Phi} p^{-1}(V), \qquad \Phi(e, y) = s_e(y).$$

It's not hard to prove that Φ is open and bijective, hence a homeomorphism.

Exercises

12.6 Let X, F be topological spaces, with X connected. Prove that the projection $X \times F \to X$ is a covering space if and only if F has the discrete topology.

12.7 Let $p \colon E \to X$ be a covering space. Prove that if X is Hausdorff, then E is Hausdorff, too.

12.8 Let $p \colon E \to X$ and $q \colon F \to Y$ be coverings. Prove that

$$p \times q \colon E \times F \to X \times Y, \qquad (p \times q)(e, f) = (p(e), q(f))$$

is a covering.

12.9 Let $p \colon E \to X$ be a covering space. Prove that for every connected Y and every continuous map $f \colon Y \to X$,

$$f^* p \colon \{(e, y) \in E \times Y \mid p(e) = f(y)\} \to Y, \qquad f^* p(e, y) = y,$$

is a covering space. (Hint: identify Y with the graph of f.)

12.10 ($\heartsuit$) For any $a, b \in \mathbb{R}$, $a < b$, the map

$$e \colon \,]a, b[\to S^1, \qquad e(t) = \cos(2\pi t) + i \sin(2\pi t)$$

is a local homeomorphism but not a covering. Why?

12.11 ($\heartsuit$) Prove that the multiplicative groups S^1 and $\mathbb{C} - \{0\}$ are isomorphic as Abelian groups but not homeomorphic as topological spaces.

12.12 ($\clubsuit$, $\heartsuit$) Let $p \colon E \to X$ be a local homeomorphism of connected Hausdorff spaces. Prove that if E is compact then p is a covering map of finite degree.

12.13 ($\clubsuit$, $\heartsuit$) Let $A = \begin{pmatrix} a & b \\ c & d \end{pmatrix}$ be a 2×2-matrix with integer coefficients, and consider

$$p_A \colon S^1 \times S^1 \to S^1 \times S^1, \qquad p_A(z_1, z_2) = (z_1^a z_2^b, z_1^c z_2^d).$$

Prove that if $\det(A) \neq 0$, then p_A is a covering map of degree $|\det(A)|$.

12.3 Quotients by Properly Discontinuous Actions

Definition 12.15 Let G be a subgroup of the group $\mathrm{Homeo}(E)$ of homeomorphisms of a space E. The group G is said to act **properly discontinuously** if every point $e \in E$ has a neighbourhood U such that $g(U) \cap U = \emptyset$ for any $g \in G$ different from the identity.

Example 12.16 The translation $t \mapsto t + 1$ generates a subgroup of Homeo($\mathbb{R}$), isomorphic to $\mathbb{Z}$, that acts properly discontinuously.

Example 12.17 The subgroup in Homeo($\mathbb{R}^2 - \{0\}$) generated by the multiplication by a number $\lambda > 1$ acts in a properly discontinuous fashion.

Theorem 12.18 *Let E be a topological space and* $G \subset$ Homeo(E) *a subgroup acting properly discontinuously. If E/G is connected, then the quotient map $p: E \rightarrow E/G$ is a covering map.*

Proof Fix $e \in E$ and choose an open set $U \subset E$ such that $e \in U$ and $g(U) \cap U = \emptyset$ for every g different from the identity.

Proposition 5.15 implies that $p: E \rightarrow E/G$ is an open map, and

$$p^{-1}(p(U)) = \cup\{g(U) \mid g \in G\}.$$

So we just need to prove that, for any $g \in G$, the open sets $g(U)$ are disjoint and that $p: g(U) \rightarrow p(U)$ is a homeomorphism.

Since $g(U) \cap h(U) = h(h^{-1}g(U) \cap U)$, it follows $g(U) \cap h(U) = \emptyset$ for every $g \neq h$. The quotient map $p: U \rightarrow p(U)$ is open and bijective hence a homeomorphism. The map $p: g(U) \rightarrow p(U)$ is the composite of the homeomorphisms $g^{-1}: g(U) \rightarrow U$ with $p: U \rightarrow p(U)$.

Proposition 12.19 *Let $G \subset$ Homeo(E) be a subgroup of homeomorphisms of a Hausdorff space E. If:*

1. *G acts freely, i.e. $g(e) \neq e$ for every $e \in E$ and every g different from the identity, and*
2. *any point $e \in E$ has an open neighbourhood U such that $g(U) \cap U \neq \emptyset$ for finitely many $g \in G$ at most (NB: this is automatic if G is finite),*

then G acts properly discontinuously.

Proof Let e and U be as in the statement; we shall find a neighbourhood V of e such that $g(V) \cap V = \emptyset$ for every g other than the identity. Write

$$\{g_1, \ldots, g_n\} = \{g \in G \mid g(U) \cap U \neq \emptyset\}.$$

As E is Hausdorff and G acts freely, by Lemma 3.72 there exist n disjoint open sets $U_1, \ldots, U_n \subset E$ such that $g_i(e) \in U_i$ for every i. Hence the point e belongs in the open set
$$V = U \cap g_1^{-1}(U_1) \cap \cdots \cap g_n^{-1}(U_n).$$

To show that $g(V) \cap V = \emptyset$ for every g different from the identity, suppose $g_1 = Id$ to fix ideas, so that $V \subset U \cap U_1$. For every $i = 2, \ldots, n$ we have $g_i(V) \subset U_i$, hence $g_i(V) \cap V = \emptyset$; if $g \notin \{g_1, \ldots, g_m\}$ then $g(V) \subset g(U)$, and therefore $g(V) \cap V = \emptyset$. $\qquad\square$

Remark 12.20 The quotient of a Hausdorff space by a group acting properly discontinuously could be non-Hausdorff. We refer to Massey's textbook [Ma67, p.167] for an example where $E = \mathbb{R}^2$, $G = \mathbb{Z}$ but the quotient E/G isn't Hausdorff.

Example 12.21 Let $G \subset \mathrm{Homeo}(\mathbb{R}^2)$ be the subgroup generated by the translations

$$a, b \colon \mathbb{R}^2 \to \mathbb{R}^2, \qquad a(x, y) = (x + 1, y), \qquad b(x, y) = (x, y + 1).$$

Notice that $ab = ba$, the map $\mathbb{Z}^2 \to G, (n, m) \mapsto a^n b^m$, is a group isomorphism, and that if $U \subset \mathbb{R}^2$ is an open set contained in a ball of radius $< 1/2$, then $g(U) \cap U = \emptyset$ for every g different from the identity.

The quotient $\mathbb{R}^2/G$ is homeomorphic to the torus $S^1 \times S^1$: in fact the map

$$\mathbb{R}^2 \to S^1 \times S^1, \qquad (x, y) \mapsto (e(x), e(y)),$$

factorises through a continuous bijection $\mathbb{R}^2/G \to S^1 \times S^1$. Now observe that $\mathbb{R}^2/G$ is compact and the product of circles is Hausdorff.

Example 12.22 Let $G \subset \mathrm{Homeo}(\mathbb{R}^2)$ denote the subgroup generated by the isometries

$$a, b \colon \mathbb{R}^2 \to \mathbb{R}^2, \qquad a(x, y) = (x + 1, 1 - y), \qquad b(x, y) = (x, y + 1).$$

Note the relation $bab = a$, and that the elements of G are the isometries of the form $g(x, y) = (x + n, (-1)^n y + m)$ for some $n, m \in \mathbb{Z}$. Hence, if $U \subset \mathbb{R}^2$ is an open set contained in a ball of radius $< 1/2$, then $g(U) \cap U = \emptyset$ for every g other than the identity. The quotient $p \colon \mathbb{R}^2 \to \mathbb{R}^2/G$ is therefore a connected covering space.

The same argument used for Example 5.18 shows here that $\mathbb{R}^2/G$ is Hausdorff and the composite of the inclusion $[0, 1]^2 \subset \mathbb{R}^2$ with p factorises through a homeomorphism between $\mathbb{R}^2/G$ and the Klein bottle.

Exercises

12.14 Let G be a group of homeomorphisms of a Hausdorff space that acts in a properly discontinuous way. Prove that the orbits of G are closed and discrete. (Hint: every point has a neighbourhood intersecting any orbit in one point at most.)

12.15 Let G be a group of homeomorphisms of a locally compact Hausdorff space E. Assume that:

1. G acts freely on E;
2. for every compact set $K \subset E$, we have $g(K) \cap K \neq \emptyset$ for finitely many $g \in G$.

Prove that the action of G is properly discontinuous and that the quotient E/G is locally compact and Hausdorff.

12.16 Let G be a group of isometries of a metric space (E, d). Prove that G acts properly discontinuously if and only if for any $e \in E$ there exists $l > 0$ such that $d(e, g(e)) \geq l$ for every $g \in G$ different from the identity.

12.17 Let $0 < p < q$ be fixed coprime integers and let $\xi \in \mathbb{C}$ be a primitive qth root of unity. Prove that the group of homeomorphisms of the sphere $S^3 = \{(z_1, z_2) \in \mathbb{C}^2 \mid \|z_1\|^2 + \|z_2\|^2 = 1\}$ generated by $(z_1, z_2) \mapsto (\xi z_1, \xi^p z_2)$ is isomorphic to the cyclic group of order q, and that it acts properly discontinuously. The corresponding quotient is called a **lens space**.

12.18 Let n be a positive integer and $\mu_n \subset \mathbb{C}$ the subgroup of nth roots of unity. For any $\xi \in \mu_n$ let

$$\Gamma_\xi = \{(z, u) \in \mathbb{C}^2 \mid |z| \leq 1, \ u = (1 - |z|)\xi\}.$$

Prove that the union $\Gamma = \cup_{\xi \in \mu_n} \Gamma_\xi$ is simply connected, and the action

$$\mu_n \times \Gamma \to \Gamma, \qquad (\xi, (z, w)) \mapsto (\xi z, \xi w)$$

is properly discontinuous.

12.19 (✥) Let $H \subset \mathbb{C}$ be the set of complex numbers z with imaginary part bigger than 0, and $G \subset \mathrm{Homeo}(H)$ the homeomorphisms of the form

$$z \mapsto \frac{az + b}{cz + d}, \qquad \text{with } a, b, c, d \in \mathbb{Z}, \quad ad - bc = 1.$$

Prove that G acts properly discontinuously on $H - (Gi \cup G\omega)$, where Gi and $G\omega$ are the orbits of the square and cubic roots of -1, i.e. $i^2 + 1 = 0$ and $\omega^2 - \omega + 1 = 0$. (Hint: it may be useful to show first that the imaginary part of $(az + b)/(cz + d)$ equals the imaginary part of z divided by $|cz + d|^2$. Use Exercise 12.15.)

12.4 Lifting Homotopies

Definition 12.23 Let $f : Y \to X$ be a continuous map and $p : E \to X$ a covering space. A continuous mapping $g : Y \to E$ is called a **lift** of f when the diagram

commutes, i.e. $f = pg$.

Lemma 12.24 *For any covering space $p : E \to X$ the diagonal $\Delta \subset E \times E$ is open and closed in the fibred product*

$$E \times_X E = \{(u, v) \in E \times E \mid p(u) = p(v)\}.$$

Proof Take $(e, e) \in \Delta$ and choose an open set $U \subset E$ such that $e \in U$ and the restriction $p: U \to X$ is 1-1. Then

$$(U \times U) \cap (E \times_X E) = U \times_X U$$

is an open neighbourhood of (e, e) in the fibred product. On the other hand

$$(U \times U) \cap (E \times_X E) = \{(u, v) \in U \times U \mid p(u) = p(v)\} \subset \Delta,$$

proving that Δ is a neighbourhood of any of its points, inside the fibred product.

Conversely, if $(e_1, e_2) \in E \times_X E - \Delta$ we pick an admissible open set V containing $p(e_1) = p(e_2)$. Since $e_1 \neq e_2$, there exist disjoint open sets $U_1, U_2 \subset p^{-1}(V)$ such that $e_1 \in U_1, e_2 \in U_2$. Therefore

$$(e_1, e_2) \in (U_1 \times U_2) \cap (E \times_X E) \subset E \times_X E - \Delta,$$

so that the diagonal is closed in the fibred product.   □

Theorem 12.25 (Uniqueness of lifts) *Let* $p: E \to X$ *be a covering space, Y a connected space and* $f: Y \to X$ *a continuous map. For any two lifts* $g, h: Y \to E$ *of f we have either* $g = h$ *or* $g(y) \neq h(y)$ *for every* $y \in Y$.

Proof Let $g, h: Y \to E$ be lifts of f, and consider the continuous map

$$\Phi: Y \to E \times_X E, \qquad \Phi(y) = (f(y), h(y)).$$

By Lemma 12.24

$$A = \{y \in Y \mid g(y) = h(y)\} = \Phi^{-1}(\Delta)$$

is open and closed in Y. Since Y is connected, $A = Y$ or $A = \emptyset$.

Corollary 12.26 *Let* $p: E \to X$ *be a covering, Y connected and* $f: Y \to X$ *continuous. Then for every* $y \in Y$ *and* $e \in E$ *such that* $p(e) = f(y)$ *there exists at most one lift* $g: Y \to E$ *of f such that* $g(y) = e$.

Proof Straightforward consequence of Theorem 12.25.   □

It's easy to see that lifts don't always exist: any lift of the identity is a continuous section, and we did remark earlier that the covering space $e: \mathbb{R} \to S^1$ doesn't possess continuous sections.

Theorem 12.27 (Lift of a path) *Let $p\colon E \to X$ be a covering space, $\alpha\colon I \to X$ a path and $e \in E$ a point such that $p(e) = \alpha(0)$. Then α lifts to a unique map $\alpha_e\colon I \to E$ such that $\alpha_e(0) = e$.*

Proof Uniqueness follows from Corollary 12.26 and it's then enough to prove existence. Since admissible open sets form a cover, by Corollary 11.24 there exist a positive integer n and n admissible open sets $V_1, \ldots, V_n \subset X$ such that the range of α, when restricted to $\left[\dfrac{i-1}{n}, \dfrac{i}{n}\right]$, is contained in V_i, for every $i = 1, \ldots, n$.

We define recursively n continuous maps

$$\gamma_i\colon \left[\frac{i-1}{n}, \frac{i}{n}\right] \to E,$$

such that $p(\gamma_i(t)) = \alpha(t)$, $\gamma_1(0) = e$ and $\gamma_i(i/n) = \gamma_{i+1}(i/n)$ for every i. The product of all γ_i will give us the required path α_e. Suppose to have defined γ_i, set $e_i = \gamma_i(i/n)$ and call $s_{i+1}\colon V_{i+1} \to p^{-1}(V_{i+1})$ a section of p such that $s_{i+1}(\alpha(i/n)) = e_i$ (Lemma 12.14). Now it is sufficient to define $\gamma_{i+1}(t) = s_{i+1}(\alpha(t))$, $i/n \le t \le (i+1)/n$, and everything follows. $\square$

Lemma 12.28 *Write*

$$L = \{(t, s) \in I^2 \mid ts = 0\}$$

for the union of two successive sides of a square and call $i\colon L \hookrightarrow I^2$ the inclusion morphism.

Let $p\colon E \to X$ be a covering space, $F\colon I^2 \to X$ and $f\colon L \to E$ continuous maps such that $pf = Fi$. Then F lifts to a map $G\colon I^2 \to E$ such that $Gi = f$.

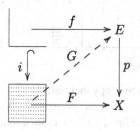

Proof For starters consider the special case in which the range of F is completely contained in an admissible open set V. From condition $pf = Fi$ follows $f(L) \subset p^{-1}(V)$, and as V is admissible we can write $p^{-1}(V)$ as disjoint union of open sets U, W such that $p\colon U \to V$ is a homeomorphism and $f(0, 0) \in U$. Therefore $L = f^{-1}(U) \cup f^{-1}(W)$. Since L is connected and $f^{-1}(U) \neq \emptyset$, $f(L)$ is contained in U. Now indicate with $s\colon V \to U$ the inverse homeomorphism to $p\colon U \to V$ and set $G = sF$. Then $pG = psF = F$, so G is a lift of F. Furthermore $f = spf = sFi = Gi$, as we wanted.

Now to the general situation. By Corollary 11.24 there exist a positive integer n and n^2 admissible sets $V_{ij} \subset X$ such that

$$F(Q(i, j)) \subset V_{ij}, \quad \text{where} \quad Q(i, j) = \left[\frac{i-1}{n}, \frac{i}{n}\right] \times \left[\frac{j-1}{n}, \frac{j}{n}\right],$$

for every $i, j = 1, \ldots, n$. Put on $\mathbb{N}^2$ the total order

$$(i, j) \leq (h, k) \quad \text{if } i + j < h + k \text{ or } i + j = h + k, \; j \leq k,$$

and observe that for every (h, k), $1 \leq h, k \leq n$, the intersection

$$Q(h, k) \cap \left(L \cup \bigcup_{(i,j)<(h,k)} Q(i, j)\right)$$

equals two consecutive sides of the square $Q(h, k)$. We can then use the previous construction of lifts on the smaller squares $Q(h, k)$, and recursively manufacture lifts

$$G_{hk} : L \cup \bigcup_{(i,j)\leq(h,k)} Q(i, j) \to E.$$

The requested lift is $G = G_{nn}$.

To understand better the recursive procedure it's useful to draw the sequence of domains of the lifts G_{hk}:

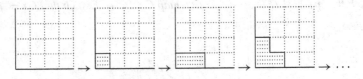

Theorem 12.29 (Covering homotopy theorem) *Let $p : E \to X$ be a covering space and $F : I^2 \to X$, $\alpha : I \to E$ continuous maps such that $F(t, 0) = p(\alpha(t))$ for any $t \in I$. Then F admits a unique lift $G : I^2 \to E$ such that $G(t, 0) = \alpha(t)$ for every t.*

$$
\begin{array}{ccc}
I \times \{0\} & \xrightarrow{\alpha} & E \\
\downarrow & \nearrow^{G} & \downarrow{p} \\
I \times [0, 1] & \xrightarrow{F} & X .
\end{array}
$$

Proof Proving existence is enough, for uniqueness is a consequence of the connectedness of I^2 and Theorem 12.25. Write $\beta : I \to E$ for the lift of the path $s \mapsto F(0, s)$ starting at $\beta(0) = \alpha(0)$; the glueing of α and β defines a continuous map

$$f: L = \{(t, s) \in I^2 \mid ts = 0\} \to E, \qquad f(t, 0) = \alpha(t), \ f(0, s) = \beta(s),$$

that, together with F, satisfies Lemma 12.28 by construction.

The existence and uniqueness theorems for lifts have important repercussions on the fundamental group.

Lemma 12.30 Let $p: E \to X$ be a covering space, $\alpha, \beta \in \Omega(X, a, b)$ two paths with the same endpoints, and $e \in E$ a point such that $p(e) = a$. Denote by $\alpha_e, \beta_e: I \to E$ the lifts of α and β beginning at $\alpha_e(0) = \beta_e(0) = e$. The following statements are equivalent:

1. α and β are path homotopic;
2. $\alpha_e(1) = \beta_e(1)$ and α_e, β_e are path homotopic.

Proof (1)$\Rightarrow$(2). Let $F: I^2 \to X$ be a path homotopy such that $F(0, t) = \alpha(t)$, $F(1, t) = \beta(t)$, $F(s, 0) = a$, $F(s, 1) = b$. By the covering homotopy theorem there exists a continuous $G: I^2 \to E$ such that $G(s, 0) = e$. Uniqueness implies $G(0, t) = \alpha_e(t)$ and $G(1, t) = \beta_e(t)$. The path $G(s, 1)$ lift the constant path 1_b, so it must be constant. Consequently $\alpha_e(1) = \beta_e(1)$, and G is a path homotopy.

(2)$\Rightarrow$(1). If α_e and β_e are path homotopic, then also $\alpha = p\alpha_e$ and $\beta = p\beta_e$ are path homotopic. $\qquad\square$

Theorem 12.31 Let $p: E \to X$ be a covering space. If X, E are path connected and X is simply connected, then p is a homeomorphism.

Proof Covering maps are open and onto, and hence homeomorphisms iff they are 1-1. Let $e, u \in E$ be such that $p(e) = p(u) = x \in X$; as E is path connected, there exists a path $\alpha: I \to E$ such that $\alpha(0) = e$ and $\alpha(1) = u$. But $\pi_1(X) = 0$, so $p\alpha$ is path homotopic to a constant, and by Lemma 12.30 the endpoints of α coincide, i.e. $u = e$. This proves that p is injective. $\qquad\square$

Corollary 12.32 The circle S^1 is not simply connected.

Proof The circle has non-trivial connected covering spaces (Example 12.8), so we can invoke Theorem 12.31. $\qquad\square$

Corollary 12.33 Let $p: E \to X$ be a covering and $f: S^2 \to X$ a continuous map. For every $y \in S^2$ and $e \in p^{-1}(f(y))$ there exists a unique lift $g: S^2 \to E$ of f such that $g(y) = e$.

Proof Without loss of generality we may assume y is $(1, 0, 0)$. We remind that there's an identification map $q: I^2 \to S^2$ that contracts the boundary of I^2 to the point $(1, 0, 0)$ and induces a homeomorphism between the open square $]0, 1[^2$ and the punctured sphere $S^2 - \{(1, 0, 0)\}$.

Theorem 12.29 warrants that there exists a continuous map $h\colon I^2 \to E$ lifting fq that assumes the value e on one side of the square:

Let F denote a fibre of q; as the restriction $h_{|F}\colon F \to E$ is the lift of a constant map, and F is connected, $h_{|F}$ must be constant. Then the universality of identifications (Lemma 5.6) forces the existence of a continuous map $k\colon S^2 \to E$ such that $kq = h$. Hence $pkq = ph = fq$, and because q is onto, $pk = f$.

Exercises

12.20 Let $p\colon E \to X$ be a covering space. Prove that:

1. every continuous map $f\colon D^2 \to X$ has a lift;
2. every continuous map $f\colon \mathbb{R}^2 \to X$ has a lift;
3. let $q\colon I^2 \to Y$ be an identification map whose fibres $q^{-1}(y)$ are all connected. Then every continuous map $f\colon Y \to X$ has a lift.

12.21 Let $f\colon \mathbb{C} \to \mathbb{C}-\{0\}$ be a continuous mapping. Prove there exists a continuous $g\colon \mathbb{C} \to \mathbb{C} - \{0\}$ such that $g^n = f$, for every $n > 0$.

12.22 Prove that any continuous map $S^2 \to S^1$ is homotopic to a constant.

12.23 ($\heartsuit$) Let $p\colon G \to H$ be a continuous homomorphism of connected, locally path connected topological groups. Prove that if p is a covering map, its kernel is contained in the centre of G, i.e. $kg = gk$ for every $g \in G, k \in \ker(p)$.

12.24 Let $p\colon E \to F$ and $q\colon F \to X$ be continuous and surjective maps between connected, locally path connected spaces. Prove the following:

1. if $qp\colon E \to X$ is a covering, also p and q are coverings;
2. if p and q are coverings and q has finite degree, qp is a covering;
3. if p and q are coverings and every point in X has a simply connected open neighbourhood, then qp is a covering.

12.5 Brouwer's Theorem and Borsuk's Theorem

Theorem 12.34 (Borsuk) *There is no continuous map* $f: S^2 \to S^1$ *such that* $f(-x) = -f(x)$ *for every* $x \in S^2$.

We will present four different proofs of this, each unveiling a different aspect of the theory.

Proof (first proof of Theorem 12.34) Let $f: S^2 \to S^1$ be a continuous mapping; we wish to show that there exists a point $x_0 \in S^2$ such that $f(-x_0) \neq -f(x_0)$. Consider the covering $e: \mathbb{R} \to S^1$. By Corollary 12.33 there's a continuous map $g: S^2 \to \mathbb{R}$ lifting f, i.e. such that $eg = f$. Lemma 4.16 provides us with an $x_0 \in S^2$ satisfying $g(x_0) = g(-x_0)$, hence $f(x_0) = f(-x_0)$: in particular $f(-x_0) \neq -f(x_0)$. $\square$

Corollary 12.35 *For any continuous map* $g: S^2 \to \mathbb{R}^2$ *there exists a point* $x \in S^2$ *such that* $g(x) = g(-x)$. $\square$

Proof By contradiction: if $g(x) - g(-x) \neq 0$ for every $x \in S^2$, the continuous map

$$f: S^2 \to S^1, \qquad f(x) = \frac{g(x) - g(-x)}{\|g(x) - g(-x)\|},$$

would satisfy $f(-x) = -f(x)$ for every x, and thus violate Borsuk's theorem. $\square$

Corollary 12.36 *Take* $n \geq 3$ *and* $A \subset \mathbb{R}^n$ *open, non-empty. Then no continuous map* $f: A \to \mathbb{R}^2$ *can be one-to-one.*

Proof Just note that A contains a subspace homeomorphic to S^2 and apply Corollary 12.35. $\square$

Corollary 12.37 *There are no continuous maps* $r: D^2 \to S^1$ *satisfying* $r(-x) = -r(x)$ *for every* $x \in S^1$.

Proof Let's fix notations and set

$$D^2 = \{(x_1, x_2) \in \mathbb{R}^2 \mid x_1^2 + x_2^2 \leq 1\}, \qquad S^1 = \{(x_1, x_2) \in \mathbb{R}^2 \mid x_1^2 + x_2^2 = 1\},$$

$$S^2 = \{(x_1, x_2, x_3) \in \mathbb{R}^3 \mid x_1^2 + x_2^2 + x_3^2 = 1\}.$$

Now suppose there was $r: D^2 \to S^1$ continuous with $r(-x_1, -x_2) = -r(x_1, x_2)$ for every $x \in S^1$. Then

$$f: S^2 \to S^1, \qquad f(x_1, x_2, x_3) = \begin{cases} r(x_1, x_2) & \text{if } x_3 \geq 0, \\ -r(-x_1, -x_2) & \text{if } x_3 \leq 0, \end{cases}$$

would contradict Theorem 12.34. $\square$

Corollary 12.38 *The circle S^1 is not a retract of the disc D^2, i.e. there is no continuous map $r: D^2 \to S^1$ such that $r(x) = x$ for every $x \in S^1$.*

Proof This is a straightforward consequence of Corollary 12.37. Another argument is as follows: if S^1 were a retract of D^2, by Example 11.18 the inclusion $S^1 \to D^2$ would induce a 1-1 homomorphism of the fundamental groups.   □

Corollary 12.39 (Brouwer's fixed-point theorem) *Any continuous map $f: D^2 \to D^2$ has at least one fixed point.*

Proof Suppose by contradiction that $f(x) \neq x$ for every x. We can define $r: D^2 \to S^1$ by

$$r(x) = x + t(x - f(x)), \quad \text{where } t \geq 0 \text{ and } \|r(x)\| = 1.$$

Geometrically this means that $r(x)$ is the point on the circle that x meets when moving away from $f(x)$ along a straight line (Fig. 12.2). The map r is continuous (see Exercise 12.31 as well) and $r(x) = x$ for every $x \in S^1$. Therefore r is a retraction of the disc onto the boundary, in breach of Corollary 12.38.   □

Corollary 12.40 *Let $f: \mathbb{R}^2 \to \mathbb{R}^2$ be a continuous map. Assume there are positive numbers $a < 1$ and b such that $\|x - f(x)\| \leq a\|x\| + b$ for every $x \in \mathbb{R}^2$. Then f is onto.*

Proof Suppose by contradiction there is a $p \in \mathbb{R}^2$ that doesn't belong in the range of f, and choose a real number R large enough to satisfy the inequality $aR + b < R - \|p\|$.

Write $D = \{x \in \mathbb{R}^2 \mid \|x\| \leq R\}$ and $S = \{x \in \mathbb{R}^2 \mid \|x\| = R\}$. As $R > \|p\|$, the point p belongs in the interior of D, and so for every $x \in S$ the triangle of vertices $x, -x, p$ has an obtuse angle at p. Consider the continuous map

$$r: D \to S, \quad r(x) = p + t(f(x) - p), \quad \text{where } t > 0, \|r(x)\| = R.$$

Notice that $r(x)$ is the intersection between S and the affine half-line through $f(x)$ starting from p. If we take $x \in S$,

$$\|x - f(x)\| \leq a\|x\| + b = aR + b < R - \|p\| = \|x\| - \|p\| \leq \|x - p\|$$

Fig. 12.2 Proof of Brouwer's fixed-point theorem

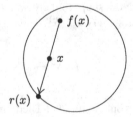

and $f(x)$ doesn't belong to the segment joining p and $-x$. Consequently $r(x) \neq -x$, so the continuous map $h: D \to D$, $h(x) = -r(x)$, has no fixed points, infringing Brouwer's theorem.

Exercises

12.25 Show that at any given moment there are on the surface of the Earth two antipodal places with the same temperature and atmospheric pressure.

12.26 Show that a continuous map $f: S^2 \to \mathbb{R}^2$ such that $f(-x) = -f(x)$ for every $x \in S^2$ must have a zero.

12.27 (Watermelon theorem) Consider a watermelon having finite volume and continuous, bounded density function. Prove that one can cut it through a point c and simultaneously separate flesh and pips in two parts of the same weight. (Hint: fix a Cartesian framing in $\mathbb{R}^3$ with origin at c. For any $x \in S^2$ consider the vector-valued function $f(x) = (f_+ - f_-, p_+ - p_-)$, where f_+ represents the flesh and p_+ the pips contained in $\{y \in \mathbb{R}^3 \mid (x \cdot y) \geq 0\}$, while f_-, p_- the respective quantities in the other half-plane.

12.28 (Ham-and-cheese sandwich theorem, ♡) Take three open bounded sets B (the bread), H (the ham) and C (the cheese) in $\mathbb{R}^3$, and assume further that B is connected.

Prove that there exists a plane in $\mathbb{R}^3$ that simultaneously divides each set B, H, C in two halves of the same volume.

12.29 (Lusternik-Schnirelmann theorem) Let $A_1, A_2, A_3 \subset S^2$ be closed sets with $A_1 \cup A_2 \cup A_3 = S^2$. Prove that there exists at least one index i such that A_i contains a pair of antipodal points. (Hint: write $B_i = \{-x \mid x \in A_i\}$ and suppose $A_1 \cap B_1 = A_2 \cap B_2 = \emptyset$. Consider

$$f: S^2 \to \mathbb{R}^2, \qquad f(x) = \left(\frac{d_{A_1}(x)}{d_{A_1}(x) + d_{B_1}(x)}, \frac{d_{A_2}(x)}{d_{A_2}(x) + d_{B_2}(x)} \right),$$

where d_Z indicates the distance from Z. Show that $f(x) = f(-x)$ implies $x \in A_3 \cap B_3$.)

12.30 Let $A_1, A_2, A_3 \subset S^2$ be closed, connected subsets such that

$$A_1 \cup A_2 \cup A_3 = S^2.$$

Prove that there's at least one i such that, for every real $0 \leq d \leq 2$, A_i contains a pair of points x, y with $\|x - y\| = d$ (Hint: Exercise 12.29).

12.31 (♡) Find the explicit expression of the function $r(x)$ introduced in the proof of Corollary 12.39.

12.32 Let $f, g: S^1 \to S^1$ be continuous and such that $f(x) \neq g(x)$ for every x. Prove that f and g are homotopic.

12.33 Prove that $\mathbb{R}^2$ is not homeomorphic to $\mathbb{R} \times [0, +\infty[$.

12.34 (☀) Prove that the group of homeomorphisms of $\mathbb{R}^2$, with the compact-open topology, is a topological group. (Hint: by Exercise 8.22 we just need to show that the transformation mapping a homeomorphism to its inverse is continuous at the identity. Argue as in the proof of Corollary 12.40.)

12.35 (☀) Prove the Ham-and-cheese sandwich theorem of Exercise 12.28 without assuming B connected. (Hint.: keeping in mind Exercise 4.34, for every n there exists an open connected set $B_n \subset \mathbb{R}^3$ such that $B \subset B_n$ and the volume of $B_n - B$ is smaller than $1/n$.)

12.6 A Non-abelian Fundamental Group

Consider the union X of two tangent circles. It is convenient to view X as a graph with one vertex e and two edges a, b. Fixing the edges' orientations defines elements $[a], [b] \in \pi_1(X, e)$ corresponding to the homotopy classes of the simple loops obtained by running along the edges in the chosen direction. We want to prove that $[a][b] \neq [b][a]$, and thence that *the fundamental group of X is not Abelian*.

Consider the connected covering space $p \colon E \to X$ of degree 3 depicted in Fig. 12.3. The nodes e_1, e_2, e_3 of the graph E (left) are mapped by p to the unique vertex e of X, while the interior of each edge of E is mapped homeomorphically to the interior of the edge of X indexed alike. Note that every node in the figure has an incoming a edge, an outgoing a edge, an incoming b edge and an outgoing b edge.

Observe that the lift of the path $a{*}b \in \Omega(X, e, e)$ starting at e_1 is the product of $a \in \Omega(E, e_1, e_1)$ and $b \in \Omega(E, e_1, e_2)$, so it has to end at e_2. Similarly, $b{*}a \in \Omega(X, e, e)$ lifts to a path with initial point e_1: the lift is the product of $b \in \Omega(E, e_1, e_2)$ and $a \in \Omega(E, e_2, e_3)$, and ends necessarily at e_3. Since the two lifts have same initial points but different final points, Lemma 12.30 tells that $a * b$ is not homotopic to $b * a$ in $\Omega(X, e, e)$.

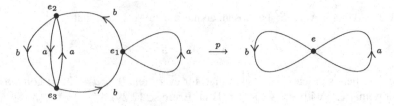

Fig. 12.3 A covering space of the 'figure of eight'

Exercises

We retain the notations introduced throughout this section.

12.36 Prove that $p_*\pi_1(E, e_1) \neq p_*\pi_1(E, e_2)$, and that $p_*\pi_1(E, e_1)$ is not a normal subgroup in $\pi_1(X, e)$.

12.37 Prove that $([a][b])^n \neq ([b][a])^n$ in $\pi_1(X, e)$ for every $n > 0$.

Reference

[Ma67] Massey, W.: Algebraic Topology: An Introduction. Harcourt Brace and World, New York (1967)

Chapter 13
Monodromy

From this point onwards we shall normally assume, to make the theory more agile and simplify both statements and proofs, that total spaces and bases of all coverings $E \to X$ are *locally path connected*. In particular, given a connected covering $E \to X$, both E, X will be path connected.

There are situations in which the fundamental group acts in a special way on sets, depending on the so-called **monodromy**. Paths with non-trivial monodromy are homotopically non-trivial. In the present chapter we shall discuss monodromies arising from coverings, and we will show that sometimes the monodromy action allows to describe the fundamental group completely.

13.1 Monodromy of Covering Spaces

Let $p \colon E \to X$ be a covering and $x, y \in X$ two points. One calls **monodromy** the map

$$\mathrm{Mon} \colon p^{-1}(x) \times \Omega(X, x, y) \to p^{-1}(y), \qquad \mathrm{Mon}(e, \alpha) = \alpha_e(1),$$

where $\alpha_e \colon I \to E$ is the unique lift of the path α with initial point $\alpha_e(0) = e$. Note that for every path $\beta \in \Omega(X, y, z)$ one has $(\alpha * \beta)_e = \alpha_e * \beta_{\alpha_e(1)}$, so

$$\mathrm{Mon}(e, \alpha * \beta) = \mathrm{Mon}(\mathrm{Mon}(e, \alpha), \beta).$$

Lemma 12.30 implies that $\mathrm{Mon}(e, \alpha)$ depends only on the homotopy class of α. In particular, $\mathrm{Mon}(e, \alpha * i(\alpha)) = \mathrm{Mon}(e, 1_x) = e$, and so for every given $\alpha \in \Omega(X, x, y)$, the map

$$p^{-1}(x) \to p^{-1}(y), \qquad e \mapsto \mathrm{Mon}(e, \alpha),$$

© Springer International Publishing Switzerland 2015
M. Manetti, *Topology*, UNITEXT - La Matematica per il 3+2 91,
DOI 10.1007/978-3-319-16958-3_13

is bijective with inverse

$$p^{-1}(y) \to p^{-1}(x), \qquad u \mapsto \mathrm{Mon}(u, i(\alpha)).$$

The monodromy's homotopy invariance has another consequence if we pass to the quotient; namely, taking $y = x$ gives

$$p^{-1}(x) \times \pi_1(X, x) \to p^{-1}(x), \qquad (e, [\alpha]) \mapsto e \cdot [\alpha] = \mathrm{Mon}(e, \alpha).$$

It follows easily from the definitions that

$$e \cdot [1_x] = e. \qquad e \cdot ([\alpha][\beta]) = (e \cdot [\alpha]) \cdot [\beta],$$

for every $[\alpha], [\beta] \in \pi_1(X, a)$ and any $e \in p^{-1}(x)$.

Theorem 13.1 *Let* $p\colon E \to X$ *be a connected covering,* $e \in E$ *a base point and* $x = p(e)$. *Then:*

1. *the group homomorphism* $p_*\colon \pi_1(E, e) \to \pi_1(X, x)$ *is injective, and*

$$p_*\pi_1(E, e) = \{[\alpha] \in \pi_1(X, x) \mid e \cdot [\alpha] = e\};$$

2. *there's a one-to-one correspondence between the fibre* $p^{-1}(x)$ *and right cosets of* $p_*\pi_1(E, e)$ *in* $\pi_1(X, x)$;
3. *for every* $[\alpha] \in \pi_1(X, x)$ *we have* $[\alpha]^{-1} p_*\pi_1(E, e)[\alpha] = p_*\pi_1(E, e \cdot [\alpha])$. *In particular the subgroups* $\{p_*\pi_1(E, u) \mid p(u) = x\}$ *are precisely the conjugates of* $p_*\pi_1(E, e)$ *in* $\pi_1(X, x)$.

Recall that a right coset of a subgroup $H \subset G$ is a subset in G of the type Hg, for $g \in G$. Equivalently, right cosets of H in G are the equivalence classes of the relation '$g_1 \sim g_2 \iff g_1 g_2^{-1} \in H$'.

Proof Let's recall what we said at the very beginning of the chapter: since we assume E, X locally path connected, if E, X are connected they become path connected.

Proof of (1). Injectivity is a straightforward consequence of Lemma 12.30 (with $a = b = x$). If $[\alpha] \in p_*\pi_1(E, e)$ there's a path $\beta\colon I \to E$ such that $\beta(0) = \beta(1) = e$ and $[\alpha] = [p\beta]$; since lifts are unique $\beta = \alpha_e$, so $\beta(1) = \alpha_e(1)$ and $e \cdot [\alpha] = e$. If, conversely, $\alpha \in \Omega(X, x, x)$ and $\mathrm{Mon}(e, \alpha) = e \cdot [\alpha] = e$, then the lift α_e is a loop and therefore $[\alpha] = [p\alpha_e] \in p_*\pi_1(E, e)$.

Proof of (2). The monodromy allows to define

$$\pi_1(X, x) \to p^{-1}(x), \qquad [\alpha] \mapsto e \cdot [\alpha] = \mathrm{Mon}(e, \alpha).$$

Let's show that $[\alpha] \mapsto e \cdot [\alpha]$ is onto, and $e \cdot [\alpha] = e \cdot [\beta]$ if and only if $[\alpha][\beta]^{-1} \in p_*\pi_1(E, e)$. If $u \in p^{-1}(x)$, since by assumption E is path connected, there's a path $\gamma\colon I \to E$ such that $\gamma(0) = e$ and $\gamma(1) = u$. The lift's uniqueness implies

$\gamma = (p\gamma)_e$, and by definition of monodromy we have $u = e \cdot [p\gamma]$. Now given homotopy classes $[\alpha], [\beta] \in \pi_1(X, x)$, we have $e \cdot [\alpha] = e \cdot [\beta]$ if and only if $e \cdot [\alpha] \cdot [\beta]^{-1} = e$ if and only if $[\alpha][\beta]^{-1} \in p_*\pi_1(E, e)$.

Proof of (3). It suffices to prove that

$$[\alpha]^{-1} p_*\pi_1(E, e)[\alpha] \subset p_*\pi_1(E, e \cdot [\alpha])$$

for every $\alpha \in \Omega(X, x, x)$. Let $\alpha_e : I \to E$ be the lift of α starting at $\alpha_e(0) = e$; then $\alpha_e(1) = e \cdot [\alpha]$ and

$$i(\alpha_e) * \Omega(E, e, e) * \alpha_e \subset \Omega(E, e \cdot [\alpha], e \cdot [\alpha]). \qquad \square$$

Corollary 13.2 *The fundamental group of $\mathbb{P}^n(\mathbb{R})$ is isomorphic to $\mathbb{Z}/2$, for any $n \geq 2$.*

Proof The quotient map $p : S^n \to \mathbb{P}^n(\mathbb{R})$ is a connected double covering. Let $e \in S^n$ be a point and $x = p(e)$ its image; Theorem 13.1 implies that the subgroup $p_*\pi_1(S^n, e) \subset \pi_1(\mathbb{P}^n(\mathbb{R}), x)$ has two right cosets. If $n \geq 2$, the sphere S^n is simply connected, so the subgroup $p_*\pi_1(S^n, e)$ is trivial and therefore $\pi_1(\mathbb{P}^n(\mathbb{R}), x)$ contains just two elements. But up to isomorphism there's only one group with two elements. $\qquad \square$

Monodromy actions commute with continuous maps:

Proposition 13.3 *Let*

$$
\begin{array}{ccc}
E & \xrightarrow{\varphi} & F \\
\downarrow{\scriptstyle p} & & \downarrow{\scriptstyle q} \\
X & \xrightarrow{f} & Y
\end{array}
$$

be a commutative diagram of continuous maps, with p, q coverings. Then

$$\varphi(e \cdot [\alpha]) = \varphi(e) \cdot f_*([\alpha])$$

for every $e \in E$ and $[\alpha] \in \pi_1(X, p(e))$.

Proof Let $\alpha : I \to X$ be a loop based at the point $p(e)$, and $\alpha_e : I \to E$ its unique lift through $\alpha_e(0) = e$. By definition of monodromy $\alpha_e(1) = e \cdot [\alpha]$, so $\varphi(e \cdot [\alpha]) = \varphi(\alpha_e(1))$.

Consider the path $\beta : I \to F$, $\beta(t) = \varphi(\alpha_e(t))$. Since $q(\beta(t)) = q\varphi\alpha_e(t) = fp\alpha_e(t) = f\alpha(t)$ for every t, it follows that β is the lift of $f\alpha$ such that $\beta(0) = \varphi(e)$. Hence

$$\varphi(e) \cdot f_*([\alpha]) = \varphi(e) \cdot [f\alpha] = \beta(1) = \varphi\alpha_e(1). \qquad \square$$

Exercises

13.1 Show that the quotient space of $\mathbb{R}^3$ by the group generated by the involution $(x, y, z) \mapsto (-x, -y, -z)$ is not homeomorphic to $\mathbb{R}^3$.

13.2 ($\clubsuit$, $\heartsuit$) Compute the fundamental group of the space of real 3×3-matrices of rank one.

13.2 Group Actions on Sets

Definition 13.4 Let G be a group and T a set. A **left action** of G on T is a map

$$G \times T \to T, \qquad (g, t) \mapsto g \cdot t,$$

such that:

1. $1 \cdot t = t$ for every $t \in T$, where $1 \in G$ is the neutral element;
2. $(gh) \cdot t = g \cdot (h \cdot t)$ for every $t \in T$, $g, h \in G$.

Example 13.5 Let E be a space and Homeo(E) the group of homeomorphisms of E with the composition product. The map

$$\text{Homeo}(E) \times E \to E, \qquad \phi \cdot e = \phi(e),$$

is a left action.

If G acts on the left on T, then for every $g \in G$ one can define

$$L_g \colon T \to T, \qquad L_g(t) = g \cdot t.$$

As $L_g L_h = L_{gh}$, it follows that for any g the map L_g is bijective with inverse $L_{g^{-1}}$.

Definition 13.6 A **right action** of G on T is a map

$$T \times G \to T, \qquad (t, g) \mapsto t \cdot g,$$

such that:

1. $t \cdot 1 = t$ for every $t \in T$, where $1 \in G$ is the neutral element;
2. $t \cdot (gh) = (t \cdot g) \cdot h$ for every $t \in T$, $g, h \in G$.

Example 13.7 Let $p \colon E \to X$ be a covering. Then for every $x \in X$ the monodromy map

$$p^{-1}(x) \times \pi_1(X, x) \to p^{-1}(x), \qquad (u, [\alpha]) \mapsto u \cdot [\alpha],$$

is a right action.

If G acts on the right on T, then for every $g \in G$ we set

$$R_g : T \to T, \qquad R_g(t) = t \cdot g.$$

From $R_g R_h = R_{hg}$ descends, in particular, that every R_g is bijective with inverse $R_{g^{-1}}$.

We remark that any right action induces a canonical left action, and conversely, by the rule $g \cdot t = t \cdot g^{-1}$.

Definition 13.8 A left action of a group G on a non-empty set T is called:

1. **faithful**, if for every $g \neq 1$ there exists a $t \in T$ such that $g \cdot t \neq t$;
2. **free**, if $g \cdot t \neq t$ for every $g \neq 1, t \in T$;
3. **transitive**, if for every $t, s \in T$ there exist a $g \in G$ such that $g \cdot t = s$.

Similar notions hold for right actions.

Observe that a free action is faithful, while if G acts freely and transitively on T, the element $g \in G$ mapping any $t, s \in T$ one to the other is unique.

Definition 13.9 Consider a set T on which a group G acts on the left and another group H acts on the right. The two actions are called **compatible** if

$$g \cdot (t \cdot h) = (g \cdot t) \cdot h$$

for every $t \in T, g \in G$ and $h \in H$.

Example 13.10 The product $G \times G \to G$ may be viewed as both a right and a left action of the group G on itself. The multiplication's associativity is equivalent to the above notion of compatible actions.

Example 13.11 Let V, W be vector spaces, $\mathrm{Hom}(V, W)$ the set of linear maps from V to W. The group $\mathrm{GL}(W)$ acts on the left on $\mathrm{Hom}(V, W)$, while $\mathrm{GL}(V)$ acts on the right, and the two are compatible.

Proposition 13.12 *Consider groups G, H and a set T. Suppose G acts freely and transitively on T on the left, while H acts on the right, compatibly with G. For any $e \in T$ define*

$$\theta_e : H \to G, \qquad \theta_e(h) = \text{the unique } g \in G \text{ such that } g \cdot e = e \cdot h.$$

The map θ_e is a group homomorphisms and $\{h \in H \mid e \cdot h = e\}$ is a normal subgroup of H.

Proof As the actions are compatible, for every $h, k \in H$

$$\theta_e(hk) \cdot e = e \cdot hk = (e \cdot h) \cdot k = (\theta_e(h) \cdot e) \cdot k$$
$$= \theta_e(h) \cdot (e \cdot k) = \theta_e(h) \cdot (\theta_e(k) \cdot e) = \theta_e(h)\theta_e(k) \cdot e,$$

whence $\theta_e(hk) = \theta_e(h)\theta_e(k)$. The set $\{h \in H \mid e \cdot h = e\}$ is the kernel of θ_e. $\qquad\square$

The homomorphism θ_e depends on the choice of $e \in T$. If we take $u \in T$, and $g \in G$ is the unique element with $g \cdot e = u$, then

$$u \cdot h = \theta_u(h) \cdot u = \theta_u(h) \cdot (g \cdot e), \qquad u \cdot h = (g \cdot e) \cdot h = g \cdot (\theta_e(h) \cdot e),$$

for every $h \in H$. Consequently $\theta_u(h)g = g\theta_e(h)$, and θ_u is then the composite of θ_e with the inner automorphism $G \to G$, $\hat{g} \mapsto g\hat{g}g^{-1}$.

Exercises

13.3 Let G be a group of homeomorphisms of a space X. Show that there is a natural right action of G on continuous maps $f: X \to \mathbb{R}$.

13.4 Let groups G and H, a set T, a left action of G on T and a compatible right action of H on T be given.

Show that G acts on the left on the quotient T/H, H acts on the right on T/G, and that there exist natural bijections

$$(T/H)/G = (T/G)/H = T/\sim,$$

where $t \sim s$ iff $t = g \cdot s \cdot h$ for some $g \in G, h \in H$. Sometimes one writes $G\backslash T/H$ to indicate the double quotient $(T/H)/G = (T/G)/H$.

13.3 An Isomorphism Theorem

Let G be a group of homeomorphisms of a space E that acts properly discontinuously. Suppose $X = E/G$ is connected, and write $p: E \to X$ for the natural quotient map. By Theorem 12.18 p is a covering map. Given $x \in X$, for every pair of points $u, v \in p^{-1}(x)$ there's a unique element $g \in G$ such that $g(u) = v$. Therefore the action

$$G \times p^{-1}(x) \to p^{-1}(x), \qquad (g, u) \mapsto g(u),$$

is free and transitive.

Lemma 13.13 *In the notation above, the left action of G on $p^{-1}(x)$ is compatible with the monodromy's right action*

$$p^{-1}(x) \times \pi_1(X, x) \to p^{-1}(x), \qquad (e, [\alpha]) \mapsto e \cdot [\alpha].$$

Proof Let $e \in p^{-1}(x)$, $g \in G$ and $\alpha: I \to X$ a loop with base point x. To prove $g(e \cdot [\alpha]) = g(e) \cdot [\alpha]$ it suffices to apply Proposition 13.3 to the commutative diagram

$$
\begin{array}{ccc}
E & \xrightarrow{\ g\ } & E \\
\Big\downarrow{\scriptstyle p} & & \Big\downarrow{\scriptstyle p} \\
X & \xrightarrow{\ Id\ } & X.
\end{array}
$$

$\qquad\qquad\qquad\qquad\qquad\qquad\qquad\qquad\qquad\qquad\qquad\qquad$ □

Now fix a point $e \in p^{-1}(x)$ and define

$$\theta_e : \pi_1(X, x) \to G, \qquad \theta_e([\alpha]) = \text{the unique } g \in G \text{ such that } g(e) = e \cdot [\alpha].$$

Put differently, if $\alpha_e : I \to E$ is the lift of α such that $\alpha_e(0) = e$, then $\theta_e([\alpha])$ is the only element of G satisfying $\theta_e([\alpha])(e) = \alpha_e(1)$.

Theorem 13.14 *Retaining the notations above, the map $\theta_e : \pi_1(X, x) \to G$ is a group homomorphisms with kernel $p_*\pi_1(E, e)$; in particular $p_*\pi_1(E, e)$ is a normal subgroup in $\pi_1(X, x)$. If E is connected, then θ_e factorises through a group isomorphism*

$$\theta_e : \frac{\pi_1(X, x)}{p_*\pi_1(E, e)} \xrightarrow{\ \sim\ } G.$$

Proof θ_e is a homomorphisms with kernel $p_*\pi_1(E, e)$ by virtue of Lemma 13.13 and Proposition 13.12.

If E is connected θ_e is onto: for every $g \in G$, in fact, there's a path β in E such that $\beta(0) = e$ and $\beta(1) = g(e)$. Then $\alpha = p\beta$ is a loop, and $g(e) = e \cdot [\alpha]$, i.e. $\theta_e([\alpha]) = g$. $\qquad\qquad\qquad\qquad\qquad\qquad\qquad\qquad\qquad\qquad\qquad\qquad$ □

Quite often it's useful to describe the inverse of θ_e: assuming E connected,

$$\theta_e^{-1} : G \xrightarrow{\ \sim\ } \frac{\pi_1(X, x)}{p_*\pi_1(E, e)}, \qquad \theta_e^{-1}(g) = [p\gamma] \text{ where } \gamma \in \Omega(E, e, g(e)).$$

So we may say that $\theta_e^{-1}(g)$ is the coset containing the homotopy class of the image under p of any path in E with initial point e and endpoint $g(e)$.

Corollary 13.15 *Let E be simply connected, G a group of homeomorphisms of E acting properly discontinuously. Then the fundamental group of E/G is isomorphic to G.*

Proof Immediate consequence of Theorem 13.14. $\qquad\qquad\qquad\qquad\qquad\qquad\qquad$ □

Example 13.16 The fundamental group of the circle is isomorphic to $\mathbb{Z}$. In fact $S^1 = \mathbb{R}/\mathbb{Z}$, where $\mathbb{Z}$ acts on the left by $n \cdot x = n + x$, and the quotient map is the exponential map e. By Corollary 13.15

$$\pi_1(S^1, 1) \to \mathbb{Z}, \qquad [\alpha] \mapsto 0 \cdot [\alpha] \in \mathbb{Z},$$

is an isomorphism.

Example 13.17 By applying Corollary 13.15 to the construction of Example 12.22 we deduce that the fundamental group of the Klein bottle is isomorphic to the group with two generators a, b subject to $bab = a$. This group, by the way, is not Abelian.

At this juncture we can use the fact that $\pi_1(S^1)$ is isomorphic to $\mathbb{Z}$ to give another proof of Borsuk's theorem.

Proof (Second proof of Theorem 12.34) Suppose by contradiction to have a continuous map $f: S^2 \to S^1$ such that $f(-x) = -f(x)$ for every x. Identify S^1 with unit complex numbers and consider the double covering $h: S^1 \to S^1$, $h(z) = z^2$.

Let $\alpha: I \to S^2$ be a path joining two antipodal points. The closed path $hf\alpha: I \to S^1$ is not homotopically trivial because its lift $f\alpha$ is not a loop. Indicate with $\beta: I \to S^2$ the path $\beta(t) = -\alpha(t)$. Then $hf\beta = hf\alpha$, and $\alpha * \beta$ is a loop in S^2 hence homotopically trivial. So we have

$$0 = [hf(\alpha * \beta)] = [hf\alpha][hf\beta] = [hf\alpha][hf\alpha],$$

which is at odds with the fact that $\pi_1(S^1)$ doesn't contain non-trivial elements of order 2.   □

Exercises

13.5 Compute the fundamental group of

$$\mathbb{R}^3 - \left(\{x = y = 0\} \cup \{z = 0, \ x^2 + y^2 = 1\} \right).$$

(Hint: consider the revolution of $\{(r, z) \in \mathbb{R}^2 \mid r > 0, (r, z) \neq (1, 0)\}$ about the axis $x = y = 0$.)

13.6 Compute the fundamental group of the Möbius strip. Prove that this surface cannot be retracted to its boundary.

13.7 Compute the fundamental group of the quotient induced by the action of Exercise 12.18.

13.8 Compute the fundamental group of

$$X = \{(x_0, \dots, x_n) \in S^n \mid x_0^2 + x_1^2 < 1\}$$

for every $n \geq 3$.

13.9 Let $f: S^1 \times S^1 \to S^2$ be a continuous map and $\Gamma \subset S^1 \times S^1 \times S^2$ its graph. Determine the fundamental group of $S^1 \times S^1 \times S^2 - \Gamma$.

13.10 Set $X = \{(x, y, z) \in \mathbb{R}^3 \mid x^2 + y^2 = z^2, \ z \neq 0\}$ and let $G \subset \text{Homeo}(X)$ be the group with generators $a, b \colon X \to X$:

$$a(x, y, z) = (-x, -y, z), \qquad b(x, y, z) = (x, y, -z).$$

Prove that G is isomorphic to $\mathbb{Z}/2 \times \mathbb{Z}/2$ and it acts in a properly discontinuous fashion. Furthermore, say whether X/G is homeomorphic to $\mathbb{R}^2 - \{(0, 0)\}$. Explain.

13.11 ($\heartsuit$) Prove that the fundamental group of $\text{GL}(n, \mathbb{C})$ is infinite.

13.12 ($\clubsuit$, $\heartsuit$) Let $\alpha, \beta \colon I \to I^2$ be paths such that $\alpha(0) = (0, 0)$, $\alpha(1) = (1, 1)$, $\beta(0) = (1, 0)$ and $\beta(1) = (0, 1)$. Prove that they have to intersect at some point, i.e. $\alpha(t) = \beta(s)$ for certain $t, s \in I$.

13.13 ($\clubsuit$, $\heartsuit$) Let $X \subset \mathbb{R}^2$ be homeomorphic to S^1. Assume there exist a point $x \in X$ and a real number $r > 0$ such that $X \cap B(x, r)$ concedes with the intersection of $B(x, r)$ with a straight line. Prove that $\mathbb{R}^2 - X$ can't be connected. (Hint: identify the plane with the sphere minus a point; for every $0 < t \leq r$ the closed set $S^2 - B(x, t)$ is homeomorphic to I^2; apply Exercise 13.12 to show that $\pi_0(B(x, r) - X) \to \pi_0(S^2 - X)$ is one-to-one.)

13.4 Lifting Arbitrary Maps

Consider a covering $p \colon E \to X$, a connected and locally path connected space Y and a continuous map $f \colon Y \to X$. We'd like to determine a necessary and sufficient condition for a lift of f to exist.

Fix a point $y_0 \in Y$ and suppose there exists $g \colon Y \to E$ such that $pg = f$; if we write $e_0 = g(y_0)$ and $x_0 = f(y_0) = p(e_0)$, the homomorphisms $f_* \colon \pi_1(Y, y_0) \to \pi_1(X, x_0)$ equals the composite

$$\pi_1(Y, y_0) \xrightarrow{g_*} \pi_1(E, e_0) \xrightarrow{p_*} \pi_1(X, x_0),$$

so the range of p_* contains the range of f_*.

Theorem 13.18 *Let $f \colon Y \to X$ be a continuous map between connected, locally path connected spaces, $p \colon E \to X$ a covering and $y_0 \in Y$, $e_0 \in E$ points such that $f(y_0) = p(e_0)$. Then there exists a continuous map $g \colon Y \to E$ such that $pg = f$ and $g(y_0) = e_0$ if and only if*

$$f_* \pi_1(Y, y_0) \subset p_* \pi_1(E, e_0).$$

Proof We already proved this condition necessary, so let's demonstrate it's sufficient, too.

Set $x_0 = f(y_0)$, and for any $y \in Y$ choose a path $\alpha \colon I \to Y$ such that $\alpha(0) = y_0$ and $\alpha(1) = y$. The composite $f\alpha \colon I \to X$ is a path starting at $f(y_0)$; consider the monodromy

$$\text{Mon} \colon p^{-1}(x_0) \times \Omega(X, x_0, f(y)) \to p^{-1}(f(y))$$

and set $g(y) = \text{Mon}(e_0, f\alpha)$. We claim that g is well defined: if $\beta \colon I \to Y$ is another path with $\beta(0) = y_0$ and $\beta(1) = y$, then $\alpha * i(\beta)$ is a loop with base point y_0, and the homotopy class of $f\alpha * i(f\beta) = f(\alpha * i(\beta))$ belongs in $f_*\pi_1(Y, y_0)$. By assumption $[f\alpha * i(f\beta)] \in p_*\pi_1(E, e_0)$ and therefore $\text{Mon}(e_0, f\alpha) = \text{Mon}(e_0, f\beta)$.

Let's show g is continuous everywhere. Take $y \in Y$ and an open neighbourhood $A \subset E$ of $g(y)$. Pick an admissible open set $V \subset X$ containing $f(y)$, then call $U \subset p^{-1}(V)$ an open set in E containing $g(y)$ and such that $p \colon U \to V$ is a homeomorphism. In particular $p(U \cap A)$ is open, and by continuity there exists a path connected open $W \subset Y$ such that $y \in W$ and $f(W) \subset p(U \cap A) \subset V$; we want to show $g(W) \subset A$. Fix a path $\alpha \in \Omega(Y, y_0, y)$. For every point $w \in W$ we consider a path $\beta \colon I \to W$ from y to w. The lift of $f\beta$ at $g(y)$ is entirely contained in $U \cap A$, and as $g(y) = \text{Mon}(e_0, f\alpha)$ and $g(w) = \text{Mon}(e_0, f(\alpha * \beta)) = \text{Mon}(g(y), f\beta)$, eventually $g(w) \in U \cap A$. ◻

Example 13.19 The topological group $\text{SU}(2, \mathbb{C})$ is homeomorphic to the sphere S^3 (Example 1.6). As such, it is compact and simply connected. Consider a continuous group homomorphism $f \colon \text{SU}(2, \mathbb{C}) \to S^1$: we shall prove that $f(A) = 1$ for every $A \in \text{SU}(2, \mathbb{C})$.

Let $Id \in \text{SU}(2, \mathbb{C})$ denote the identity matrix. As $f(Id) = 1$, Theorem 13.18 ensures that there's a unique lift $g \colon \text{SU}(2, \mathbb{C}) \to \mathbb{R}$ such that $\mathrm{e}g = f$ and $g(Id) = 0$. We want to show that g is a homomorphism, i.e. that

$$G \colon \text{SU}(2, \mathbb{C}) \times \text{SU}(2, \mathbb{C}) \to \mathbb{R}, \qquad G(A, B) = g(AB) - g(B) - g(A),$$

is the zero map. As e is a homomorphism,

$$\mathrm{e}G(A, B) = f(AB)f(B)^{-1}f(A)^{-1} = 1,$$

so G is the lift of a constant map. But $\text{SU}(2, \mathbb{C})$ is connected, implying that G is constant and $G(Id, Id) = 0$.

The range of g is compact in $\mathbb{R}$, so bounded; at the same time, for every $A \in \text{SU}(2, \mathbb{C})$ and every $n \in \mathbb{Z}$ we have $g(A^n) = ng(A)$, whence $g(A) = 0$. The latter means g is constant.

Proof (Third proof of Theorem 12.34) Suppose there existed a continuous map $f \colon S^2 \to S^1$ such that $f(-x) = -f(x)$ for every x, and let us show we'll end up with a contradiction.

Write $q \colon S^2 \to \mathbb{P}^2(\mathbb{R})$ and $p \colon S^1 \to \mathbb{P}^1(\mathbb{R})$ for the canonical quotients, then consider the commutative diagram of continuous maps

$$\begin{array}{ccc} S^2 & \xrightarrow{f} & S^1 \\ {\scriptstyle q}\downarrow & & \downarrow{\scriptstyle p} \\ \mathbb{P}^2(\mathbb{R}) & \xrightarrow{g} & \mathbb{P}^1(\mathbb{R}) \end{array} \qquad \text{where} \quad g([x]) = [f(x)].$$

The fundamental group of $\mathbb{P}^2(\mathbb{R})$ is isomorphic to $\mathbb{Z}/2$, while the projective line $\mathbb{P}^1(\mathbb{R})$ is homeomorphic to S^1 and has thus fundamental group isomorphic to $\mathbb{Z}$. As any homomorphism from $\mathbb{Z}/2$ to $\mathbb{Z}$ is trivial, by Theorem 13.18 g lifts to a continuous map $h\colon \mathbb{P}^2(\mathbb{R}) \to S^1$ such that $ph = g$.

Choose $x \in S^2$: then $hq(x) = \pm f(x)$, and $hq(x) = f(x)$ or $hq(-x) = hq(x) = -f(x) = f(-x)$. In either case hq and f agree at one point at least. But being lifts of $pf = gq$, f and hq coincide everywhere, in violation of $f(x) \neq f(-x)$.

Exercises

13.14 Prove that any continuous map $\mathbb{P}^3(\mathbb{R}) \to S^1$ is homotopic to a constant map.

13.15 ($\heartsuit$) Consider a commutative diagram of continuous maps and connected, locally path connected spaces

$$\begin{array}{ccc} A & \xrightarrow{f} & E \\ {\scriptstyle q}\downarrow & & \downarrow{\scriptstyle p} \\ Y & \xrightarrow{F} & X. \end{array}$$

Suppose p is a covering and q induces a surjective homomorphism of the relative fundamental groups. Prove that there exists a continuous map $G\colon Y \to E$ such that $pG = F$, $Gq = f$.

13.16 ($\heartsuit$) Let $\alpha\colon [0, 1] \to \mathbb{C} - \{0\}$ be a loop with base point 1 and such that $\alpha(t) = \alpha(t + 1/2)$ for every $0 \leq t \leq 1/2$. Prove that $[\alpha]$ is an even multiple of the generator of $\pi_1(\mathbb{C} - \{0\}, 1)$.

13.17 Let $\alpha\colon [0, 1] \to \mathbb{C} - \{0\}$ be a loop based at 1 and such that $\alpha(t) = -\alpha(t + 1/2)$ for every $0 \leq t \leq 1/2$. Prove that $[\alpha]$ is an odd multiple of the generator of $\pi_1(\mathbb{C} - \{0\}, 1)$. (Hint: consider the covering $\mathbb{C} - \{0\} \to \mathbb{C} - \{0\}$, $z \mapsto z^2$, and the lift α_1 of α with $\alpha_1(0) = 1$. Prove that either $\alpha_1(t + 1/2) = i\alpha_1(t)$ or $\alpha_1(t + 1/2) = -i\alpha_1(t)$ for every $0 \leq t \leq 1/2$.)

13.18 (Fourth proof of Theorem 12.34) Use Exercise 13.17 to show that there cannot be any continuous map $f\colon D^2 \to S^1$ such that $f(-x) = -f(x)$ for every $x \in S^1$. Deduce Borsuk's theorem.

13.19 Determine all continuous group homomorphisms $S^1 \to S^1$. (Hint: first, classify continuous homomorphisms from $(\mathbb{R}, +)$ to itself and from $(\mathbb{R}, +)$ to S^1.)

13.20 (✋) Let $p\colon E \to G$ be a connected covering of a topological group G with neutral element $1 \in G$. Given a point $e \in E$ such that $p(e) = 1$, prove that there's a unique multiplication $E \times E \to E$ rendering E a topological group with neutral element e and p a group homomorphism. Show, moreover, that E is Abelian if G is Abelian.

13.21 (✋) Let G be a connected, locally path connected, compact group with neutral element e. Prove that every continuous homomorphism $f\colon G \to S^1$ is uniquely determined by the homomorphism of Abelian groups

$$f_*\colon \pi_1(G, e) \to \pi_1(S^1, 1).$$

13.5 Regular Coverings ↻

We continue assuming that all spaces are locally path connected.

Lemma 13.20 *Let X be path connected, $x_1, x_2 \in X$ points and $f\colon X \to Y$ a continuous mapping. There exists an isomorphism $\gamma_\sharp\colon \pi_1(Y, f(x_1)) \to \pi_1(Y, f(x_2))$ such that $\gamma_\sharp(f_*\pi_1(X, x_1)) = f_*\pi_1(X, x_2)$. In particular $f_*\pi_1(X, x_1)$ is a normal subgroup in $\pi_1(Y, f(x_1))$ if and only if $f_*\pi_1(X, x_2)$ is normal in $\pi_1(Y, f(x_2))$.*

Proof Let $\delta\colon [0, 1] \to X$ be a path with $\delta(0) = x_1$, $\delta(1) = x_2$, and write $\gamma = f\delta\colon [0, 1] \to Y$. By Lemma 11.13 we have isomorphisms:

$$\delta_\sharp\colon \pi_1(X, x_1) \to \pi_1(X, x_2), \qquad \delta_\sharp[\beta] = [i(\delta) * \beta * \delta],$$

$$\gamma_\sharp\colon \pi_1(Y, f(x_1)) \to \pi_1(Y, f(x_2)), \qquad \gamma_\sharp[\alpha] = [i(\gamma) * \alpha * \gamma].$$

Then $\gamma_\sharp$ has the requested properties, for if $\beta \in \Omega(X, x_1, x_1)$ then

$$\gamma_\sharp f_*[\beta] = \gamma_\sharp[f\beta] = [i(f\delta) * f\beta * f\delta] = [f(i(\delta) * \beta * \delta)] = f_*\delta_\sharp[\beta]. \qquad \square$$

Definition 13.21 A connected covering $p\colon E \to X$ is **regular** whenever $p_*\pi_1(E, e)$ is a normal subgroup of $\pi_1(X, p(e))$, where e is an arbitrary point in E.

Lemma 13.20 implies that 13.21 is a proper definition.

Example 13.22

1. If E is a simply connected space, any covering $p\colon E \to X$ is regular.
2. If X is connected and $\pi_1(X)$ Abelian, every connected covering $p\colon E \to X$ is regular.

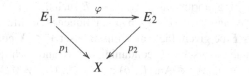

3. Every connected double covering is regular: it's a well-known algebraic fact that any subgroup of index two is normal.
4. The three-fold covering of Sect. 12.6 is not regular.

A consequence of Theorem 13.14 is that if a group G acts properly discontinuously on a connected space E, the covering $E \rightarrow E/G$ is regular. In this section we'll prove that the converse is true, too: regular coverings are precisely those that arise as quotients of properly discontinuous actions.

Definition 13.23 Let $p_1 \colon E_1 \rightarrow X$, $p_2 \colon E_2 \rightarrow X$ be coverings of a space X. A continuous map $\varphi \colon E_1 \rightarrow E_2$ is called a **covering morphism** if

$$
\begin{array}{ccc}
E_1 & \xrightarrow{\;\varphi\;} & E_2 \\
& \searrow{\scriptstyle p_1} \quad \swarrow{\scriptstyle p_2} & \\
& X &
\end{array}
$$

is commutative.

We can also say that covering morphisms $\varphi \colon E_1 \rightarrow E_2$ are lifts of p_1. If E_1 is connected, in particular, two distinct covering morphisms $\varphi, \psi \colon E_1 \rightarrow E_2$ must satisfy $\varphi(e) \neq \psi(e)$ for every $e \in E_1$. The composite of two covering morphisms is a covering morphism.

Having introduced the notion of morphism, it makes sense to talk about the category of coverings of a given space; from this will descend a natural notion of isomorphism.

Definition 13.24 Let $p_1 \colon E_1 \rightarrow X$, $p_2 \colon E_2 \rightarrow X$ be coverings of the same space X. A covering morphism $\varphi \colon E_1 \rightarrow E_2$ is a **covering isomorphism** if there is a covering morphism $\psi \colon E_2 \rightarrow E_1$ such that $\varphi\psi$ and $\psi\varphi$ are the identity morphisms.

Definition 13.25 Denote by $\mathrm{Aut}(E, p)$ the group of automorphisms of a covering $p \colon E \rightarrow X$, i.e. the set of covering isomorphisms of $\varphi \colon E \rightarrow E$ equipped with the composition product.

The group $\mathrm{Aut}(E, p)$ acts on E: as p is constant on orbits, the universal property of quotients provides us with the factorisation

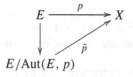

where $\tilde{p}$ is open and onto. The consequence is that $\tilde{p}$ is a homeomorphism if and only if $\mathrm{Aut}(E, p)$ acts transitively on the fibres of p.

Theorem 13.26 *Let $p\colon E \to X$ be a connected covering.*

1. *Given $e_1, e_2 \in E$, there exists an automorphism $\varphi \in \mathrm{Aut}(E, p)$ such that $\varphi(e_1) = e_2$ if and only if*

$$p(e_1) = p(e_2), \qquad p_*\pi_1(E, e_1) = p_*\pi_1(E, e_2).$$

2. $\mathrm{Aut}(E, p)$ *acts on E properly discontinuously.*

Proof (1) If there is an automorphism $\varphi \in \mathrm{Aut}(E, p)$ such that $\varphi(e_1) = e_2$, then $p(e_2) = p\varphi(e_1) = p(e_1)$ and $p_*\pi_1(E, e_2) = p_*\varphi_*\pi_1(E, e_1) = p_*\pi_1(E, e_1)$.

Conversely, if $p(e_1) = p(e_2)$ and $p_*\pi_1(E, e_1) = p_*\pi_1(E, e_2)$, by Theorem 13.18 there exist covering morphisms $\varphi, \psi\colon E \to E$ such that $\varphi(e_1) = e_2$ and $\psi(e_2) = e_1$. As $\psi\varphi(e_1) = e_1$ and $\varphi\psi(e_2) = e_2$, by the uniqueness of lifts $\varphi\psi = \psi\varphi = Id$.

(2) Let $e \in E$ be given. Pick an admissible set $V \subset X$ containing $p(e)$ and call $U \subset p^{-1}(V)$ an open set in E containing $e \in U$ and such that $p\colon U \to V$ is 1-1. We want to prove that $\varphi \in \mathrm{Aut}(E, p)$ and $U \cap \varphi(U) \neq \emptyset$ imply $\varphi = Id$. Choose $u \in U \cap \varphi(U)$, so $u, \varphi^{-1}(u) \in U$; then $u = \varphi^{-1}(u)$ as p is 1-1 on U. Therefore $\varphi = Id$ because lifts are unique. $\qquad\square$

Corollary 13.27 *A connected covering $p\colon E \to X$ is regular if and only if $\mathrm{Aut}(E, p)$ acts transitively on the fibres of p. In that case $E/\mathrm{Aut}(E, p) \simeq X$, and for every $e \in E$ there's an isomorphism*

$$\frac{\pi_1(X, p(e))}{p_*\pi_1(E, e)} \simeq \mathrm{Aut}(E, p).$$

Proof The first part descends directly from Theorem 13.26 and the fact that for any given $e_1 \in E$, the conjugate subgroups to $p_*\pi_1(E, e_1)$ are exactly $p_*\pi_1(E, e_2)$, with $e_2 \in E$ such that $p(e_2) = p(e_1)$.

If $\mathrm{Aut}(E, p)$ acts transitively on fibres we know that $E/\mathrm{Aut}(E, p) \simeq X$, and $\dfrac{\pi_1(X, p(e))}{p_*\pi_1(E, e)} \simeq \mathrm{Aut}(E, p)$ follows from Theorem 13.14. $\qquad\square$

Remark 13.28 There's a suggestive analogy between the theories of field extensions and covering spaces, according to which Galois extensions correspond to regular coverings: thus, regular coverings are called **Galois coverings** at times.

Exercises

13.22 Always assuming that every space is locally path connected, prove that a covering morphism is an isomorphism if and only if it is bijective.

13.23 Let $\varphi\colon E_1 \to E_2$ be a covering morphism. Prove that if E_2 is connected, φ is a covering map.

13.24 Given groups $H \subset G$, the **normaliser** of H in G is

$$N(H) = \{g \in G \mid gHg^{-1} = H\}.$$

1. Prove that $N(H)$ is a subgroup of G, H is a normal subgroup in $N(H)$ and $N(H)$ is the largest subgroup of G containing H as a normal subgroup. In particular $N(H) = G$ if and only if H is normal in G.
2. Let $p: E \to X$ be a connected covering and $e \in E$. Then for every homotopy class $[\alpha] \in N(p_*\pi_1(E, e))$ there's a unique automorphism $\varphi \in \text{Aut}(E, p)$ such that $\varphi(e) = e \cdot [\alpha]$. Prove that the map $N(p_*\pi_1(E, e)) \to \text{Aut}(E, p)$ just defined is onto, and descends to an isomorphism

$$\frac{N(p_*\pi_1(E, e))}{p_*\pi_1(E, e)} \simeq \text{Aut}(E, p).$$

13.6 Universal Coverings ∿

Definition 13.29 A covering $u: \widetilde{X} \to X$ is said to be **universal** if the total space $\widetilde{X}$ is connected and simply connected.

We saw in Example 13.22 that universal coverings are regular. In particular, if $u: \widetilde{X} \to X$ is universal then $\text{Aut}(\widetilde{X}, u)$ acts freely and transitively on fibres, and $\pi_1(X) \simeq \text{Aut}(\widetilde{X}, u)$ are isomorphic.

Proposition 13.30 (Universal property of universal coverings) *Let $u: \widetilde{X} \to X$ be a universal covering. For every covering $p: E \to X$ and any points $\tilde{x} \in \widetilde{X}$, $e \in E$ such that $u(\tilde{x}) = p(e)$, there exists a unique covering morphism $\phi: \widetilde{X} \to E$ such that $\phi(\tilde{x}) = e$. In particular, all universal coverings of a space X are isomorphic to one another.*

Proof Since $0 = u_*\pi_1(\widetilde{X}, \tilde{x}) \subset p_*\pi_1(E, e)$, ϕ exists by virtue of Theorem 13.18. Additionally, if $p: E \to X$ is universal then the previous arguments show that there's a covering morphism $\psi: E \to \widetilde{X}$ with $\psi(e) = \tilde{x}$. But $\widetilde{X}$ and E are connected by definition, so the lift's uniqueness forces $\phi\psi$ and $\psi\phi$ to be identity maps. $\square$

This proves the uniqueness of universal coverings. The remaining part of the section is devoted entirely to the issue of existence. We begin with a simple necessary condition.

Definition 13.31 A space X is **semi-locally simply connected** if any point $x \in X$ has a path connected neighbourhood V such that $i_*\pi_1(V, x) = 0$ in $\pi_1(X, x)$, where $i: V \to X$ is the inclusion map.

Hence, a locally path connected space X is semi-locally simply connected if and only if each point lies in an open set V where any loop is homotopically trivial in X.

Lemma 13.32 *Let X be connected and locally path connected. If X admits a universal covering, then X is semi-locally simply connected.*

Proof Let $u \colon \widetilde{X} \to X$ denote the universal covering, and $U \subset X$ an admissible set. Call $i \colon U \to X$ the inclusion. Since we have a continuous section $s \colon U \to \widetilde{X}$,

$$i_* \pi_1(U, x) = u_* s_* \pi_1(U, x) \subset u_* \pi_1(\widetilde{X}, s(x)) = 0$$

for every $x \in U$. $\square$

Example 13.33 There are connected and locally path connected space that aren't semi-locally simply connected, and hence don't admit a universal covering. One such example is

$$X = \mathbb{R}^2 - \{(2^{-n}, 0) \mid n \in \mathbb{N}\}.$$

Take a neighbourhood $U \subset X$ of 0, fix an irrational number $r > 0$ such that

$$\{x \in X \mid \|x\| \le 2r\} \subset U$$

and choose an integer N satisfying $2^{-N} < 2r$. Consider

$$Y = \{x \in \mathbb{R}^2 \mid \|x - (r, 0)\| = r\}, \qquad Z = \mathbb{R}^2 - \{(2^{-N}, 0)\}.$$

The inclusions $0 \in Y \subset U \subset X \subset Z$ induce an isomorphism $\pi_1(Y, 0) \to \pi_1(Z, 0)$, so the morphism $\pi_1(U, 0) \to \pi_1(X, 0)$ cannot be trivial.

Let $u \colon \widetilde{X} \to X$ be the universal covering of X. Any point $\tilde{x} \in \widetilde{X}$ determines a bijective map

$$\Phi \colon \widetilde{X} \xrightarrow{\;\simeq\;} \bigcup_{y \in X} \pi(X, x, y),$$

where $x = u(\tilde{x})$ and $\pi(X, x, y) = \Omega(X, x, y)/ \sim$ are the homotopy classes of paths joining x and y. Given $\tilde{y} \in \widetilde{X}$, in fact, there is a unique class $\xi \in \pi(\widetilde{X}, \tilde{x}, \tilde{y})$, whence $\Phi(\tilde{y}) = u_*(\xi)$ is well defined. The inverse to Φ is the covering's monodromy map.

Theorem 13.34 *A connected and locally path connected space admits a universal covering if and only if it is semi-locally simply connected.*

Proof One implication follows from Lemma 13.32, so we'll prove that a connected, locally path connected and semi-locally simply connected space X admits a universal covering $u \colon \widetilde{X} \to X$.

The previous argument suggests we choose a point $x \in X$ and consider the set

$$\widetilde{X} = \bigcup_{y \in X} \pi(X, x, y), \qquad \text{where} \quad \pi(X, x, y) = \frac{\Omega(X, x, y)}{\text{path homotopies}}.$$

Path homotopies preserve endpoints, so

$$u: \widetilde{X} \to X, \qquad u([\alpha]) = \alpha(1)$$

is a well-defined onto map. There is a natural 1-1 correspondence between $u^{-1}(x)$ and $\pi_1(X, x)$. Let's define a topology on $\widetilde{X}$ making u a universal covering. Given $\alpha \in \Omega(X, x, y)$ and an open neighbourhood $U \ni y$, define $W(\alpha, U) \subset \widetilde{X}$ to be homotopy classes of paths of the form $\alpha * \alpha'$, with α' in U starting at y. Clearly $W(\alpha, U) = W(\alpha, U')$, where U' is the connected component of U containing $\alpha(1)$. Moreover, if $\beta \in W(\alpha, U)$ and $V \subset U$ open contains $\beta(1)$, then $W(\beta, V) \subset W(\alpha, U)$.

The subsets $W(\alpha, U)$ form a basis for a topology on the set $\widetilde{X}$, because any $[\alpha] \in \pi(X, x, y)$ belongs in $W(\alpha, X)$, and if $[\gamma] \in W(\alpha, U) \cap W(\beta, V)$, then $\gamma(1) \in U \cap V$ and also

$$[\gamma] \in W(\gamma, U \cap V) \subset W(\alpha, U) \cap W(\beta, V).$$

In this topology $u: \widetilde{X} \to X$ is continuous and open, since

$$u^{-1}(U) = \bigcup_{\alpha(1) \in U} W(\alpha, U), \qquad u(W(\alpha, U)) = U$$

for every open connected $U \subset X$.

Suppose now $U \subset X$ is open, connected, and such that any loop in it is homotopically trivial in X. Fix a path $\alpha: I \to X$ starting at x and such that $\alpha(1) = u([\alpha]) \in U$. The projection $u: W(\alpha, U) \to U$ is a homeomorphism: if $u([\alpha * \beta]) = u([\alpha * \gamma])$ then $i(\beta) * \gamma$ is a loop in U, so β and γ are homotopic in X and $[\alpha * \beta] = [\alpha * \gamma]$ in $\widetilde{X}$, i.e. u is one-to-one on $W(\alpha, U)$.

Given α, β in X with initial point x and endpoint in U, we have $W(\alpha, U) = W(\beta, U)$ or $W(\alpha, U) \cap W(\beta, U) = \emptyset$. That's true because if γ in X satisfies $[\gamma] \in W(\alpha, U) \cap W(\beta, U)$, then $[\alpha] = [\gamma * \alpha']$, $[\beta] = [\gamma * \beta']$, and further, $[\beta] = [\alpha * i(\alpha') * \beta'] \in W(\alpha, U)$. Therefore $W(\beta, U) \subset W(\alpha, U)$, and by symmetry $W(\alpha, U) \subset W(\beta, U)$.

At this stage we have that u is a covering. Given $\alpha \in \Omega(X, x, y)$, consider

$$\alpha_s: [0, 1] \to X, \qquad \alpha_s(t) = \alpha(ts), \qquad s \in [0, 1],$$

and define

$$\hat{\alpha}: [0, 1] \to \widetilde{X}, \qquad \hat{\alpha}(s) = [\alpha_s] \in \pi(X, x, \alpha(s)).$$

Fix $s \in [0, 1]$ and an open set U containing the point $\alpha(s)$. Then a positive real number ε exists, such that $\alpha(t) \in U$ for every $s - \varepsilon \leq t \leq s + \varepsilon$. By Corollary 11.7 we have $[\alpha_t] \in W(\alpha_s, U)$ for every $s - \varepsilon \leq t \leq s + \varepsilon$. Therefore the map $\hat{\alpha}$ is continuous, and $\widetilde{X}$ is path connected. Notice that $u(\hat{\alpha}(s)) = \alpha(s)$, so $\hat{\alpha}$ is the lift of α

with $\hat{\alpha}(0) = [1_x]$. This eventually proves $[\alpha] = \hat{\alpha}(1) = \text{Mon}([1_x], \alpha)$; in particular the monodromy action of the covering u is free, so $u_*\pi_1(\widetilde{X}, [1_x]) = 0$.   □

Exercises

13.25 Let $p: E \to F$ and $q: F \to X$ be continuous, onto maps between connected spaces. Prove that if p, q are coverings and X has a universal covering, then qp is a covering.

13.26 Prove that every continuous map $X \to Y$ lifts to a continuous map $\widetilde{X} \to \widetilde{Y}$ between the universal coverings (provided these exist).

13.27 Let X be the sphere S^2 with North pole and South pole identified. Determine the universal covering and the fundamental group of X. (Hint: consider the quotient of the sausage space of Exercise 11.22 by some properly discontinuous action.)

13.7 Coverings with Given Monodromy ↻

Consider a covering $p: E \to X$, a point $x \in X$ and the monodromy action

$$p^{-1}(x) \times \pi_1(X, x) \to p^{-1}(x).$$

It's not hard to show that E is connected if and only if the monodromy action is transitive. In fact if E is connected, for every pair $a, b \in p^{-1}(x)$ we can find a path $\alpha \in \Omega(E, a, b)$ and hence $b = a \cdot [p\alpha]$. Conversely, if the monodromy is transitive, the fibre $p^{-1}(x)$ is contained in a path component. Given any point $a \in E$ we choose a path $\alpha: I \to X$ such that $\alpha(0) = p(a)$, $\alpha(1) = x$. The lift $\alpha_a: I \to E$ joins a to some point in $p^{-1}(x)$, so that a belongs in the same connected component where $p^{-1}(x)$ lies.

We saw already, in Theorem 13.1, that the stabiliser of any $e \in p^{-1}(x)$, i.e. the subgroup

$$Stab(e) = \{a \in \pi_1(X, x) \mid e \cdot a = e\},$$

coincides with $p_*\pi_1(E, e)$. In particular the covering $p: E \to X$ is universal if and only if the monodromy is free and transitive. Moreover, the covering is regular if and only if the monodromy acts transitively and all stabilisers are normal subgroups.

Theorem 13.35 *Let X be connected, locally path connected and semi-locally simply connected. For every non-empty set T and every right action*

$$T \times \pi_1(X, x) \xrightarrow{\ \bullet\ } T$$

there exists a covering $p: E \to X$ *and a bijection* $\phi: T \to p^{-1}(x)$ *such that* $\phi(t \bullet a) = \phi(t) \cdot a$, *for every* $t \in T$ *and* $a \in \pi_1(X, x)$. *The pair* (p, ϕ) *is unique up to isomorphism.*

Before we go to the proof let's clarify the last statement. The pair (p, ϕ) being unique up to isomorphism means that if $q: F \to X$ and $\psi: T \to q^{-1}(x)$ are another covering-bijection pair, there is a covering isomorphism $E \to F$ extending $\psi\phi^{-1}$.

Proof We'll begin with uniqueness; the argument itself will eventually show existence as well.

So let $p: E \to X$ be a covering and $\phi: T \to p^{-1}(x)$ a bijection such that $\phi(t \bullet a) = \phi(t) \cdot a$ for every $t \in T, a \in \pi_1(X, x)$. Write $u: \widetilde{X} \to X$ for the universal covering and fix a point $\tilde{x} \in u^{-1}(x)$. By Proposition 13.30, for every $e \in p^{-1}(x)$ there is a unique covering morphism $\eta_e: \widetilde{X} \to E$ such that $\eta_e(\tilde{x}) = e$. We also have an isomorphism

$$\theta: \mathrm{Aut}(\widetilde{X}, u) \to \pi_1(X, x), \qquad \tilde{x} \cdot \theta(g) = g(\tilde{x}).$$

The monodromy commutes with covering morphisms, so for any $e \in p^{-1}(x)$ and $g \in \mathrm{Aut}(\widetilde{X}, u)$ we obtain

$$e \cdot \theta(g) = \eta_e(\tilde{x}) \cdot \theta(g) = \eta_e(\tilde{x} \cdot \theta(g)) = \eta_e(g(\tilde{x}))$$

whence $\eta_e g = \eta_{e \cdot \theta(g)}$. Put on T the discrete topology and define $q: T \times \widetilde{X} \to X$ as the composite of u with the projection $T \times \widetilde{X} \to \widetilde{X}$. Then

$$\eta: T \times \widetilde{X} \to E, \qquad \eta(t, \tilde{y}) = \eta_{\phi(t)}(\tilde{y})$$

is a covering morphism of X. Note that η is onto, and for every $g \in \mathrm{Aut}(\widetilde{X}, u)$

$$\eta(t, g(\tilde{y})) = \eta_{\phi(t)}(g(\tilde{y})) = \eta_{\phi(t) \cdot \theta(g)}(\tilde{y}) = \eta(t \bullet \theta(g), \tilde{y}).$$

Therefore we can view $\mathrm{Aut}(\widetilde{X}, u)$ as a subgroup of homeomorphisms of $T \times \widetilde{X}$ that commute with η, just by mapping $g \in \mathrm{Aut}(\widetilde{X}, u)$ to the function

$$T \times \widetilde{X} \to T \times \widetilde{X}, \qquad (t, \tilde{y}) \mapsto (t \bullet \theta(g)^{-1}, g(\tilde{y})).$$

The group $\mathrm{Aut}(\widetilde{X}, u)$ acts properly discontinuously on $\widetilde{X}$, so it also acts in a properly discontinuous way on $T \times \widetilde{X}$. The action of $\mathrm{Aut}(\widetilde{X}, u)$ on the fibre $q^{-1}(x) = T \times u^{-1}(x)$ is free, and every orbit intersects $T \times \{\tilde{x}\}$ in one point. On the other hand $\eta(t, \tilde{x}) = \phi(t)$, so η induces on quotients a bijection

$$q^{-1}(x)/\mathrm{Aut}(\widetilde{X}, u) \to p^{-1}(x).$$

Consequently η descends to the quotient to give an isomorphism

$$\frac{T \times \widetilde{X}}{\text{Aut}(\widetilde{X}, u)} \simeq E.$$

The covering $h : T \times \widetilde{X}/\text{Aut}(\widetilde{X}, u) \to X$ was built using only X and the action $\bullet$, without any mention to E. This proves uniqueness, and at the same time existence. □

Corollary 13.36 *Let X be connected, locally path connected and semi-locally simply connected. Given $x \in X$ and a subgroup $H \subset \pi_1(X, x)$, there exist a covering $p : E \to X$ and a point $e \in p^{-1}(x)$ such that $p_*\pi_1(E, e) = H$.*

Proof Let $T = \{Ha \mid a \in \pi_1(X, x)\}$ denote right H-cosets. There is an obvious right action

$$T \times \pi_1(X, x) \xrightarrow{\;\bullet\;} T, \qquad Ha \bullet b = Hab,$$

and by Theorem 13.35 also a covering $p : E \to X$ such that $p^{-1}(x) = T$, whereby $\bullet$ is the monodromy action. If we define $e \in T \subset E$ to be the point corresponding to the trivial coset, i.e. $e = H$,

$$p_*\pi_1(E, e) = \{a \in \pi_1(X, x) \mid H \bullet a = H\} = H$$

follows. □

Exercises

13.28 Let $p : E \to X$ be a connected covering, $A \subset X$ an open subset and $e \in p^{-1}(A)$ a point. Prove that $p^{-1}(A)$ is connected if and only if the range of the map $\pi_1(A, p(e)) \to \pi_1(X, p(e))$ intersects every right coset of $p_*\pi_1(E, e)$.

13.29 Let X be connected, locally connected and semi-locally simply connected. Suppose X contains a retract homeomorphic to S^1. Prove that X admits regular connected coverings of degree d, for any integer $d > 0$.

13.30 (☙, ♡) Consider the topological group G of real unipotent 3×3-matrices (matrices having 1s on the diagonal and 0s below), and the discrete subgroup Γ of matrices with integer entries. Is G/Γ or some covering space of it homeomorphic to the product $(S^1)^3$ of three circles? Explain your answer.

Chapter 14
van Kampen's Theorem

The notation for this chapter will be as follows: if G is a group and $S \subset G$ a subset we will write $\langle\langle S \rangle\rangle \subset G$ or $\langle\langle s \mid s \in S \rangle\rangle \subset G$ for the **normal** subgroup generated by S, which is the intersection of all normal subgroups in G containing S. It's easy to show that $\langle\langle S \rangle\rangle$ coincides with the subgroup generated by all elements of the type gsg^{-1}, for $s \in S$ and $g \in G$.

14.1 van Kampen's Theorem, Universal Version

Let A, B be open in a space X with $X = A \cup B$; suppose A, B and $A \cap B$ are **path connected**, $A \cap B \neq \emptyset$ and then fix a point $x_0 \in A \cap B$. The inclusions $A \subset X$, $B \subset X$, $A \cap B \subset A$ and $A \cap B \subset B$ induce a commutative diagram of group homomorphisms:

$$
\begin{array}{ccc}
\pi_1(A \cap B, x_0) & \xrightarrow{\ \alpha_* \ } & \pi_1(A, x_0) \\
\Big\downarrow{\beta_*} & & \Big\downarrow{f_*} \\
\pi_1(B, x_0) & \xrightarrow[\ g_* \ ]{} & \pi_1(X, x_0) .
\end{array}
$$

We showed in Sect. 11.4 that when A and B are simply connected, then also X is simply connected. If we pay a little attention it's not hard to see that the argument of Theorem 11.25 implies that if α_* is onto then g_* is onto.

In general, van Kampen's theorem asserts that the fundamental group of X is determined, up to isomorphism, by the fundamental groups of A, B, $A \cap B$ and the homomorphisms α_*, β_*. In a convenient formulation of the theorem $\pi_1(X, x_0)$ is the solution to a universal problem. Later we'll give an equivalent description in terms of free products of groups (Fig. 14.1).

Theorem 14.1 (van Kampen, universal version) *Retaining the previous setup, for every group G and every pair of homomorphisms $h : \pi_1(A, x_0) \to G, k : \pi_1(B, x_0) \to$*

© Springer International Publishing Switzerland 2015
M. Manetti, *Topology*, UNITEXT - La Matematica per il 3+2 91,
DOI 10.1007/978-3-319-16958-3_14

G such that $h\alpha_* = k\beta_*$, there exists a unique homomorphism $\varphi\colon \pi_1(X, x_0) \to G$ making the diagram

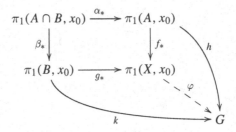

commutative.

Proof (partial) The proof of Theorem 11.25 told us that any element in $\Omega(X, x_0, x_0)$ is path homotopic to the product of a finite number of loops contained either in A or in B. This means that for any $\gamma \in \pi_1(X, x_0)$ we may find two finite sequences $a_1, \ldots, a_n \in \pi_1(A, x_0)$, $b_1, \ldots, b_n \in \pi_1(B, x_0)$ such that

$$\gamma = f_*(a_1)g_*(b_1) \cdots f_*(a_n)g_*(b_n).$$

If we are seeking φ as in the statement, we must require

$$\varphi(\gamma) = h(a_1)k(b_1) \cdots h(a_n)k(b_n).$$

We've thus settled the uniqueness of φ and also provided a construction for it; now we should check the definition of φ makes sense, i.e. that $\varphi(\gamma)$ shouldn't depend on the a_i, b_j. This is the complicate part of the proof, for which we refer the reader to either Massey [Ma67, Ma91] or Vick [Vi94]. But it's the consequences of van Kampen's theorem, as for Zorn's lemma, that are far more telling than the actual proof.

In Sect. 15.7 we'll show that Theorem 14.1 descends easily from Theorem 13.35, under mild additional hypotheses.

Corollary 14.2 *Let's keep the previous notions and suppose the homomorphism* $\beta_*\colon \pi_1(A \cap B, x_0) \to \pi_1(B, x_0)$ *is onto. Then the homomorphism* $f_*\colon \pi_1(A, x_0) \to \pi_1(X, x_0)$ *is onto and has as kernel the normal subgroup generated by* $\alpha_*(\ker \beta_*)$:

$$\pi_1(X, x_0) \cong \frac{\pi_1(A, x_0)}{\langle \alpha_*(\ker \beta_*) \rangle}.$$

Proof Call $h\colon \pi_1(A, x_0) \to \dfrac{\pi_1(A, x_0)}{\langle \alpha_*(\ker \beta_*) \rangle}$ the quotient map. The homomorphism $h\alpha_*$ kills the kernel of β_*, and so there is a homomorphism

$$k\colon \pi_1(B, x_0) \simeq \frac{\pi_1(A \cap B, x_0)}{\ker \beta_*} \to \frac{\pi_1(A, x_0)}{\langle \alpha_*(\ker \beta_*) \rangle}$$

such that $k\beta_* = h\alpha_*$. By Theorem 14.1 there is a unique homomorphism $\varphi \colon \pi_1(X, x_0)$ $\to \dfrac{\pi_1(A, x_0)}{\langle \alpha_*(\ker \beta_*) \rangle}$ rendering

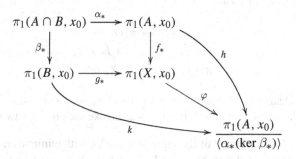

a commutative diagram. As h is onto, also φ is onto. But $f_*\alpha_* = g_*\beta_*$, so the kernel of $f_*\alpha_*$ contains the kernel of β_* and then $\alpha_*(\ker \beta_*) \subset \ker f_*$. Consequently f_* factorises through a homomorphism $\psi \colon \dfrac{\pi_1(A, x_0)}{\langle \alpha_*(\ker \beta_*) \rangle} \to \pi_1(X, x_0)$. Hence $f_* = \psi h$,

$$\psi k \beta_* = \psi h \alpha_* = f_* \alpha_* = g_* \beta_*$$

and the surjectivity of β_* implies $\psi k = g_*$. That is to say,

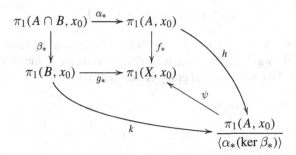

commutes. By uniqueness the composite map $\psi\varphi$ must be the identity, proving that φ is 1-1.

Corollary 14.3 *With the above notations, if* $\beta_* \colon \pi_1(A \cap B, x_0) \to \pi_1(B, x_0)$ *is an isomorphism, then* $f_* \colon \pi_1(A, x_0) \to \pi_1(X, x_0)$ *is an isomorphism.*

Example 14.4 Let $X \subset \mathbb{R}^n$ be open, connected and $p, q \in X$ distinct points. The homomorphism $f_* \colon \pi_1(X - \{p\}, q) \to \pi_1(X, q)$ induced by the inclusion $X - \{p\} \subset X$ is:

1. onto for $n = 2$;
2. one-to-one and onto for $n \geq 3$.

For the purposes of the proof consider a positive real number r such that the ball $B = \{x \in \mathbb{R}^n \mid \|x - p\| < r\}$ is contained in X. Set $A = X - \{p\}$. If $n \geq 2$ the sets $A, B, A \cap B$ are connected and $X = A \cup B$. Choose $x_0 \in A \cap B$ and a path $\gamma \in \Omega(X - \{p\}, q, x_0)$. Then

$$\begin{array}{ccc} \pi_1(A, q) & \xrightarrow{\gamma_\sharp} & \pi_1(A, x_0) \\ \downarrow f_* & & \downarrow f_* \\ \pi_1(X, q) & \xrightarrow{\gamma_\sharp} & \pi_1(X, x_0) \end{array}$$

commutes, and since the homomorphism $\pi_1(A \cap B, x_0) \to \pi_1(B, x_0)$ is surjective for $n \geq 2$ and bijective for $n \geq 3$, it suffices to invoke the previous two corollaries.

Example 14.5 The argument of Example 14.4 works, with minimal changes that we leave to the reader, in the more general case where X is a connected topological manifold of dimension n.

Example 14.6 Let $X \subset \mathbb{R}^3$ be the complement of a circle. Example 1.15 shows that X is homeomorphic to $\mathbb{R}^3 - (\text{line} \cup \text{point})$. By Example 14.4 the inclusion

$$\mathbb{R}^3 - (\text{line} \cup \text{point}) \subset \mathbb{R}^3 - (\text{line})$$

induces an isomorphism of the fundamental groups, so

$$\pi_1(X) \simeq \pi_1(\mathbb{R}^3 - (\text{line})) .$$

On the other hand the complement of a line arises also by rotating an open half-plane about the boundary axis, so the space $\mathbb{R}^3 - (\text{line})$ is homeomorphic to $S^1 \times \{(x, y) \in \mathbb{R}^2 \mid y > 0\}$, and the fundamental group is isomorphic to $\mathbb{Z}$.

Example 14.7 The isomorphism

$$\pi_1(\mathbb{R}^3 - (\text{circle})) \cong \mathbb{Z}$$

can be proved using Corollary 14.3, directly. Suppose K is the circle of equation

$$y = 0, \qquad (x - 4)^2 + z^2 = 1,$$

in $\mathbb{R}^3$ and write $\mathbb{R}^3 - K = A \cup B$, where:

1. B is the complement of the closed disc bounded by K.

$$B = \mathbb{R}^3 - \{(x, y, z) \mid y = 0, \ (x - 4)^2 + z^2 \leq 1\};$$

2. A is the solid torus of revolution obtained by rotating about the z-axis the open ball centred at $(4, 0)$ with radius 1, i.e.

$$A = \{(r - 4)^2 + z^2 < 1\}, \qquad x = r\cos(\theta), \ y = r\sin(\theta)$$

Fig. 14.1 Exercise 14.2: what's the fundamental group?

in cylindrical coordinates (r, θ, z).

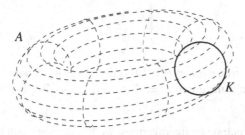

Clearly, B has the homotopy type of S^2 so it's simply connected, while $A \cap B$ is contractible. Hence by Corollary 14.3 the inclusion $A \subset \mathbb{R}^3 - K$ induces an isomorphism of the fundamental groups.

Exercises

14.1 ($\heartsuit$) Compute the fundamental group of the complement in $\mathbb{R}^3$ of the set

$$\{(x, y, z) \mid y = 0, \ x^2 + z^2 = 1\} \cup \{(x, y, z) \mid y = z = 0, \ x \geq 1\}.$$

14.2 Compute the fundamental group of the union of three mutually tangent spheres S^2 (Fig. 14.1).

14.3 Compute the fundamental group of the complement of

$$\{(x, y, z) \mid y = 0, \ x^2 + z^2 = 1\} \cup \{(x, y, z) \mid z = 0, \ (x - 1)^2 + y^2 = 1\}$$

in $\mathbb{R}^3$.

14.4 Compute the fundamental group of the union of a sphere $S^2 \subset \mathbb{R}^3$ centred at the origin with a finite number of planes through its centre.

14.2 Free Groups

Definition 14.8 Let S be a set. A **free group generated by** S is a group F together with a map $\phi\colon S \to F$ with the following universal property: for every group H and every map $\psi\colon S \to H$ there exists a unique homomorphism $\eta\colon F \to H$ such that $\psi = \eta\phi$.

$$
\begin{array}{ccc}
S & \xrightarrow{\ \psi\ } & H \\
{\scriptstyle\phi}\big\downarrow & \nearrow & \\
& {\scriptstyle \exists!\,\eta} & \\
F & &
\end{array}
$$

Example 14.9 The trivial group $F = \{0\}$ is the free group generated by $\emptyset$. If $S = \{*\}$ has one element only, the group $F = \mathbb{Z}$ with $\phi\colon \{*\} \to \mathbb{Z}$, $\phi(*) = 1$, defines a free group generated by S.

In order to concentrate on their features, we'll prove the existence of free groups in a while, and for the moment take it for granted.

Lemma 14.10 *Let* $\phi\colon S \to F$ *be a free group generated by* S. *The map* ϕ *is 1-1 and its range* $\phi(S)$ *generates* F.

Proof Let $a, b \in S$ be distinct elements and consider a map $\psi\colon S \to \mathbb{Z}$ such that $\psi(a) = 0, \psi(b) = 1$. By definition of free group there's a homomorphism $\eta\colon F \to \mathbb{Z}$ such that $\psi = \eta\phi$, and this implies $\phi(a) \neq \phi(b)$.

Write $G \subset F$ for the subgroup generated by $\phi(S)$ and $i\colon G \to F$ for the inclusion morphism; the claim is that i is onto. As $\phi(S) \subset G$, we can decompose $\phi = i\psi$ for some $\psi\colon S \to G$, and by universality there is only one homomorphism $\eta\colon F \to G$ such that $\eta\phi = \psi$. Consider now the homomorphisms

$$\mathrm{Id}\colon F \to F, \qquad i\eta\colon F \to F.$$

Since $\mathrm{Id}\,\phi = i\eta\phi = \phi$, by uniqueness they must be the same, i.e. $\mathrm{Id} = i\eta$, so i is onto.

Lemma 14.11 *Free groups generated by the same set are canonically isomorphic. More precisely, if* $\phi\colon S \to F$ *and* $\phi'\colon S \to F'$ *are free groups generated by* S, *there exists a unique isomorphism* $\eta\colon F \to F'$ *such that* $\eta\phi = \phi'$.

Proof Because $\phi\colon S \to F$ is a free group, there's a homomorphism $\eta\colon F \to F'$ such that $\eta\phi = \phi'$, and similarly, $\phi'\colon S \to F'$ is free so there's $\eta'\colon F' \to F$ such that $\eta'\phi' = \phi$. Hence $\eta\eta'\phi' = \phi'$, and by uniqueness $\eta\eta' = \mathrm{Id}$. In the same way $\eta'\eta = \mathrm{Id}$ can be proved, so η is an isomorphism.

Lemma 14.12 *Let* F *and* G *be the free groups generated by two sets* S *and* T. *If at least one of* S, T *is finite and* F *is isomorphic to* G, *then* S *and* T *have the same cardinality.*

Proof By definition there is a natural bijective map between the set of maps $S \to \mathbb{Z}/2$, of cardinality $2^{|S|}$, and homomorphisms $F \to \mathbb{Z}/2$. Hence the free group F determines $2^{|S|}$, and for the same reasons the free group G determines $2^{|T|}$. If F and G are isomorphic, then $2^{|S|} = 2^{|T|}$.

Remark 14.13 Lemma 14.12 holds even when S, T are both infinite, but this case requires a different proof, see Exercise 14.8.

Theorem 14.14 *For any set there exists a free group generated by it.*

Proof The argument presented here is not the simplest available, but has the advantage of describing the free group explicitly. Given any S we will construct a larger set $S \subset F_S$ and a group structure on F_S; this F_S, together the inclusion $S \hookrightarrow F_S$, will satisfy the universal property of Definition 14.8.

We define F_S to consist of reduced words on the alphabet $S \cup S^{-1}$. This means that F_S is made by the empty word 1 and all finite expressions of the type

$$s_1^{a_1} s_2^{a_2} \cdots s_n^{a_n}$$

where $s_i \in S$, $a_i \in \mathbb{Z} - \{0\}$ and $s_i \neq s_{i+1}$ for every i. For example, if $S = \{a, b\}$ the set F_S consists of $a, a^2, ab, aba, ab^2a^{-1}b, \ldots$; on the other hand the words $aa^{-1}, aab^2, \ldots$, not being reduced, don't belong in F_S.

The group structure on F_S is defined as follows: first, 1 is the neutral element. The inversion is defined through

$$(s_1^{a_1} s_2^{a_2} \cdots s_n^{a_n})^{-1} = s_n^{-a_n} \cdots s_2^{-a_2} s_1^{-a_1},$$

while the multiplication is given by the recursive rule

$$(s_1^{a_1} \cdots s_n^{a_n})(t_1^{b_1} \cdots t_m^{b_m}) = \begin{cases} s_1^{a_1} \cdots s_n^{a_n} t_1^{b_1} \cdots t_m^{b_m} & \text{if } s_n \neq t_1, \\ s_1^{a_1} \cdots s_n^{a_n + b_1} t_2^{b_2} \cdots t_m^{b_m} & \text{if } s_n = t_1, \ a_n + b_1 \neq 0, \\ (s_1^{a_1} \cdots s_{n-1}^{a_{n-1}})(t_2^{b_2} \cdots t_m^{b_m}) & \text{if } s_n = t_1, \ a_n + b_1 = 0. \end{cases}$$

It's not hard to understand that F_S is a group, at least at an intuitive level. As a matter of fact, the rigorous proof for the associativity is rather boring, so several algebra books adopt van der Waerden's trick (Exercise 14.6), or use an argument of Artin [Ar47].

It is often useful to reason by induction on the length of words. The length $l(u)$ of a reduced word $u \in F_S$ is defined by the formula

$$l(1) = 0, \qquad l(s_1^{a_1} \cdots s_n^{a_n}) = |a_1| + \cdots + |a_n|.$$

Clearly S generates F_S. Moreover, any map $\psi \colon S \to G$ with values in a group G extends to a homomorphism

$$\eta\colon F_S \to G, \qquad \eta(s_1^{a_1} s_2^{a_2} \cdots s_n^{a_n}) = \psi(s_1)^{a_1}\psi(s_2)^{a_2} \cdots \psi(s_n)^{a_n}. \qquad\qquad \square$$

Example 14.15 The free group generated by two elements a, b is infinite, although there are only finitely many elements of given length.

Example 14.16 Let $G \subset SO(3, \mathbb{R})$ be the subgroup generated by matrices

$$A = \frac{1}{3} \begin{pmatrix} 1 & -2\sqrt{2} & 0 \\ 2\sqrt{2} & 1 & 0 \\ 0 & 0 & 3 \end{pmatrix}, \qquad B = \frac{1}{3} \begin{pmatrix} 3 & 0 & 0 \\ 0 & 1 & -2\sqrt{2} \\ 0 & 2\sqrt{2} & 1 \end{pmatrix}.$$

It's a free group generated by A, B; equivalently, the induced homomorphism $F_{\{A,B\}} \to SO(3, \mathbb{R})$ is 1-1. We will only sketch the rough idea, and let the interested reader work out the details.

Let $g = g_1 g_2 \cdots g_k$ be a reduced word of length $k \geq 2$ where every g_i belongs to $\{A, A^{-1}, B, B^{-1}\}$ and $g_i \neq g_{i+1}^{-1}$. As A and B are similar matrices, to prove that g is not the identity it suffices to show that if $g_k = A^{\pm 1}$, then $g(1, 0, 0) = (a, b\sqrt{2}, c)/3^k$ with a, b, c integer and b not divisible by 3. By induction on k

$$g_3 g_4 \cdots g_k(1, 0, 0) = \frac{(a'', b''\sqrt{2}, c'')}{3^{k-2}}, \qquad g_2 g_3 \cdots g_k(1, 0, 0) = \frac{(a', b'\sqrt{2}, c')}{3^{k-1}}$$

with b' not divisible by 3. There are 12 conditions to verify since $g_1 g_2$ can be one of

$$A^{\pm 2}, \quad B^{\pm 2}, \quad A^{\pm 1} B^{\pm 1}, \quad B^{\pm 1} A^{\pm 1}.$$

Here are two cases in detail (the remaining ten are completely similar). If $g_1 = g_2 = A$ then

$$b = 2a' + b' = 4a'' - 7b'', \quad b' = 2a'' + b'';$$

eliminating a'' yields $b = 2b' - 9b''$, and as 3 is not a factor in b', also b is not divisible by 3.

If $g_1 = B$, $g_2 = A$ then $b = b' - 2c'$ and $c' = 3c''$, hence $b = b' - 6c''$.

Exercises

14.5 ($\heartsuit$) Let F_S be the free group generated by a set S. Given elements $a \neq b$ in S, prove that $ab \neq ba$.

14.6 Given a set S define $A = S \cup S^{-1}$, where S^{-1} denotes the set of formal inverses of S. Write $P(A)$ for the set of finite reduced sequences $a_1, a_2, \ldots, a_n$ in A: reduced means that $a_{i-1}, a_{i+1} \neq s^{-1}$ in case $a_i = s \in S$. Note that $P(A)$ contains the empty sequence $\emptyset$. Let Σ be the group of bijections $\sigma\colon P(A) \to P(A)$ and $\phi\colon S \to \Sigma$ the injective map defined by the rules

$$\phi(s)(\emptyset) = s, \qquad \phi(s)(a_1, a_2, \ldots) = \begin{cases} s, a_1, a_2, \ldots & \text{if } a_1 \neq s^{-1}, \\ \\ a_2, a_3, \ldots & \text{if } a_1 = s^{-1}. \end{cases}$$

Check that ϕ is well defined and 1-1, then prove that the subgroup of Σ generated by $\phi(S)$ is in a natural one-to-one correspondence with F_S.

14.7 Show that a free group cannot contain elements of finite order except 1. (Hint: write $u^p = 1$ and consider the conjugate of u with smallest length.)

14.8 Given a group G write $G(x^2) \subset G$ to denote the subgroup generated by $\{a^2 \mid a \in G\}$. Prove that:

1. $G(x^2)$ is normal in G, the quotient $G/G(x^2)$ is Abelian and every coset has order 2;
2. if $\phi: G \to H$ is a homomorphism, $\phi(G(x^2)) \subset H(x^2)$. In particular, G isomorphic to H implies $G/G(x^2)$ isomorphic to $H/H(x^2)$;
3. $G/G(x^2)$ has a natural structure of $\mathbb{Z}/2$-vector space and its dimension (cardinality of a basis) depends only on the isomorphism class of G;
4. let F_S be the free group generated by S. The map $S \to F_S/F_S(x^2)$ is 1-1 and its image is a basis of a vector space over $\mathbb{Z}/2$.

14.9 (Hausdorff paradox) Follow the outline given and show that the axiom of choice implies the existence of a countable subset $D \subset S^2$, four disjoint sets $M_1, M_2, M_3, M_4 \subset S^2 - D$ and two rotations $A, B \in SO(3, \mathbb{R})$ such that

$$M_1 \cap A(M_2) = \emptyset, \qquad M_1 \cup A(M_2) = S^2 - D,$$

$$M_3 \cap B(M_4) = \emptyset, \qquad M_3 \cup B(M_4) = S^2 - D.$$

Let A, B be the transformations given in Example 14.16 and call G the free group they generate. For any $\phi \in \{A, A^{-1}, B, B^{-1}\}$ we indicate with G_ϕ the subset of elements in G corresponding to reduced words starting with ϕ. We have three partitions:

1. $G = \{Id\} \cup G_A \cup G_{A^{-1}} \cup G_B \cup G_{B^{-1}}$;
2. $G = G_A \cup AG_{A^{-1}}$;
3. $G = G_B \cup BG_{B^{-1}}$.

Let $D \subset S^2$ consists of the intersection points of S^2 with the axes of the rotations in G. The set D is countable and G acts freely on $S^2 - D$. By the axiom of choice there exists $K \subset S^2 - D$ meeting every G-orbit in one point. For any subset $H \subset G$ we write $H(K) = \{g(x) \mid g \in H, x \in K\}$. Then $G(K) = S^2 - D$, and if H, L are disjoint in G it follows that $H(K) \cap L(K) = \emptyset$. Now take $M_1 = G_A(K)$, $M_2 = G_{A^{-1}}(K)$, $M_3 = G_B(K)$ and $M_4 = G_{B^{-1}}(K)$.

14.3 Free Products of Groups

The notion of free product of groups is very similar to that of free group, and in some sense it generalises the latter.

Definition 14.17 The **free product** of a family of groups $\{G_s \mid s \in S\}$ is a group F together with homomorphisms $\phi_s : G_s \to F$, for every $s \in S$, so that the following universal property holds: for every group H and every family of homomorphisms $\psi_s : G_s \to H$, $s \in S$, there is a unique homomorphism $\eta : F \to H$ such that $\psi_s = \eta \phi_s$ for every s.

$$
\begin{array}{ccc}
G_s & \xrightarrow{\psi_s} & H \\
{\scriptstyle \phi_s}\downarrow & \nearrow & \\
F & {\scriptstyle \exists! \, \eta} &
\end{array}
$$

Rather easily, the homomorphisms ϕ_s are all one-to-one: for every given $s \in S$ we consider $H = G_s$, $\psi_s = Id$ and $\psi_i = 0$ for $i \neq s$, and deduce that ϕ_s is injective.

Example 14.18 The free product of two groups G_1, G_2 is a group $G_1 * G_2$ together with homomorphisms $\phi_i : G_i \to G_1 * G_2$, $i = 1, 2$, such that for every group H and every pair of homomorphisms $\psi_i : G_i \to H$ there's a unique homomorphism $\eta : G_1 * G_2 \to H$ that makes

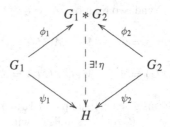

a commutative diagram.

Proposition 14.19 *Any family of groups admits the free product.*

Proof Let $\{G_s \mid s \in S\}$ be a family of groups, and for simplicity of notation suppose every pair has only the neutral element in common, i.e. $G_s \cap G_h = \{1\}$ for every $s \neq h$. On $\cup_s G_s - \{1\}$ consider the equivalence relation

$$g \sim h \iff g, h \in G_s \text{ for some } s \in S.$$

Define F to be the set of reduced words on the alphabet $\cup_s G_s$, meaning that F is made by the empty word 1 and all finite sequences

$$g_1 \times g_2 \times \cdots \times g_n$$

with $g_i \in \cup_s G_s$. We demand further that $g_i \neq 1$ and $g_i \not\sim g_{i+1}$ for every i. Let's describe the group structure: 1 is the neutral element, the law

$$(g_1 \times g_2 \cdots g_n)^{-1} = g_n^{-1} \cdots g_2^{-1} \times g_1^{-1}$$

determines inverse elements and the product is defined recursively via

$$(g_1 \cdots g_n) \times (t_1 \cdots t_m) = \begin{cases} g_1 \cdots g_n \times t_1 \cdots \times t_m & \text{if } g_n \not\sim t_1, \\ g_1 \cdots g_{n-1} \times g_n t_1 \times t_2 \cdots t_m & \text{if } g_n \sim t_1, \ g_n t_1 \neq 1, \\ (g_1 \cdots g_{n-1}) \times (t_2 \cdots t_m) & \text{if } g_n \sim t_1, \ g_n t_1 = 1. \end{cases}$$

As seen in the case of free groups, F is the free product of the family $\{G_s \mid s \in S\}$. □

Exercises

14.10 Let G_s, $s \in S$, be groups isomorphic to $\mathbb{Z}$. Prove that the free product of $\{G_s \mid s \in S\}$ is isomorphic to the free group generated by S.

14.11 Prove that the free product of a family of free groups is free.

14.12 Prove that the free product of two non-trivial groups is infinite.

14.13 Show that the group of permutations on n elements is a quotient of the free product of $n - 1$ copies of $\mathbb{Z}/2$.

14.14 Prove that the kernel of the natural homomorphism

$$\mathbb{Z}/2 * \mathbb{Z}/2 \to \mathbb{Z}/2 \times \mathbb{Z}/2$$

is isomorphic to $\mathbb{Z}$.

14.15 Show that the free product of two groups coincides with the coproduct in the category of groups (Exercise 10.29).

14.4 Free Products and van Kampen's Theorem

There is a certain similarity between Theorem 14.1 and the universal property of the free product of groups. It is possible—and this is exactly what we'll do here—to replace the universal property characterising van Kampen's theorem with a more explicit construction, that uses suitable quotients of a free product of groups.

Let's return to the situation of Sect. 14.1, where a space X was the union of open sets A, B with A, B and $A \cap B$ **path connected**, then fix a point $x_0 \in A \cap B$. The inclusions $A \subset X$, $B \subset X$, $A \cap B \subset A$ and $A \cap B \subset B$ induce a commutative diagram of homomorphisms between fundamental groups:

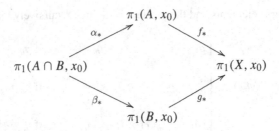

By definition of free product there is a unique homomorphism

$$h \colon \pi_1(A, x_0) * \pi_1(B, x_0) \to \pi_1(X, x_0)$$

so that

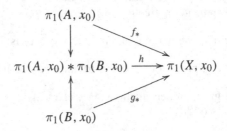

becomes commutative. The vertical arrows are the inclusions in the free product. The range of h coincides with the subgroup of $\pi_1(X, x_0)$ generated by the ranges of f_* and g_*. As we assumed A, B, $A \cap B$ path connected, the universal version of van Kampen's theorem implies that h is onto.

In general h is not 1-1, though: write $\hat{\alpha}$ and $\hat{\beta}$ for the composites of α_* and β_* with the inclusion in the free product:

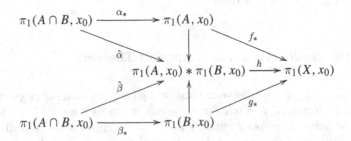

The kernel of h contains all the elements of the form $\hat{\alpha}(\gamma) \times \hat{\beta}(\gamma^{-1})$, where $\gamma \in \pi_1(A \cap B, x_0)$. In fact $h\hat{\alpha} = f_*\alpha_* = g_*\beta_* = h\hat{\beta}$, so

$$h(\hat{\alpha}(\gamma) \times \hat{\beta}(\gamma^{-1})) = h(\hat{\alpha}(\gamma))h(\hat{\beta}(\gamma^{-1})) = (h\hat{\alpha}(\gamma))(h\hat{\beta}(\gamma))^{-1} = 1.$$

Theorem 14.20 (van Kampen) *Retaining the previous notation, if A, B and $A \cap B$ are path connected, the kernel of h is the smallest normal subgroup containing elements $\hat{\alpha}(\gamma) \times \hat{\beta}(\gamma^{-1})$, for any $\gamma \in \pi_1(A \cap B, x_0)$.*

Proof Let $N \subset \pi_1(A, x_0) * \pi_1(B, x_0)$ be the normal subgroup generated by $\hat{\alpha}(\gamma)\hat{\beta}(\gamma^{-1})$ as γ varies in $\pi_1(A \cap B, x_0)$. We saw that $N \subset \ker(h)$, and also that $\hat{\alpha} \equiv \hat{\beta} \pmod{N}$, by construction. Hence a commutative diagram arises:

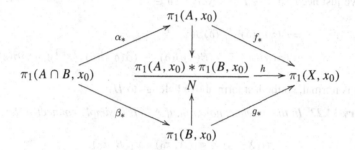

By Theorem 14.1 there's a unique homomorphism

$$\varphi : \pi_1(X, x_0) \to \frac{\pi_1(A, x_0) * \pi_1(B, x_0)}{N}$$

such that also

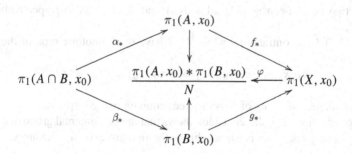

commutes.

By uniqueness $h\varphi$ must be the identity, making φ injective. On the other hand $\dfrac{\pi_1(A, x_0) * \pi_1(B, x_0)}{N}$ is generated by the images of $\pi_1(A, x_0)$ and $\pi_1(B, x_0)$, so φ is onto.

Corollary 14.21 *In the above notation, suppose A, B, A ∩ B are path connected and S is a set of generators of* $\pi_1(A \cap B, x_0)$. *Then*

$$\pi_1(X, x_0) = \frac{\pi_1(A, x_0) * \pi_1(B, x_0)}{\langle \hat{\alpha}(s)\hat{\beta}(s^{-1}) \mid s \in S \rangle}.$$

Proof Write

$$H = \langle \hat{\alpha}(s)\hat{\beta}(s^{-1}) \mid s \in S \rangle, \qquad T = \{\gamma \in \pi_1(A \cap B, x_0) \mid \hat{\alpha}(\gamma)\hat{\beta}(\gamma)^{-1} \in H\}.$$

By Theorem 14.20 it's enough to show $T = \pi_1(A \cap B, x_0)$. As $S \subset T$, and S generates $\pi_1(A \cap B, x_0)$, we can reduce to proving that T is a subgroup, and since $T \neq \emptyset$ we just need $\gamma\delta^{-1} \in T$ for every $\gamma, \delta \in T$:

$$\hat{\alpha}(\gamma\delta^{-1})\hat{\beta}(\delta\gamma^{-1}) = \hat{\alpha}(\gamma)\hat{\alpha}(\delta)^{-1}\hat{\beta}(\delta)\hat{\beta}(\gamma)^{-1}$$
$$= (\hat{\alpha}(\gamma)\hat{\beta}(\gamma)^{-1})\,(\hat{\beta}(\gamma)\hat{\beta}(\delta)^{-1})\,(\hat{\alpha}(\delta)\hat{\beta}(\delta)^{-1})^{-1}\,(\hat{\beta}(\gamma)\hat{\beta}(\delta)^{-1})^{-1}.$$

Now, H is normal, so the last term above belongs to H.   □

Corollary 14.22 *In the previous notation, if A ∩ B is simply connected then*

$$\pi_1(X, x_0) = \pi_1(A, x_0) * \pi_1(B, x_0).$$

Proof Straightforward consequence of Theorem 14.20.

Example 14.23 Let X denote two tangent circles and let's find its fundamental group. We may think of X as a graph in the plane formed by one node x_0 and two edges a, b. Choosing orientations on the latter defines two loops, whence two elements in $\pi_1(X, x_0)$.

Take two small open neighbourhoods A and B of a and b respectively, as in Fig. 14.2.

Then $A \cap B$ is contractible, while A, B have the homotopy type of the circle. Hence

$$\pi_1(X, x_0) = \pi_1(A, x_0) * \pi_1(B, x_0) = (\mathbb{Z}a) * (\mathbb{Z}b)$$

and the fundamental group of X is the free group on two generators.

Using induction on n, the same idea shows that the fundamental group of n circles touching at only one point is isomorphic to the free group on n generators.

Fig. 14.2 Computing the fundamental group of two tangent circles using van Kampen's theorem

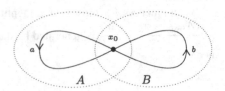

Exercises

14.16 Let X_1, X_2 be disjoint path connected spaces. Call X the space obtained by connecting X_1 and X_2 with a path γ.

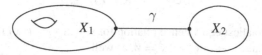

Prove $\pi_1(X) \simeq \pi_1(X_1) * \pi_1(X_2)$.

14.17 Compute the fundamental group of the complement in $\mathbb{R}^3$ of the set

$$\{(x, y, z) \mid y = 0, \ (x + 2)^2 + z^2 = 1\} \cup \{(x, y, z) \mid y = 0, \ (x - 2)^2 + z^2 = 1\}.$$

14.18 ($\heartsuit$) Compute the fundamental group of the complement in $\mathbb{R}^3$ of the three coordinate semi-axes

$$\{z = y = 0, \ x \geq 0\} \cup \{z = x = 0, \ y \geq 0\} \cup \{y = x = 0, \ z \geq 0\}.$$

14.19 Compute the fundamental group of the complement of n points on S^2.

14.20 Let $K \subset S^2 = \{(x, y, z) \in \mathbb{R}^3 \mid x^2 + y^2 + z^2 = 1\}$ be a closed subset such that:

1. $S^2 - K$ is connected;
2. $\{(x, y, z) \in K \mid x > 0\}$ is consists of n points.

Prove that $\pi_1(S^2 - K)$ is not zero for $n \geq 2$, and not Abelian if $n \geq 3$.

14.21 Consider a circle in $\mathbb{R}^3$ and n parallel straight lines going through the middle. Let X be the complement of their union. Prove that $\pi_1(X)$ is not free if $n \geq 1$, and not Abelian when $n \geq 2$.

14.22 ($\heartsuit$) Let F be the free group on two elements. Prove that for every $n > 0$ there exists a subgroup $G \subset F$ of index n that is free on $n + 1$ generators.

14.5 Attaching Spaces and Topological Graphs

The vague ideas of cutting and pasting seen in Chap. 1 can be formalised using quotients, according to the following general scheme.

Take disjoint spaces X, Y, a subspace $K \subset X$ and a continuous map $f: K \to Y$. Put on $X \cup Y$ the disjoint-union topology (cf. Example 3.10), the one for which the inclusions $X \to X \cup Y$ and $Y \to X \cup Y$ are open immersions. Now define a new space

$$X \cup_f Y = (X \cup Y)/\sim,$$

where $\sim$ is the smallest equivalence relation for which $x \sim f(x)$ for every $x \in K$. This definition makes sense when $K = \emptyset$, as well.

The space $X \cup_f Y$ is said to be obtained by **attaching** X **to** Y **via** f; **the map** f **is called attaching map**.

Example 14.24 In case the attaching map $f: K \to Y$ is onto we have $X \cup_f Y = X/\approx$, where $\approx$ is the smallest equivalence relation on X such that $x \approx y$ if $x, y \in K$ and $f(x) = f(y)$. In particular, $X \cup_f Y = X/K$ when Y consists of one point and $K \neq \emptyset$.

Lemma 14.25 *In the previous notation, if $H \subset X$ and $g: H \cap K \to Y$ is the restriction of f, the natural inclusion*

$$\Phi: H \cup_g Y \hookrightarrow X \cup_f Y$$

is continuous. In case $H \cup K = X$ and H, K are both closed or both open, Φ is a homeomorphism. When $K = X$, $X \cup_f Y = Y$.

Proof The inclusion $H \cup Y \subset X \cup Y$ passes to the quotient as the inclusion Φ, which is continuous by the universal property of identifications. Write

$$q: X \to X \cup_f Y, \qquad i: Y \to X \cup_f Y$$

for the restrictions to X and Y of the quotient map $p: X \cup Y \to X \cup_f Y$. Then i is 1-1, and a subset $A \subset X \cup_f Y$ is open (or closed) if and only if $i^{-1}(A)$ and $q^{-1}(A)$ are open (closed).

Indicate with

$$j: Y \to H \cup_g Y, \qquad r: H \to H \cup_g Y$$

the natural inclusions. Then for every subset $A \subset H \cup_g Y$

$$i^{-1}(\Phi(A)) = j^{-1}(A), \qquad q^{-1}(\Phi(A)) = r^{-1}(A) \cup f^{-1}(j^{-1}(A)).$$

Consequently if H, K are open, the map Φ is open, while if H and K are both closed, Φ is closed. To finish the proof it suffices to observe that $H \cup K = X$ implies Φ bijective.

Attaching spaces commutes with (deformation) retracts:

Proposition 14.26 *With the same notation as before, if $K \subset Z \subset X$ and Z is a deformation retract of X, then $Z \cup_f Y$ is a deformation retract of $X \cup_f Y$.*

Proof Let $R: X \times I \to X$ be a deformation of X into Z, and consider the continuous map

$$g: (X \cup Y) \times I \to X \cup Y, \qquad \begin{cases} g(x,t) = R(x,t), & x \in X, \\ g(y,t) = y, & y \in Y. \end{cases}$$

From Corollary 5.27 we know that $(X \cup Y) \times I \to (X \cup_f Y) \times I$ is an identification, and by the universal property g descends to the quotients as a continuous map

$$G: (X \cup_f Y) \times I \to X \cup_f Y.$$

This is a deformation of $X \cup_f Y$ into $Z \cup_f Y$.

Example 14.27 We continue using the notation introduced thus far. Consider the case in which X is a square minus an interior point, and Z, K are the perimeter and one side respectively. Just to fix ideas set $X = [-1,1]^2 - \{(0,0)\}$ and $K = \{-1\} \times [-1,1]$. Since Z is a deformation retract of X, Proposition 14.26 implies that $Z \cup_f Y$ is a deformation retract of $X \cup_f Y$.

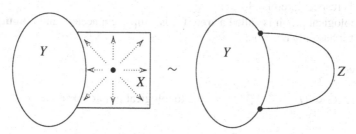

Now take two real numbers $-1 < b < a < 0$ and open sets in $X \cup_f Y$:

$$A = ([-1,a[\times [-1,1]) \cup_f Y, \qquad B =]b,1] \times [-1,1].$$

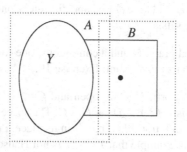

As $A \cap B$ is simply connected Corollary 14.22 applies, and we deduce

$$\pi_1(X \cup_f Y) = \pi_1(A) * \pi_1(B) = \pi_1(A) * \mathbb{Z}.$$

On the other hand K is a deformation retract of $[-1,a[\times [-1,1]$, hence $Y = K \cup_f Y$ is a deformation retract of A. Altogether

$$\pi_1(Z \cup_f Y) = \pi_1(X \cup_f Y) \simeq \pi_1(Y) * \mathbb{Z}.$$

Example 14.28 From the combinatorial viewpoint a (possibly infinite) graph is given by a set of edges L, a set of nodes V and a map $f \colon L \times \{0, 1\} \to V$ that assigns to an edge l its endpoints $f(l, 0)$, $f(l, 1)$.

If we wish to pass from combinatorics to topology we need to match nodes to points and edges to 'regular' arcs. We define a **topological graph** to be the space

$$G = (L \times [0, 1]) \cup_f V,$$

where L and V have the discrete topology. Notice that for any edge l the natural map $\{l\} \times [0, 1] \to G$ determines a path $q_l \colon [0, 1] \to G$ with endpoints $q_l(0) = f(l, 0)$, $q_l(1) = f(l, 1)$, whose image is exactly what we intended in Chap. 1 by an edge.

Since L and V are discrete spaces, a subset $A \subset G$ is open (closed) if and only if $q_l^{-1}(A)$ is open (closed) for every $l \in L$. In particular, a map $\phi \colon G \to X$ is continuous if and only if the composites ϕq_l are all continuous. This implies that the topology of a topological graph is finer than those arising from its possible embeddings in $\mathbb{R}^n$: see in this respect Exercise 14.24.

A topological graph is called a **tree** if it is simply connected, and a **bouquet of circles** if it has only one node.

Exercises

14.23 Let $G = (L \times [0, 1]) \cup_f V$ be a topological graph (Example 14.28). Prove that:

1. G is a Hausdorff space, and every point has a local basis of contractible neighbourhoods;
2. every subgraph $(H \times [0, 1]) \cup_f S$, where $H \subset L$ and $f(H \times \{0, 1\}) \subset S \subset V$, is closed in G;
3. the subspaces V and $L \times \{\frac{1}{2}\}$ are closed and discrete. Conclude that G is compact iff V and L are finite sets.

14.24 Find a topological graph G and a continuous, one-to-one map $G \to \mathbb{R}^2$ that is not homeomorphic on its range (Hint: Exercise 10.16)

14.25 Let X, Y be Hausdorff, $U \subset X$ open and $C \subset U$, $D \subset Y$ compact, then consider a homeomorphism $f \colon Y - D \to U - C$. The space $Y \cup_f (X - C)$ arises by 'carving out' C from X and 'transplanting' D in its place. Prove that $Y \cup_f (X - C)$ is Hausdorff. Show by an example that $Y \cup_f X$ is not Hausdorff in general.

14.26 Set $Y = S^1 = \{(x, y) \in \mathbb{R}^2 \mid x^2 + y^2 = 1\}$, $X = Y \times [0, 1]$ and $K = Y \times \{0, 1\}$. Consider maps $f, g \colon K \to Y$

$$f(x, y, t) = (x, y), \qquad g(x, y, t) = (x, (-1)^t y).$$

Show that $X \cup_f Y$ is homeomorphic to the torus $S^1 \times S^1$, whereas $X \cup_g Y$ is homeomorphic to the Klein bottle.

14.27 (☕) Prove that any point in a topological tree is a deformation retract of the whole space.

14.6 Cell Complexes

Definition 14.29 Let Y be a space and $f: S^{n-1} \to Y$ a continuous map. We say that $D^n \cup_f Y$ is obtained by **attaching an n-cell** to Y.

Example 14.30 Any finite topological graph arises by attaching finitely many 1-cells (the edges) to a finite discrete set (the nodes).

Example 14.31 The sphere S^n can be obtained by attaching an n-cell to a space with one point: $D^n \cup_f \{*\}$, namely, coincides with the quotient D^n/S^{n-1}.

Example 14.32 The real projective space $\mathbb{P}^n(\mathbb{R})$ arises by attaching an n-cell to $\mathbb{P}^{n-1}(\mathbb{R})$. In fact if $f: S^{n-1} \to \mathbb{P}^{n-1}(\mathbb{R})$ is the usual quotient map, $D^n \cup_f \mathbb{P}^{n-1}(\mathbb{R})$ coincides with the quotient of D^n under the relation that identifies antipodal boundary points.

Example 14.33 The sewing operations described in Sect. 1.2 attach a 2-cell (the polygon's interior) to a finite graph (the boundary modulo the equivalence relation).

We wish to understand the effect that attaching an n-cell to a path connected space has on its fundamental group. We shall treat the cases $n = 1, n = 2$ and $n \geq 3$ separately.

14.6.1 Attaching 1-cells

Let Y be a path connected space: attaching a 1-cell to Y means choosing two points y_0, y_1 (possibly equal) and build the union of Y with the interval $[0, 1]$ upon identifying 0 with y_0 and 1 with y_1 (Fig. 14.3).

As Y is path connected we can find a path $\gamma: [-1, 1] \to Y$ such that $\gamma(-1) = y_0$, $\gamma(1) = y_1$. Consider the punctured square $X = [-1, 1]^2 - \{(0, 0)\}$, the side $K = \{-1\} \times [-1, 1]$ and the attaching map

Fig. 14.3 Two 1-cells attached to the sphere S^2

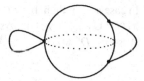

$$f: K \to Y, \qquad f(-1, t) = \gamma(t).$$

From Example 14.27 we know that $X \cup_f Y$ has the same homotopy type of the space obtained by attaching to Y a 1-cell with endpoints y_0, y_1. Therefore

$$\pi_1(D^1 \cup_f Y) = \pi_1(Y) * \mathbb{Z}.$$

14.6.2 Attaching 2-cells

Let us attach a 2-cell D^2 to Y via the function $f: S^1 = \partial D^2 \to Y$, and write $y_0 = f(1)$. Then f defines a loop

$$\gamma \in \Omega(Y, y_0, y_0), \qquad \gamma(t) = f(\cos 2\pi t, \sin 2\pi t).$$

We claim that the inclusion $Y \subset D^2 \cup_f Y$ induces a *surjective* homomorphism $\pi_1(Y, y_0) \twoheadrightarrow \pi_1(D^2 \cup_f Y, y_0)$ whose kernel is the normal subspace spanned by the homotopy class of γ. We can, in fact, write $D^2 \cup_f Y = A \cup B$, where

$$A = (D^2 - \{0\}) \cup_f Y, \qquad B = D^2 - S^1.$$

The open set B is simply connected, while the intersection $A \cap B$ is connected and its fundamental group is isomorphic to $\mathbb{Z}$. By Corollary 14.2

$$\pi_1(D^2 \cup_f Y) = \frac{\pi_1(A)}{\langle a \rangle},$$

where a is the image in $\pi_1(A)$ of the generator of $\pi_1(A \cap B)$. Now it suffices to note that Y is a deformation retract of A, and that under the induced isomorphism $\pi_1(Y) \simeq \pi_1(A)$, a corresponds to the class of γ.

14.6.3 Attaching n-cells, n ≥ 3

Attaching to Y the n-cell D^n allows to write $D^n \cup_f Y = A \cup B$, where

$$A = (D^n - \{0\}) \cup_f Y, \qquad B = D^n - S^{n-1}.$$

In case $n \geq 3$, then, both B and $A \cap B$ are simply connected, so Corollary 14.3 forces $\pi_1(D^n \cup_f Y) = \pi_1(A)$. As already noticed, moreover, Y is a deformation retract of A, so $\pi_1(D^n \cup_f Y) = \pi_1(Y)$.

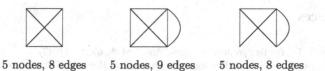

5 nodes, 8 edges   5 nodes, 9 edges   5 nodes, 8 edges

Fig. 14.4 Finite topological graphs in $\mathbb{R}^2$

Exercises

14.28 Prove that Y Hausdorff implies $D^n \cup_f Y$ Hausdorff, irrespective of the attaching function f.

14.29 Compute the fundamental groups of the graphs in Fig. 14.4.

14.30 ($\heartsuit$) Let X be a connected graph with v nodes, l edges and Euler-Poincaré characteristic $e(X) = v - l$. Prove that the fundamental group of X is a free group on $1 - e(X)$ generators.

14.31 (Euler formula for polyhedra) Let $G \subset S^2$ denote a finite graph of l edges and v nodes lying on the sphere, and $F \subset S^2 - G$ a non-empty finite set made of f distinct points. Prove that if G is a deformation retract of $S^2 - F$, necessarily $v - l + f = 2$. (Hint: $S^2 - F$ has the homotopy type of a bouquet of $f - 1$ circles.)

14.32 (Euler formula for doughnuts) Let $G \subset S^1 \times S^1$ be a finite graph of l edges and v nodes lying on the torus, $F \subset S^1 \times S^1 - G$ a non-empty finite set of f distinct points. Prove that if G is a deformation retract of $S^1 \times S^1 - F$, then $v - l + f = 0$. (Hint: $S^1 \times S^1 - F$ has the homotopy type of a bouquet of $f + 1$ circles.)

14.33 ($\clubsuit$) Let $L \subset \mathbb{P}^2(\mathbb{R})$ be a projective line and X a connected space whose fundamental group is finite cyclic of odd order. Prove that if $f : L \to X$ induces a surjective homomorphism between the corresponding fundamental groups, the space $\mathbb{P}^2(\mathbb{R}) \cup_f X$ is simply connected.

14.34 ($\clubsuit$, $\heartsuit$) Use the theory of covering spaces and van Kampen's theorem to show that the kernel of the natural homomorphism

$$\mathbb{Z}/3 * \mathbb{Z}/2 \to \mathbb{Z}/3 \times \mathbb{Z}/2$$

is a free group on two generators. More precisely: let a be the generator of $\mathbb{Z}/2$ and b the generator of $\mathbb{Z}/3$; then the kernel is the free group generated by $abab^2$, ab^2ab.

References

[Ar47] Artin, E.: The free product of groups. Am. J. Math. **69**, 1–4 (1947)
[Ma67] Massey, W.: Algebraic Topology: An Introduction. Harcourt, Brace and World, New York (1967)
[Ma91] Massey, W.: Basic Course in Algebraic Topology. Springer, Berlin (1991)
[Vi94] Vick, J.W.: Homology Theory. Springer, Berlin (1994)

Chapter 15
Selected Topics in Algebraic Topology ↻

15.1 Natural Transformations and Equivalence of Categories

Among the most riveting concepts in category theory are natural transformation and equivalence of categories; much of the mathematics of the last 60 years wouldn't exist without them. Rephrasing MacLane [ML71], one may say that functors were introduced for the purpose of defining natural transformations, and categories in order to define functors.

Let $\mathbf{A}, \mathbf{B}$ be two categories and $F, G: \mathbf{A} \to \mathbf{B}$ two functors. A **natural transformation** $\gamma: F \to G$ assigns to every object X of $\mathbf{A}$ a morphism $\gamma_X \in \mathrm{Mor}_{\mathbf{B}}(F(X), G(X))$ such that, for every morphism $f \in \mathrm{Mor}_{\mathbf{A}}(X, Y)$, the diagram

$$
\begin{array}{ccc}
F(X) & \xrightarrow{F(f)} & F(Y) \\
\downarrow{\gamma_X} & & \downarrow{\gamma_Y} \\
G(X) & \xrightarrow{G(f)} & G(Y)
\end{array}
$$

commutes.

A natural transformation $\gamma: F \to G$ is called an **isomorphism of functors** if γ_X is an isomorphism in the category $\mathbf{B}$, for every $X \in \mathbf{A}$.

Example 15.1 For any functor $F: \mathbf{A} \to \mathbf{B}$ the identity on F is by definition the natural transformation $1_F: F \to F$ such that $(1_F)_X = 1_{F(X)}$, for every $X \in \mathbf{A}$.

Example 15.2 Given sets X, Y of the same cardinality, the functors

$$F, G: \mathbf{Set} \to \mathbf{Set}, \qquad F(Z) = Z \times X, \quad G(Z) = Z \times Y,$$

© Springer International Publishing Switzerland 2015
M. Manetti, *Topology*, UNITEXT - La Matematica per il 3+2 91,
DOI 10.1007/978-3-319-16958-3_15

are isomorphic. If $f : X \to Y$ is an invertible map, the natural transformation

$$\gamma : F \to G, \qquad \gamma_Z(z, x) = (z, f(x)),$$

is an isomorphism.

Example 15.3 Given an arbitrary space X we define a category $\pi(X)$ with the points in X as objects and

$$\mathrm{Mor}_{\pi(X)}(x, y) = \frac{\Omega(X, x, y)}{\text{path homotopy}}$$

as morphisms. The composition of morphisms is the product of paths: any path α is thus invertible when seen as morphism in the category $\pi(X)$, and has inverse $i(\alpha)$. For every $x \in X$ we have $\pi_1(X, x) = \mathrm{Mor}_{\pi(X)}(x, x)$, and every continuous map $f : X \to Y$ determines a functor $f_* : \pi(X) \to \pi(Y)$ in a natural manner.

Consider now X, Y and a homotopy $F : X \times I \to Y$ from f to g. There is an isomorphism of functors $\gamma : f_* \to g_*$, whereby for every point $x \in X$ the morphism $\gamma_x \in \mathrm{Mor}_{\pi(Y)}(f(x), g(x))$ is the homotopy class of the path $t \mapsto F(x, t)$.

Remark 15.4 The category $\pi(X)$ of Example 15.3 has these properties:

1. the class of objects is a set;
2. every morphism is invertibile, hence an isomorphism.

Generally speaking, a category satisfying these conditions is called a **groupoid**. The category $\pi(X)$ is called the **fundamental groupoid** of the space X.

Natural transformations can be composed: if $\gamma : F \to G$ and $\delta : G \to H$ denote natural transformations, their composite $\delta\gamma : F \to H$ is defined by the rule

$$\delta\gamma_X = \delta_X\gamma_X, \qquad \text{for every} \quad X \in \mathbf{A}.$$

Natural transformations are also referred to as **morphisms of functors**.

Lemma 15.5 *Let $F, G : \mathbf{A} \to \mathbf{B}$ be functors. A natural transformation $\gamma : F \to G$ is an isomorphism if and only if there exists a natural transformation $\delta : G \to F$ such that $\delta\gamma = 1_F$, $\gamma\delta = 1_G$.*

Proof If a natural transformation $\delta : G \to F$ such that $\delta\gamma = 1_F$ and $\gamma\delta = 1_G$ exists, then $\gamma_X\delta_X = 1_{G(X)}$, $\delta_X\gamma_X = 1_{F(X)}$ for every $X \in \mathbf{A}$. This means γ_X is an isomorphism.

If γ is an isomorphism it is enough to set $\delta_X = \gamma_X^{-1}$ for every object $X \in \mathbf{A}$. ☐

Definition 15.6 A functor $F : \mathbf{A} \to \mathbf{B}$ is called:

1. **fully faithful** if the map

$$F : \mathrm{Mor}_{\mathbf{A}}(X, Y) \to \mathrm{Mor}_{\mathbf{B}}(F(X), F(Y))$$

is bijective for every pair of objects $X, Y \in \mathbf{A}$;

2. **essentially surjective on objects** if every object $Z \in \mathbf{B}$ is isomorphic to $F(X)$ for some $X \in \mathbf{A}$.

Example 15.7 Let $f: X \to Y$ be a continuous map. The functor $f_*: \pi(X) \to \pi(Y)$ is essentially surjective iff $f_*: \pi_0(X) \to \pi_0(Y)$ is onto, and fully faithful iff $f_*: \pi_0(X) \to \pi_0(Y)$ is one-one and $f_*: \pi_1(X, x) \to \pi_1(Y, f(x))$ is bijective for every $x \in X$.

Definition 15.8 A functor $F: \mathbf{A} \to \mathbf{B}$ is an **equivalence of categories** if there exists a functor $G: \mathbf{B} \to \mathbf{A}$ such that GF and FG are isomorphic to the identity functors (in $\mathbf{A}$ and $\mathbf{B}$ respectively). Two categories are said **equivalent** if there is an equivalence between them.

Example 15.9 If two spaces X, Y have the same homotopy type, their fundamental groupoids $\pi(X), \pi(Y)$ are equivalent categories.

Theorem 15.10 *A functor $F: \mathbf{A} \to \mathbf{B}$ is an equivalence of categories if and only if it is fully faithful and essentially surjective on objects.*

Proof Suppose $F: \mathbf{A} \to \mathbf{B}$ is an equivalence of categories; there exist a functor $G: \mathbf{B} \to \mathbf{A}$ and two isomorphisms of functors $\gamma: 1_\mathbf{A} \to GF$, $\delta: 1_\mathbf{B} \to FG$. Let $Z \in \mathbf{B}$ be an object: then $\delta_Z: Z \to FG(Z)$ is an isomorphism, proving that F is essentially surjective. Given objects $X, Y \in \mathbf{A}$, for every morphism $f: X \to Y$ we have $GF(f) = \gamma_Y f \gamma_X^{-1}$. In particular the map

$$GF: \mathrm{Mor}_\mathbf{A}(X, Y) \to \mathrm{Mor}_\mathbf{A}(GF(X), GF(Y))$$

is invertible, i.e. GF is fully faithful. Similarly, FG is fully faithful and the argument used for Lemma 11.21 shows that F and G are fully faithful.

Now assume F fully faithful and essentially surjective. For every object $Z \in \mathbf{B}$ choose[1] an object $G(Z) \in \mathbf{A}$ and an isomorphism $\delta_Z: Z \to F(G(Z))$. As F is fully faithful, given a morphism $h: Z \to W$ in $\mathbf{B}$ there's a well-defined morphism $G(h): G(Z) \to G(W)$ such that $FG(h) = \delta_W h \delta_Z^{-1}$. The easy task of checking that G is a functor from $\mathbf{B}$ to $\mathbf{A}$, and that $\delta: 1_\mathbf{B} \to FG$ is an isomorphism, is left to the reader. For every $X \in \mathbf{A}$ let $\gamma_X: X \to GF(X)$ be the only isomorphism such that $F(\gamma_X) = \delta_{F(X)}: F(X) \to FGF(X)$; as F is fully faithful, γ is an isomorphism. $\square$

Example 15.11 The existence and uniqueness of covering spaces with prescribed monodromy (Theorem 13.35) can be understood as an equivalence of categories.

Take a connected, semi-locally simply connected space X and fix a point $x_0 \in X$. Let's look at the category $\mathbf{Cov}_X$, whose objects are covering spaces of X, and whose morphisms are covering morphisms. Let us denote by $\mathbf{B}$ the category having objects $(T, \bullet)$, where T is a set and $\bullet: T \times \pi_1(X, x_0) \to T$ a right action. An arrow $f: (T, \bullet) \to (\hat{T}, \hat{\bullet})$ defines a morphism in $\mathbf{B}$ if $f: T \to \hat{T}$ is a map of sets such

---

[1] This requires the axiom of choice, in its version for classes.

that $f(t \bullet a) = f(t) \hat{\bullet} a$ for every $t \in T$, $a \in \pi_1(X, x_0)$. Then we can rephrase Theorem 13.35 in a completely equivalent way, by saying that the functor

$$\mathbf{Cov}_X \to \mathbf{B}, \quad \text{covering} \mapsto \text{monodromy},$$

is an equivalence of categories.

Exercises

15.1 Let $F: \mathbf{A} \to \mathbf{B}$ be a fully faithful functor. Prove that F is essentially injective: $X, Y \in \mathbf{A}$ are isomorphic provided $F(X)$ is isomorphic to $F(Y)$.

15.2 (Yoneda's lemma) Let $\mathbf{A}$ be a category. For any object $X \in \mathbf{A}$ we denote by $h_X: \mathbf{A} \to \mathbf{Set}$ the following functor: on objects is equals

$$h_X(Y) = \mathrm{Mor}_{\mathbf{A}}(X, Y),$$

while if f is a morphism in $\mathbf{A}$, $h_X(f) = $ composition with f. Make the definition of h_X on morphisms more precise, and check that it really defines a functor.

Given a functor $K: \mathbf{A} \to \mathbf{Set}$ we write $\mathrm{Nat}(h_X, K)$ for the collection of natural transformations $\gamma: h_X \to K$. Show that for every object $X \in \mathbf{A}$ the map

$$\mathrm{Nat}(h_X, K) \to K(X), \quad \gamma \mapsto \gamma_X(1_X),$$

is bijective. Then prove that h_X is isomorphic to h_Y iff X is isomorphic to Y.

15.3 Let $\mathbf{Vect}_{\mathbb{K}}$ be the category of vector spaces over a field $\mathbb{K}$ and denote by $D: \mathbf{Vect}_{\mathbb{K}} \to \mathbf{Vect}_{\mathbb{K}}$ the functor sending a vector space V to its double dual $D(V) = V^{**}$.

Prove that the natural inclusions $i_V: V \hookrightarrow V^{**}$ induce a natural transformation between the identity functor and D.

15.4 Interpret van Kampen's theorem (Theorem 14.1) as an equivalence of categories.

15.2 Inner and Outer Automorphisms

For any group G one indicates with $\mathrm{Aut}(G)$ the set of its **automorphisms**, i.e. the isomorphisms $G \to G$. Map-composition turns $\mathrm{Aut}(G)$ in a group, actually a subgroup of the group of permutations of G.

Example 15.12 The group $\mathrm{Aut}(\mathbb{Z}^n)$ is isomorphic to the group of $n \times n$-matrices with determinant ± 1 and integer coefficients.

Given a group G there's a homomorphism

$$\text{Inn}: G \to \text{Aut}(G), \qquad g \mapsto \text{Inn}_g,$$

where with Inn_g we denote the conjugation by g: $\text{Inn}_g(a) = gag^{-1}$, $a \in G$. For any $g \in G$, in fact, $a \mapsto gag^{-1}$ is an automorphism of G.

Definition 15.13 Automorphisms of the form Inn_g are called **inner automorphisms**; automorphisms that are not of this sort are called **outer automorphisms**. The subgroup $\text{Inn}(G) \subset \text{Aut}(G)$, image of the homomorphism Inn, is the **group of inner automorphisms** of G.

The subgroup $\text{Inn}(G) \subset \text{Aut}(G)$ is normal: take an automorphism ϕ, so that

$$\phi\,\text{Inn}_g\,\phi^{-1}(a) = \phi(g\phi^{-1}(a)g^{-1}) = \phi(g)a\phi(g)^{-1} = \text{Inn}_{\phi(g)}(a)$$

for every $g, a \in G$. Then in $\text{Aut}(G)$ one has $\phi\,\text{Inn}_g\,\phi^{-1} = \text{Inn}_{\phi(g)}$. The corresponding quotient group is written

$$\text{Out}(G) = \frac{\text{Aut}(G)}{\text{Inn}(G)}.$$

Exercises

15.5 If G is an Abelian group, $\text{Inn}(G) = \{0\}$. Prove, more generally, that the kernel of the homomorphism Inn coincides with the centre of G. We remind that the centre of a group G is the set $\{a \in G \mid ab = ba \ \forall b \in G\}$ of elements that commute with any other.

15.6 Let X be a path connected space and $x \in X$ a base point. Show that there's a well-defined natural homomorphism

$$r: \text{Homeo}(X) \to \text{Out}(\pi_1(X, x)).$$

Prove that homeomorphisms homotopic to the identity lie in the kernel of r.

15.7 A subgroup H of a group G is called **characteristic** if $\phi(H) = H$ for every $\phi \in \text{Aut}(G)$. Prove that:

1. any characteristic subgroup is normal;
2. the centre of a group is characteristic;
3. the commutator subgroup $G' \subset G$, i.e. the subgroup generated by elements $aba^{-1}b^{-1}$, $a, b \in G$, is characteristic;
4. for any group G there exists a natural homomorphism

$$\text{Out}(G) \to \text{Aut}(G/G'),$$

where G' is the commutator of G.

15.3 The Cantor Set and Peano Curves

We denote by X the space of sequences $a : \mathbb{N} \to \{0, 1\}$. We may think of X as the product of $|\mathbb{N}|$ copies of the discrete space $\{0, 1\}$, and with the product topology we obtain a compact Hausdorff space.

Given a sequence $a \in X$, the sets

$$U(N, a) = \{b \in X \mid b_n = a_n \text{ for every } n \leq N\}, \qquad N \in \mathbb{N},$$

give a local basis of neighbourhoods around a. In particular X is first countable, and hence sequentially compact.

Lemma 15.14 *In the above notations:*

1. *for any positive integer n and any real $r > 1$, the map*

$$f : X \to \mathbb{R}^n, \qquad f(a) = \sum_{k=0}^{\infty} \frac{1}{r^k}(a_{kn+1}, a_{kn+2}, \ldots, a_{kn+n}),$$

  is continuous;
2. *for any positive integer n*

$$p : X \to [0, 1]^n, \qquad p(a) = \sum_{k=0}^{\infty} \frac{1}{2^{k+1}}(a_{kn+1}, a_{kn+2}, \ldots, a_{kn+n}),$$

  is continuous and onto;
3. *the map*

$$g : X \to [0, 1], \qquad g(a) = \sum_{k=1}^{\infty} \frac{2}{3^k} a_k,$$

  is continuous and $1-1$.

Proof Consider on $\mathbb{R}^n$ the distance $d(x, y) = \sum_i |x_i - y_i|$. If $b \in U(nN, a)$,

$$d(f(b), f(a)) \leq \sum_{k=N}^{\infty} \frac{n}{r^k} = \frac{nr}{r^N(r-1)}.$$

The surjectivity of p is a consequence of the fact that any real number $x \in [0, 1]$ can be written (not uniquely) as $x = \sum_{k>0} a_k/2^k$ for a suitable sequence a_k with values in $\{0, 1\}$.

Let $a, b \in X$ be distinct sequences and call N the largest integer such that $a_i = b_i$ for every $i < N$. Suppose, to fix ideas, that $a_N = 0$ and $b_N = 1$. Then

$$g(b) - g(a) = \frac{2}{3^N} + \sum_{k>N} \frac{2}{3^k}(b_k - a_k) \geq \frac{2}{3^N} - \sum_{k>N} \frac{2}{3^k} = \frac{1}{3^N} > 0.$$

□

Definition 15.15 The image $C = g(X) \subset [0, 1]$ of the mapping g from Lemma 15.14 is called **Cantor set**.

The Cantor set is the set of numbers between 0 and 1 whose decimal expansion in base 3 doesn't contain the digit 1. As X is compact and $[0, 1]$ Hausdorff, Corollary 4.52 implies that C is a closed subset in $[0, 1]$.

Actually, it can be proved directly that C is closed without invoking the compactness of X. For any positive integer n let $U_n \subset [0, 1]$ be the subset consisting of the 2^n numbers

$$\sum_{k=1}^{n} \frac{1}{3^k} a_k, \qquad a_k = 0, 2.$$

Distinct points in U_n are at least $2 \cdot 3^{-n}$ apart, and since $\sum_{k>n} 2/3^k = 3^{-n}$, for every $x \in C$ and every $n > 0$ there is a unique $x_n \in U_n$ such that $x_n \leq x \leq x_n + 3^{-n}$. Conversely, take the decimal expansion of $x \in [0, 1] - C$ in base 3

$$x = \sum_{k>0} \frac{b_k}{3^k}, \qquad b_k = 0, 1, 2,$$

and the biggest integer n such that $x_n = \sum_{k=1}^{n} \frac{b_k}{3^k} \in U_n$. Then $b_{n+1} = 1$ and $x \in [x_n + 2 \cdot 3^{-n-1}, x_n + 3^{-n}]$, so $C = \cap_n C_n$, where

$$C_n = \bigcup_{x \in U_n} [x, x + 3^{-n}]$$

is closed. Notice, eventually, that C is the closure of the countable set $\cup_n U_n$.

Theorem 15.16 (Peano curves) *There exists a continuous and onto map* $[0, 1] \to [0, 1]^n$ *for any integer $n > 0$.*

Proof Let p and g be the maps introduced in Lemma 15.14. As X is compact and $C = g(X) \subset [0, 1]$ Hausdorff, the invertible map $g \colon X \to C$ is a homeomorphism, so $pg^{-1} \colon C \to [0, 1]^n$ is continuous and surjective. Tietze's extension theorem (Theorem 8.30), applied on the components of pg^{-1}, warrants that pg^{-1} extends to a continuous map $[0, 1] \to [0, 1]^n$. ⊓

For a different approach to Peano curves, based on the complete metric space $C(I, I^n)$, we refer readers to [Mu00, Theorem 44.1].

Exercises

15.8 For any real $x \in \mathbb{R}$ define the following recursive sequence $\{p_n(x)\}$: $p_1(x) = x$ and

$$p_{n+1}(x) = \frac{3}{2} - \left| 3 p_n(x) - \frac{3}{2} \right|.$$

Write A for the set of numbers x such that $\{p_n(x)\}$ is bounded. Prove:

1. A is closed in $[0, 1]$;
2. $x \in A \iff x = \sum_{n=1}^{+\infty} \frac{a_n}{3^n}$, where each a_n is either 0 or 2;
3. A coincides with the Cantor set C.

15.9 Use the homeomorphism $S^n \simeq I^n / \partial I^n$ to show that there exists a surjective loop $\alpha \colon I \to S^n$, for any $n > 0$.

15.4 The Topology of SO(3, $\mathbb{R}$)

The object of concern of this section is the topological group $\mathrm{SO}(3, \mathbb{R})$, the importance of whom, especially in solid-state mechanics and particle physics, cannot go amiss. We already know that $\mathrm{SO}(3, \mathbb{R})$ is a compact, connected and Hausdorff space. In a moment we will show it is a 3-dimensional topological manifold, and we shall find its fundamental group.

Each $A \in \mathrm{SO}(3, \mathbb{R})$, $A \neq Id$, is a rotation about an axis in $\mathbb{R}^3$, giving us a well-defined map

$$r \colon \mathrm{SO}(3, \mathbb{R}) - \{Id\} \longrightarrow \mathbb{P}^2(\mathbb{R}), \qquad r(A) = \text{axis of } A.$$

The symmetry s_L with respect to a straight line $L \subset \mathbb{R}^3$ belongs in $\mathrm{SO}(3, \mathbb{R})$, so we also have a map

$$s \colon \mathbb{P}^2(\mathbb{R}) \longrightarrow \mathrm{SO}(3, \mathbb{R}) - \{Id\}, \qquad L \mapsto s_L,$$

such that $rs = Id$.

Proposition 15.17 *Retaining the previous notation, r and s are continuous maps. Moreover, there is no continuous mapping*

$$f \colon \mathrm{SO}(3, \mathbb{R}) - \{Id\} \longrightarrow \mathbb{R}^3 - \{0\}$$

such that $A f(A) = f(A)$ for every $A \in \mathrm{SO}(3, \mathbb{R})$, $A \neq Id$.

Proof Let's first prove that r is continuous; consider

$$Z = \{(A, x) \in (SO(3, \mathbb{R}) - \{Id\}) \times S^2 \mid Ax = x\}$$

and let $p: Z \to SO(3, \mathbb{R}) - \{Id\}$, $q: Z \to S^2 \to \mathbb{P}^2(\mathbb{R})$ be the projections on the factors. For any closed $C \subset \mathbb{P}^2(\mathbb{R})$ we have $r^{-1}(C) = p(q^{-1}(C))$. As S^2 is compact and Z closed, p is closed and also $r^{-1}(C)$ is closed.

Take $A \in SO(3, \mathbb{R})$: we can find an orthonormal basis in which A reads

$$\begin{pmatrix} 1 & 0 & 0 \\ 0 & \cos(\alpha) & -\sin(\alpha) \\ 0 & \sin(\alpha) & \cos(\alpha) \end{pmatrix}.$$

Hence A belongs in the range of s if and only if $\mathrm{tr}(A) = -1$. Therefore $s(\mathbb{P}^2(\mathbb{R}))$ coincides with the compact set $W = \{A \in SO(3, \mathbb{R}) \mid \mathrm{tr}(A) = -1\}$, and s is the inverse of the bijection $r_{|W}: W \to \mathbb{P}^2(\mathbb{R})$. As W is compact and $\mathbb{P}^2(\mathbb{R})$ Hausdorff, $r_{|W}$ is a homeomorphism.

Suppose there existed an f as stated; possibly dividing by $\|f(A)\|$ we may assume $f: SO(3, \mathbb{R}) - \{Id\} \to S^2$. But then fs would be a section of the non-trivial covering space $S^2 \to \mathbb{P}^2(\mathbb{R})$. $\qquad\square$

Theorem 15.18 *The topological group* $SO(3, \mathbb{R})$ *is homeomorphic to the projective space* $\mathbb{P}^3(\mathbb{R})$. *The map* $r: SO(3, \mathbb{R}) - \{Id\} \to \mathbb{P}^2(\mathbb{R})$ *is a homotopy equivalence.*

Proof Take $o \in \mathbb{P}^3(\mathbb{R})$. Any projective 2-plane not passing through o is a deformation retract of $\mathbb{P}^3(\mathbb{R}) - \{o\}$, so to prove the claim we need to find a homeomorphism $SO(3, \mathbb{R}) \cong \mathbb{P}^3(\mathbb{R})$ mapping the space W of symmetries to a plane.

Define a surjective map $f: S^2 \times \mathbb{R} \to SO(3, \mathbb{R})$ as follows: open your right hand and align the thumb with a unit vector $x \in S^2$, then define $f(x, \alpha)$ to be the rotation around the axis $\mathbb{R}x$ by an angle α oriented in the way the fist closes. As $f(x, \alpha) = f(-x, 2\pi - \alpha) = f(x, 2\pi + \alpha)$, the restriction

$$f: S^2 \times [0, \pi] \to SO(3, \mathbb{R})$$

is continuous and onto, and since $S^2 \times [0, \pi]$ is compact and $SO(3, \mathbb{R})$ Hausdorff, f is a closed identification. It's easy to see that $f(x, \alpha) = f(y, \beta)$ iff either $\alpha = \beta$ and $x = y$, or $\alpha = \beta = 0$, or $\alpha = \beta = \pi$ and $x = -y$. In this respect observe that $f(x, 0)$ is the identity matrix, for every $x \in S^2$, whereas $f(x, \pi)$ is the symmetry with respect to the vector subspace spanned by x. Consider

$$g: S^2 \times [0, \pi] \to \mathbb{P}^3(\mathbb{R}), \qquad g((x_1, x_2, x_3), \alpha) = [\alpha x_1, \alpha x_2, \alpha x_3, \pi - \alpha].$$

It's not difficult to show that g is onto, and $g(x) = g(y)$ if and only if $f(x) = f(y)$. The universal property of identifications now implies that there's a homeomorphism $\phi: SO(3, \mathbb{R}) \to \mathbb{P}^3(\mathbb{R})$ such that $g = \phi f$. What is more, $g(Id) = [0, 0, 0, 1]$ and $g(W) = f(\{\alpha = \pi\}) = \{[x_1, x_2, x_3, 0]\}$. $\qquad\square$

The group $PSU(n, \mathbb{C})$ is by definition the quotient of $SU(n, \mathbb{C})$ by the finite cyclic subgroup of multiples of the identity with determinant 1. For instance, $PSU(2, \mathbb{C}) = \dfrac{SU(2, \mathbb{C})}{\pm Id}$. With the quotient topology, $PSU(n, \mathbb{C})$ becomes a Hausdorff, compact, connected group.

Proposition 15.19 *There exists an isomorphism* $PSU(2, \mathbb{C}) \simeq SO(3, \mathbb{R})$ *that is a homeomorphism as well.*

Proof It suffices to find a continuous and surjective homomorphism of groups $\rho \colon SU(2, \mathbb{C}) \to SO(3, \mathbb{R})$ such that $\ker \rho = \{\pm Id\}$. As $SU(2, \mathbb{C})$ and $SO(3, \mathbb{R})$ are compact Hausdorff, the quotient map $\dfrac{SU(2, \mathbb{C})}{\pm Id} \to SO(3, \mathbb{R})$ will be a homeomorphism.

To define ρ we introduce the real vector space $H \subset M_{2,2}(\mathbb{C})$ of 2×2 traceless Hermitian matrices. This is naturally isomorphic to $\mathbb{R}^3$ under

$$\phi \colon \mathbb{R}^3 \to H, \qquad \phi(t_1, t_2, t_3) = \begin{pmatrix} t_1 & t_2 + it_3 \\ t_2 - it_3 & -t_1 \end{pmatrix}.$$

This map also shows that the bilinear form

$$H \times H \to \mathbb{R}, \qquad (A, B) \mapsto \frac{1}{2} \operatorname{tr}(AB),$$

coincides with the dot product of $\mathbb{R}^3$. Furthermore, for every $U \in SU(2, \mathbb{C})$

$$\rho(U) \colon H \to H, \qquad \rho(U)(A) = UAU^{-1},$$

is a linear isometry that, through the linear isomorphism ϕ, yields a homomorphism $\rho \colon SU(2, \mathbb{C}) \to O(3, \mathbb{R})$. Topologically, ρ is continuous and so its range lies in the connected component at the identity, i.e. $SO(3, \mathbb{R})$.

Now we will prove that

$$\rho \colon SU(2, \mathbb{C}) \to SO(3, \mathbb{R})$$

is a surjective homomorphism with kernel $\{\pm Id\}$. In Example 1.16 we saw that special unitary 2×2-matrices (elements of $SU(2, \mathbb{C})$) have the form

$$\begin{pmatrix} a & -b \\ \bar{b} & \bar{a} \end{pmatrix}, \qquad \text{where } a, b \in \mathbb{C} \text{ satisfy } |a|^2 + |b|^2 = 1.$$

For simplicity we write

$$\rho(a, b) = \rho \begin{pmatrix} a & -b \\ \bar{b} & \bar{a} \end{pmatrix}$$

and call $p\colon$ SO(3, $\mathbb{R}$) $\to S^2$ the projection on the first column vector. The identity

$$\begin{pmatrix} a & -b \\ \bar{b} & \bar{a} \end{pmatrix} \begin{pmatrix} 1 & 0 \\ 0 & -1 \end{pmatrix} \begin{pmatrix} \bar{a} & b \\ -\bar{b} & a \end{pmatrix} = \begin{pmatrix} |a|^2 - |b|^2 & 2ab \\ 2\bar{a}\bar{b} & |b|^2 - |a|^2 \end{pmatrix}$$

implies that $p\rho(a, b) = (1, 0, 0)$ iff $b = 0$, and a pinch of trigonometry produces

$$p\rho(\cos(\alpha), \sin(\alpha)e^{i\beta}) = (\cos(2\alpha), \sin(2\alpha)\cos(\beta), \sin(2\alpha)\sin(\beta)).$$

In particular $p\rho$ is onto, and $\rho(a, b) = Id$ forces $b = 0$. An easy computation (omitted) shows

$$\rho(e^{i\alpha}, 0) = \begin{pmatrix} 1 & 0 & 0 \\ 0 & \cos(2\alpha) & -\sin(2\alpha) \\ 0 & \sin(2\alpha) & \cos(2\alpha) \end{pmatrix},$$

whence $\rho(a, b) = Id$ iff $b = 0$, $a = \pm 1$. Now it's easy to prove that ρ is onto. Take $A \in$ SO(3, $\mathbb{R}$), choose a matrix $U \in$ SU(2, $\mathbb{C}$) such that $p(A) = p\rho(U)$ and set $B = \rho(U)^{-1}A$. Then $p(B) = (1, 0, 0)$, so $B = \rho(e^{i\alpha}, 0)$ for some α. $\qquad\square$

Proposition 15.19 can be used to give another proof of the homeomorphism SO(3, $\mathbb{R}$) $\cong \mathbb{P}^3(\mathbb{R})$. First, SU(2, $\mathbb{C}$) is homeomorphic to the sphere S^3, and secondly, the multiplication in SU(2, $\mathbb{C}$) by $-Id$ corresponds to the antipodal involution in S^3. Therefore PSU(2, $\mathbb{C}$) $= \dfrac{\text{SU}(2, \mathbb{C})}{\pm Id}$, with the quotient topology, is homeomorphic to $\mathbb{P}^3(\mathbb{R})$.

Exercises

15.10 If you know what Euler angles are, use them to give yet another proof that $\rho\colon$ SU(2, $\mathbb{C}$) $\to$ SO(3, $\mathbb{R}$) is onto.

15.11 Let $G \subset$ SO(3, $\mathbb{R}$) be the subgroup of matrices with $(1, 0, 0)$ as first row vector. Prove that G is not a retract of SO(3, $\mathbb{R}$).

15.12 Prove that any continuous homomorphism SO(3, $\mathbb{R}$) $\to S^1$ is trivial.

15.13 Find, for every $n > 2$, a homomorphism PSU(n, $\mathbb{C}$) $\to$ SO($n^2 - 1$, $\mathbb{R}$) that is continuous and injective.

15.14 ($\clubsuit$) Let n be a positive integer and $p\colon$ SO($n+1$, $\mathbb{R}$) $\to S^n$ the projection on the first column vector, i.e. $p(A) = Ae_1$ where e_1 is the first vector in the canonical basis.

For any $\alpha \in \mathbb{R}^{n+1}$, $\alpha \neq 0$, we call $S_\alpha \in$ SO($n + 1$, $\mathbb{R}$) the reflection with respect to the hyperplane orthogonal to α:

$$S_\alpha(x) = x - 2\frac{(x \cdot \alpha)}{(\alpha \cdot \alpha)}\alpha.$$

Note how $S_{u-v}(u) = v$ when $u, v \in S^n$.

1. Prove that for any $u \in S^n$ there's a continuous map

$$s: S^n - \{u\} \to \mathrm{SO}(n+1, \mathbb{R})$$

such that $ps(v) = v$ for every $v \in S^n - \{u\}$.
2. Prove that for every $u \in S^n$ there's a homeomorphism

$$\phi: (S^n - \{u\}) \times \mathrm{SO}(n, \mathbb{R}) \to p^{-1}(S^n - \{u\})$$

such that $p\phi$ is the projection on the first factor.
3. Use Corollary 14.3 to show that $\pi_1(\mathrm{SO}(n, \mathbb{R})) = \mathbb{Z}/2$ for every $n \geq 3$.
4. Repeat the previous argument—in the complex case—to show that $\mathrm{SU}(n, \mathbb{C})$ is simply connected for every $n \geq 2$.

15.15 (Cayley transform) Let A, B be square matrices with B invertible. Prove that $AB = BA$ implies $AB^{-1} = B^{-1}A$, in which case we'll write

$$\frac{A}{B} = AB^{-1} = B^{-1}A.$$

Now let n be a given positive integer and take the subset $U \subset M_{n,n}(\mathbb{R})$ of matrices A with $\det(I + A) \neq 0$. Prove that $\dfrac{I - A}{I + A} \in U$ for every $A \in U$, and

$$f: U \to U, \qquad f(A) = \frac{I - A}{I + A}$$

is an involutive homeomorphism (i.e. $f^2 = Id$). Show moreover:

1. $f(A)$ is orthogonal $\iff$ A is skew-symmetric;
2. if an orthogonal matrix E belongs in U, then $\det(E) = 1$.

Deduce that the group $\mathrm{SO}(n, \mathbb{R})$ is a topological manifold of dimension $n(n - 1)/2$.

15.16 (♨, ♡) Take $n \geq 3$ and the subset $D \subset \mathrm{SO}(n, \mathbb{R}) \times \mathrm{SO}(n, \mathbb{R})$ of pairs $(A, B) \in D$ such that A, B generate a free group. Prove that D is dense.

15.5 The Hairy Ball Theorem

Definition 15.20 One says that a sphere S^n can be **combed** if there exists a continuous map $f: S^n \to S^n$ such that $f(x)$ is orthogonal to x, for every $x \in S^n$.

The curious terminology becomes all the more clear if we imagine the ball centred at the origin, so that a boundary point $x \in S^n$ is identified with a position vector, a vector orthogonal to x is tangent to the sphere at the point x, and a normal vector at $x \in S^n$ can be imagined as a hair on the hypersurface.

Example 15.21 Suppose $n = 2m - 1$ is odd: then S^n can be combed. Consider $S^{2m-1} \subset \mathbb{C}^m$ and take the multiplication by the imaginary unit as f.

When n is even, instead, S^n cannot be combed without bald spots or parting the hair: for $n > 2$, as usual, the proof requires more algebraic topology than we can afford here (namely, the first applications of homology theory).

Theorem 15.22 (Hairy ball theorem) *The sphere S^2 cannot be combed.*

Proof Suppose, by contradiction, there existed $f : S^2 \to S^2$ continuous and such that $(x \cdot f(x)) = 0$ for every $x \in S^2$. Consider the continuous map

$$g : S^2 \to S^2, \qquad g(x) = x \wedge f(x),$$

where $\wedge$ indicates the cross product:

$$\begin{pmatrix} a \\ b \\ c \end{pmatrix} \wedge \begin{pmatrix} a' \\ b' \\ c' \end{pmatrix} = \begin{pmatrix} bc' - cb' \\ ca' - ac' \\ ab' - ba' \end{pmatrix}.$$

Unit vectors orthogonal to x are precisely those of the form $\cos(\alpha)f(x) + \sin(\alpha)g(x)$. Writing x, $f(x)$, $g(x)$ as column vectors let's define a continuous bijection $F : S^2 \times S^1 \to SO(3, \mathbb{R})$ by

$$F(x, e^{i\alpha}) = (x, \cos(\alpha)f(x) + \sin(\alpha)g(x), -\sin(\alpha)f(x) + \cos(\alpha)g(x)).$$

As domain and target are compact and Hausdorff, F would be a homeomorphism. But this can't be, for the fundamental group of $SO(3, \mathbb{R})$ is isomorphic to $\mathbb{Z}/2$, while the fundamental group of $S^2 \times S^1$ is isomorphic to $\mathbb{Z}$. $\square$

Corollary 15.23 *Let $g : S^2 \to \mathbb{R}^3$ be a continuous map. There exists a point $x \in S^2$ such that $g(x)$ is a multiple of x.*

Proof By contradiction, assume that the vectors $g(x)$ and x were linearly independent, for any $x \in S^2$. The map

$$f : S^2 \to S^2, \qquad f(x) = \frac{g(x) - (g(x) \cdot x)x}{\|g(x) - (g(x) \cdot x)x\|},$$

obtained by orthonormalisation, would violate the hairy ball theorem. $\square$

A layman's version of Corollary 15.23 says that at any given time there's a hurricane somewhere, i.e. there's always at least one point on the surface of the Earth where the wind blows vertically.

Exercises

15.17 (♡) Prove that any continuous map $\mathbb{P}^2(\mathbb{R}) \to \mathbb{P}^2(\mathbb{R})$ has a fixed point.

15.18 Prove that if S^n can be combed, the antipodal map on S^n is homotopic to the identity.

15.19 (✷) Let $p\colon SO(3, \mathbb{R}) \to S^2$ map a matrix to its first column vector. Prove that for every path $\alpha\colon I \to S^2$ there exists a path $\beta\colon I \to SO(3, \mathbb{R})$ with $p\beta = \alpha$.

15.6 Complex Polynomial Functions

Lemma 15.24 *Let $U \subset \mathbb{C}$ be a neighbourhood of 0 and $f\colon U \to \mathbb{C}$ the map $f(z) = z^n g(z)$, where n is a positive integer and $g\colon U \to \mathbb{C}$ a continuous map with $g(0) \neq 0$. Then $f(U)$ is a neighbourhood of 0.*

Proof Without loss of generality we can assume U is a closed ball with centre 0, i.e. $U = \{z \mid |z| \leq r\}$, $r > 0$; in particular the path $t \mapsto re(t)$ is homotopically trivial in U. We may also suppose, possibly taking a smaller U, that the real part of $\frac{g(z)}{g(r)}$ is bigger than $1/2$ for every $z \in U$, and—up to multiplication by a constant— also that $g(r) = 1$. We claim that $f(U)$ contains an open ball centred at 0 with radius $\dfrac{r^n}{2}$. By contradiction, let $p \in \mathbb{C}$ be such that $|p| < \dfrac{r^n}{2}$ and $p \notin f(U)$. Then $f(U) \subset \mathbb{C} - \{p\}$, and the path $\alpha(t) = f(re(t))$ is homotopically zero in $\mathbb{C} - \{p\}$.

We'll show that α is path homotopic to $\beta(t) = (re(t))^n$, where the path homotopy is the convex combination

$$F(t, s) = s\alpha(t) + (1 - s)\beta(t).$$

To show that p does not lie in the image of F write

$$F(t, s) = (re(t))^n (sg(re(t)) + (1 - s)) :$$

for any $s \in [0, 1]$ the complex number $sg(re(t)) + (1 - s)$ has real part larger than or equal to $1/2$, so the modulus of $F(t, s)$ is always bigger than or equal to $r^n/2$.

Now observe that the homotopy class of β in $\pi_1(\mathbb{C} - \{p\}, f(r))$ is non-trivial, being the nth power of the generator. But this contradicts the hypothesis. □

Theorem 15.25 *Let $f(z) \in \mathbb{C}[z]$ be a complex polynomial of positive degree. The associated polynomial function $f\colon \mathbb{C} \to \mathbb{C}$ is continuous, open and closed.*

Proof Continuity is clear, and openness follows directly from Lemma 15.24. As for closure, note that f extends to a continuous map $\hat{f}\colon \mathbb{P}^1(\mathbb{C}) \to \mathbb{P}^1(\mathbb{C})$ by setting $\hat{f}(\infty) = \infty$. Since $\mathbb{C} = \hat{f}^{-1}(\mathbb{C})$ and $\hat{f}$ is closed (it goes from compact to Hausdorff), by the Projection formula we infer that f is closed. □

Corollary 15.26 *Every complex polynomial of positive degree has a complex root.*

Proof Let f be a non-constant polynomial. By Theorem 15.25 the map $f : \mathbb{C} \to \mathbb{C}$ is open and closed, and in particular $f(\mathbb{C})$ is an open and closed set. Then f is onto, because $\mathbb{C}$ is connected. ☐

15.7 Grothendieck's Proof of van Kampen's Theorem

Here we shall prove a slightly weaker version of van Kampen's theorem 14.1 using the theory of covering spaces. According to reliable sources [Fu95, Go71] the argument is due to Grothendieck.

Let A, B be open sets in a space X such that $X = A \cup B$, and suppose that A, B and $A \cap B$ are **path connected**. Fix a point $x_0 \in A \cap B$. The inclusions $A \subset X$, $B \subset X, A \cap B \subset A, A \cap B \subset B$ induce a commutative diagram of homomorphisms.

$$
\begin{array}{ccc}
\pi_1(A \cap B, x_0) & \xrightarrow{\ \alpha_*\ } & \pi_1(A, x_0) \\
{\scriptstyle \beta_*}\big\downarrow & & \big\downarrow{\scriptstyle f_*} \\
\pi_1(B, x_0) & \xrightarrow[\ g_*\ ]{} & \pi_1(X, x_0)
\end{array}
$$

Theorem 15.27 *Assume, in the previous notations, that every point in X admits a local basis of simply connected neighbourhoods.*

Then for any group T and any homomorphisms $h : \pi_1(A, x_0) \to T$, $k : \pi_1$ $(B, x_0) \to T$ such that $h\alpha_* = k\beta_*$, there exists a unique homomorphism $\pi : \pi_1$ $(X, x_0) \to T$ that renders

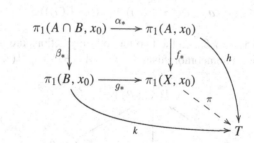

a commutative diagram.

Remark 15.28 Theorem 15.27 requires, in contrast to Theorem 14.1, the additional fact that X be locally simply connected (in the weak sense). This hypothesis is, luckily, satisfied by the vast majority of spaces of common use.

Proof We know already that $\pi_1(X, x_0)$ is generated by the images of f_* and g_*, and this implies immediately that π is unique.

Every open subset in X is locally path connected and semi-locally simply connected, so we have at our disposal the full theory of covering spaces, including the theorem of existence and uniqueness of covering spaces with assigned monodromy.

The homomorphisms h, k determine right actions

$$T \times \pi_1(A, x_0) \to T, \qquad (t, a) \mapsto t \bullet a = th(a),$$

$$T \times \pi_1(B, x_0) \to T, \qquad (t, b) \mapsto t \bullet b = tk(b),$$

compatible with the left action of the multiplication $T \times T \to T$. By Theorem 13.35 there exist covering spaces

$$p \colon E \to A, \qquad q \colon F \to B,$$

and bijections $\varphi \colon T \to p^{-1}(x_0)$, $\psi \colon T \to q^{-1}(x_0)$ that map those actions to the monodromy actions. Consider the monodromy action associated to the covering space

$$p \colon p^{-1}(A \cap B) \to A \cap B.$$

For any $a \in \pi_1(A \cap B, x_0), t \in T$ we clearly have

$$\varphi(t) \cdot a = \varphi(t \bullet \alpha_*(a)) = \varphi(th(\alpha_*(a))).$$

Analogously, from the monodromy associated to the covering

$$q \colon q^{-1}(A \cap B) \to A \cap B$$

we obtain

$$\psi(t) \cdot a = \psi(t \bullet \beta_*(a)) = \psi(tk(\beta_*(a))).$$

By assumption $h\alpha_* = k\beta_*$, so the two monodromy actions are isomorphic; by Theorem 13.35 there's a homeomorphism $\Phi \colon p^{-1}(A \cap B) \to q^{-1}(A \cap B)$ such that

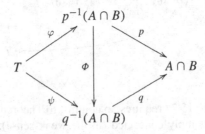

is commutative. Now let's attach E to F using the attaching map Φ: this produces a covering $E \cup_\Phi F \to X$ whose fibre over x_0 is isomorphic to T, and therefore we have a monodromy action

$$T \times \pi_1(X, x_0) \to T.$$

Let $1 \in T$ be the neutral element, and set $\pi \colon \pi_1(X, x_0) \to T$ to be $\pi(a) = 1 \cdot a$. If $a \in \pi_1(A, x_0)$ then $1 \cdot f_*(a) = 1 \bullet a = h(a)$, while if $b \in \pi_1(B, x_0)$ we have $1 \cdot g_*(b) = 1 \bullet b = k(b)$.

The left action given by the multiplication of T is compatible with the monodromy restricted to the subgroups $f_* \pi_1(A, x_0)$ and $g_* \pi_1(B, x_0)$. As the latter generate $\pi_1(X, x_0)$, the monodromy action $T \times \pi_1(X, x_0) \to T$ is compatible with the multiplication $T \times T \to T$. Proposition 13.12 then ensures that π is a homomorphism, ending the proof. $\square$

15.8 A Long Exercise: The Poincaré-Volterra Theorem

The exercises of this section will furnish, if solved in the given order, the proofs of the following facts:

Theorem 15.29 (classical Poincaré-Volterra) *Let E be a connected Hausdorff space and $p \colon E \to \mathbb{R}^n$ a local homeomorphism. Then every fibre of p is at most countable and E is second countable.*

Theorem 15.30 (Poincaré-Volterra for covering maps) *Let $p \colon E \to X$ be a covering space. If X is a topological manifold then E is a topological manifold.*

Exercises

15.20 If you haven't already done so, solve Exercises 12.2, 12.3, 12.7.

15.21 Let $\mathcal{B}$ be a countable basis in a locally connected space. Prove that the connected components of open sets in $\mathcal{B}$ form a countable family.

15.22 Let E be a separable Hausdorff space, X locally connected and second countable, and $p \colon E \to X$ a local homeomorphism. Prove that E is second countable. (Hint: consider $S \subset E$ dense and countable, $\mathcal{A}$ the family of connected components of open sets of a countable basis of X, and $\mathcal{B}$ the family of open sets $U \subset E$ such that $p(U) \in \mathcal{A}$ and $p \colon U \to p(U)$ is a homeomorphism. Prove that $\mathcal{B}$ is a basis in E, and that for every $(V, s) \in \mathcal{A} \times S$ there's at most one open set $U \in \mathcal{B}$ such that $p(U) = V$ and $s \in U$.)

15.23 Let E be path connected and $p \colon E \to \mathbb{R}^n$ a local homeomorphism. Prove that for any two points $e_1, e_2 \in p^{-1}(0)$ there exists a path $\alpha \in \Omega(E, e_1, e_2)$ such that $p\alpha$ is an edge-path with vertices in $\mathbb{Q}^n$. (Hint: take any path joining e_1, e_2 and cover it with finitely many open sets U_i for which $p(U_i)$ are convex.)

15.24 Prove Theorem 15.29.

15.25 Find an example showing that the classical Poincaré-Volterra theorem is false if we remove the Hausdorff hypothesis.

15.26 Prove that a connected manifold admits a cover by countably many contractible open sets.

15.27 Let X be a connected manifold, $S \subset X$ a countable dense subset and $\mathcal{A}$ a countable cover of X by simply connected open sets. Consider the countable set

$$P = \{(U, x, y) \mid U \in \mathcal{A}, \; x, y \in S \cap U\}$$

and choose, for every $\xi = (U, x, y) \in P$, a path $\alpha_\xi : [0, 1] \to U$ such that $\alpha(0) = x$, $\alpha(1) = y$.

Prove that any path in X with endpoints in S is path homotopic to a finite product of paths α_ξ, $\xi \in P$. Deduce that the fundamental group of X is countable.

15.28 Let X be a connected manifold and $E \to X$ a connected covering space. Prove that E is second countable.

15.29 Prove Theorem 15.30.

15.30 (👉) Under the assumptions of Theorem 15.29, denote by $R(e)$ ($e \in E$) the set of positive real numbers r for which there exists an open neighbourhood $e \in U$ such that $p : U \to B(p(e), r)$ is a homeomorphism. Prove that if p is not invertible, every $R(e)$ is bounded from above, and

$$f : E \to \mathbb{R}, \qquad f(e) = \sup R(e),$$

is continuous.

References

[Fu95] Fulton, W.: Algebraic Topology. Springer, Berlin, Heidelberg, New York (1995)
[Go71] Godbillon, C.: Éléments de topologie algébrique. Hermann, Paris (1971)
[ML71] Mac Lane, S.: Categories for the Working Mathematician. Springer, Berlin, Heidelberg, New York (1971)
[Mu00] Munkres, J.R.: Topology, 2nd edn. Prenctice-Hall, Upper Saddle River (2000)

Chapter 16
Hints and Solutions

This final chapter contains the solutions to the exercises marked with ♡ and a number of suggestions and tips.

16.1 Chapter 1

1.1 The answer to Problem 1.1 is yes, independently from the number of bridges joining 'mainland' areas. What does depend on the integers p, q, r are the ride's possible starting and arrival zones. For example, if p, q, r are all even, or all odd, the tour will have to start from the Colosseum and finish at the Vatican.

1.4 Let $p, q \in \Gamma$ be the endpoints of an edge l and set $\Gamma' = \Gamma - l$: as Γ is connected we just need to prove that p, q are the endpoints of some walk in Γ'. When $p = q$ there is nothing to prove; otherwise, p, q are the only odd-degree nodes of Γ', and as such they belong to the same connected component. More precisely, we can write Γ' as a disjoint union of connected graphs, each having an even number of odd nodes. Hence p, q lie on one connected subgraph of Γ'.

1.7 By cutting the square along the diagonal we obtain

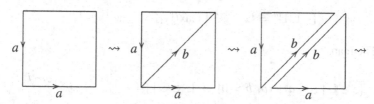

and then glueing the other pair of identified edges

© Springer International Publishing Switzerland 2015
M. Manetti, *Topology*, UNITEXT - La Matematica per il 3+2 91,
DOI 10.1007/978-3-319-16958-3_16

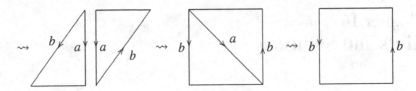

gives back the Möbius strip.

1.17 For example

$$S^n \times \mathbb{R} \to \mathbb{R}^{n+1} - \{0\}, \qquad (x, t) \mapsto e^t x,$$

with inverse $v \mapsto (v/\|v\|, \log \|v\|)$.

1.18 Suppose we have a continuous map $\Phi \colon \mathbb{R}^n \to \mathbb{R}$ such that:

1. $\Phi(tx) = t\, \Phi(x)$ for any $t \geq 0$;
2. there exist constants $m, M > 0$ for which:

$$m\|x\| \leq \Phi(x) \leq M\|x\|, \qquad \text{for every} \quad x \in \mathbb{R}^n.$$

Then $f, g \colon \mathbb{R}^n \to \mathbb{R}^n$:

$$f(x) = x\frac{\Phi(x)}{\|x\|}, \qquad g(y) = y\frac{\|y\|}{\Phi(y)},$$

are continuous and one inverse of the other. They induce a homeomorphism from $\{x \mid \Phi(x) \leq 1\}$ to the unit ball $\{y \mid \|y\| \leq 1\}$.

As for any $x = (x_1, \ldots, x_n) \in \mathbb{R}^n$ we have

$$\frac{\|x\|}{\sqrt{n}} \leq \max(|x_1|, \ldots, |x_n|) \leq \|x\|,$$

if we consider the map $\Phi(x) = \max(|x_i|)$ the previous construction gives a homeomorphism between the unit ball and the hypercube

$$[-1, 1]^n = \{x \in \mathbb{R}^n \mid \max(|x_1|, \ldots, |x_n|) \leq 1\}.$$

1.26 The maps

$$\{z \in \mathbb{C} \mid z = a + ib,\ b > 0\} \to \{z \in \mathbb{C} \mid |z| < 1\}, \qquad z \mapsto \frac{z - i}{z + i},$$

$$\{z \in \mathbb{C} \mid |z| < 1\} \to \{z \in \mathbb{C} \mid z = a + ib,\ b > 0\}, \qquad z \mapsto i\frac{1 + z}{1 - z},$$

are continuous and inverse of one other.

16.2 Chapter 2

2.4 Formulas 2, 4, 6 are always true, while 1,3 and 5 are usually false: consider for example $f : \mathbb{R} \to \mathbb{R}$, $f(x) = x^2$, $A = \{x \le 0\}$, $B = \{x \ge 0\}$.

2.6 For instance

$$3^{\sqrt{2}} = \sup \left\{ x \in \mathbb{R} \mid x^n \le 3^{\min\{m \in \mathbb{N} \mid m^2 \ge 2n^2\}} \text{ for any } n \in \mathbb{N} \right\}.$$

2.7 Denoting by $P, D \subset \mathbb{N}$ the subsets of even and odd numbers respectively, we define recursively $g : \mathbb{N} \to \mathbb{N}$ by $g(1) = 1$ and

$$g(n+1) = \begin{cases} \min(P - \{g(1), g(2), \ldots, g(n)\}) & \text{if } \sum_{i=1}^{n} \frac{(-1)^{g(i)}}{g(i)} \le x, \\[2mm] \min(D - \{g(1), g(2), \ldots, g(n)\}) & \text{if } \sum_{i=1}^{n} \frac{(-1)^{g(i)}}{g(i)} > x. \end{cases}$$

As

$$\sum_{n \in P} \frac{1}{n} = \sum_{n \in D} \frac{1}{n} = +\infty,$$

the map g is bijective and

$$\lim_{n \to \infty} g(n) = +\infty, \qquad \lim_{n \to \infty} \left| \sum_{i=1}^{n} \frac{(-1)^{g(i)}}{g(i)} - x \right| = 0.$$

2.11 Hint: call $\{p_n\}$ the sequence of primes: $p_1 = 2$, $p_2 = 3$, $p_3 = 5$ and so on. To each finite sequence $a_1, \ldots, a_n$ of natural numbers we can associate the product $p_1^{a_1} p_2^{a_2} \cdots p_n^{a_n}$.

2.12 Hint: consider, for any real $x \in [1, 10[$, the map

$$f_x : \mathbb{N} \to \mathbb{N}, \qquad f_x(n) = \lfloor 10^n x \rfloor,$$

where $\lfloor - \rfloor$ indicates the integer part: $\lfloor t \rfloor = \max\{n \in \mathbb{Z} \mid n \le t\}$. Prove that $\lim_n f_x(n)/10^n = x$, and since $10^n \le f_x(n) < 10^{n+1}$, show that $x \in [1, 10[$ is uniquely determined by the range of f_x. Then show that $f_x(n) < f_y(n)$ implies $f_x(m) < f_y(m)$ for any $m \ge n$.

2.14 By the axiom of choice we may choose, for any $i \in I$, a bijection $f_i : X_i \to Y_i$. Now consider the map

$$f : X \to Y, \qquad f(x) = f_i(x) \text{ if } x \in X_i.$$

2.16 Hint: consider the family $\{(U, V) \in \mathcal{B} \times \mathcal{A} \mid U \subset V\}$.

2.23 By Zorn's lemma it suffices to show that each chain $C \subset B$ is upper bounded: to do that it's enough to show

$$\cup\{A \mid A \in C\} \in B.$$

If $B \subset \cup\{A \mid A \in C\}$ is a finite subset, there exists $C \in C$ such that $B \subset C$, and since $C \in B$ we have $B \in B$. This proves that any finite subset of $\cup\{A \mid A \in C\}$ belongs to B, whence $\cup\{A \mid A \in C\} \in B$.

2.31 Hint: consider first the case where $\mathbb{K}$ is countable, using Exercise 2.30. If F is a subfield in $\mathbb{K}$, by the natural inclusion $F^I \subset \mathbb{K}^I$ every subset $A \subset F^I$ that is linearly independent over F is also linearly independent over $\mathbb{K}$.

16.3 Chapter 3

3.1 1, 2 and 4 are true, 3 is false.

3.6 We need to show that the family of progressions $N_{a,b}$ satisfies Theorem 3.7: but this is an immediate consequence of the formulas

$$N_{0,1} = \mathbb{Z}, \qquad N_{a,b} \cap N_{c,d} = \cup\{N_{s,bd} \mid s \in N_{a,b} \cap N_{c,d}\}.$$

As $N_{a,b}$ is the complement in $\mathbb{Z}$ of the open union

$$N_{a+1,b} \cup N_{a+2,b} \cup \ldots \cup N_{a+b-1,b},$$

the open set $N_{a,b}$ is also closed. Note that any non-empty open set contains at least one arithmetic progression, and hence it must be infinite.

3.7 For example, $A = [0, 1[$ and $B = \mathbb{R} - A$.

3.8 From $A \subset A \cup B$ follows $\overline{A} \subset \overline{A \cup B}$; similarly $\overline{B} \subset \overline{A \cup B}$ implies $\overline{A} \cup \overline{B} \subset \overline{A \cup B}$. On the other hand $\overline{A} \cup \overline{B}$ is closed and contains $A \cup B$, so $\overline{A \cup B} \subset \overline{A} \cup \overline{B}$.

3.9 As $A \cap U \subset U$, we have $\overline{U \cap A} \subset \overline{U}$; therefore it suffices to prove that $\overline{U} \subset \overline{U \cap A}$, or equivalently, that the closed set $\overline{U \cap A}$ contains U. But the set $V = U \cap (X - \overline{U \cap A})$ is open, and since $A \cap V = \emptyset$ by the density of A we obtain $V = \emptyset$.

3.14 Define a topology T on X by declaring open the subsets that are neighbourhoods of all of their points, i.e. $A \in T$ iff $A \in \mathcal{I}(x)$ for any $x \in A$. Let us check that in this way (A1), (A2), (A3) from Definition 3.1 hold true.

The empty set, having no elements, is a neighbourhood for all its points, while condition 1 implies that X is open. Let $\{A_i \mid i \in I\}$ be a family of open sets and call $A = \cup\{A_i \mid i \in I\}$ their union. If $x \in A$ then $x \in A_j$ for some j, so A_j is a neighbourhood of x and by condition 3 it follows that A is also a neighbourhood of x. If A and B are open sets and $x \in A \cap B$, then $A, B \in \mathcal{I}(x)$ and $A \cap B \in \mathcal{I}(x)$ by condition 4.

Now that we know $\mathcal{T}$ is a topology let's denote, just for the moment, by $J(x)$ the family of neighbourhoods of x in $\mathcal{T}$. We claim $\mathcal{I}(x) = J(x)$. If $U \in J(x)$ there is an open set A such that $x \in A$ and $A \subset U$. By definition $A \in \mathcal{I}(x)$, so condition 3 implies $U \in \mathcal{I}(x)$. Vice versa, take $U \in \mathcal{I}(x)$ and V as in condition 5. Then $V \in \mathcal{T}$ and so $U \in J(x)$.

3.24 Suppose $f(A)$ is open in Y for any element A in the basis $\mathcal{B}$. If $U \subset X$ is open, we can find a subfamily $\{A_i\}$ of open sets in $\mathcal{B}$ such that $U = \cup_i A_i$. Therefore $f(U) = \cup_i f(A_i)$ is a union of open sets in Y.

3.29 For instance,

$$f(x) = \begin{cases} x & \text{if } x \in \mathbb{Q}, \\ 0 & \text{if } x \notin \mathbb{Q}. \end{cases}$$

3.30 Write A as union of a countable family of finite sets $\{A_n\}$, $n \in \mathbb{N}$, and define

$$f: \mathbb{R} \to [0, 1], \qquad f(x) = \begin{cases} 0 & \text{if } x \notin A, \\ \dfrac{1}{\min\{n \mid x \in A_n\}} & \text{if } x \in A. \end{cases}$$

3.31 Hint: for any $n > 0$ consider the set $A_n \subset X$ of points x admitting a neighbourhood U such that $|f(y) - f(z)| < 1/n$ for any $y, z \in U$. Then take the set B_n of points x having a neighbourhood U such that $|f(y) - f(x)| < 1/n$ for any $y \in U$. Show that every A_n is open, $A_n \subset B_{2n}$ and $\cap A_n = \cap B_n$.

3.32 To fix ideas suppose $d(x, y) \geq d(z, w)$. By the triangle inequality

$$d(x, y) \leq d(x, z) + d(z, y) \leq d(x, z) + d(z, w) + d(y, w),$$

and so

$$d(x, y) - d(z, w) \leq d(x, z) + d(y, w).$$

3.33 If X contains just one point there's nothing to be proven. Assume then X is finite, with at least two points. Take $x \in X$ and set

$$r = \min\{d(x, y) \mid y \in X, \ y \neq x\}.$$

Clearly $r > 0$ and $B(x, r) = \{x\}$, from which the topology on X is discrete, and any subset is open and closed. Observe that there exists $y \in X$ such that $d(x, y) = r$, whence

$$\{x\} = \overline{B(x, r)} \neq \{y \in X \mid d(x, y) \leq r\}.$$

3.39 By the continuity of f at 0 it follows that the topology induced by d is finer than the one induced by h.

Conversely, note that f is non-decreasing and $f(1/n) \geq f(1)/n$ for any positive integer n. So we may consider, for any $\varepsilon > 0$, an integer n such that $1/n < \varepsilon$. If $h(x, y) < f(1)/n$, then $f(d(x, y)) \leq f(1/n)$ and so $d(x, y) < \varepsilon$; by Corollary 3.48 the topology induced by h is finer than the topology induced by d.

3.40 The function $f(t) = \frac{t}{1+t}$ satisfies $f^{-1}(0) = 0$. Moreover, f is concave and increasing when $t \geq 0$, so $\alpha f(t) \leq f(\alpha t)$ for any $t > 0$, $\alpha \in [0, 1]$. In particular,

$$f(c) \leq f(a + b) = \frac{a}{a+b} f(a+b) + \frac{b}{a+b} f(a+b) \leq f(a) + f(b)$$

for every $0 \leq c \leq a + b$, which implies $\delta = f \circ d$ is a distance. To show its equivalence to d just notice

$$d(x, y) \geq \delta(x, y) \geq \overline{d}(x, y)/2,$$

where $\overline{d}$ is the standard bound of d, and apply Corollary 3.50.

3.42 A full proof is contained in Proposition 7.31.

3.43 Both assertions are false. One counterexample is the discrete subspace $\{1/n \mid n \in \mathbb{N}\} \subset \mathbb{R}$.

3.44 Since $A \cap \overline{B}$ is closed in A and contains $A \cap B$, we have $Z \subset A \cap \overline{B}$. If $C \subset X$ is a closed subset such that $Z = A \cap C$, then $D = C \cup (X - A)$ is closed and $D \cap A = Z$; but $B \subset D$, so $\overline{B} \subset D$ and then $A \cap \overline{B} \subset A \cap D = Z$.

3.45 Let Z be locally closed and take $z \in Z$; by assumption there is an open set $U \subset X$ such that $Z \cap U$ is closed in U. From Exercise 3.44 we know $Z \cap U = \overline{Z} \cap U$, i.e. $\overline{Z} \cap U \subset Z$, proving that Z is open in $\overline{Z}$. The latter means that there's an open set $U \subset X$ such that $Z = \overline{Z} \cap U$, so Z is the intersection of a closed and an open set. The implication $(3) \Rightarrow (1)$ is obvious.

3.48 We want to prove that any discrete subspace $X \subset \mathbb{R}$ is countable. Given $x \in X$ choose a positive real $h(x)$ such that

$$X \cap B(x, h(x)) = \{x\},$$

and for any integer $n > 0$ take the subset

$$X_n = \{x \in X \mid |x| < n, \ h(x) > 1/n\}.$$

Since $X = \cup_n X_n$ it's enough to show that every X_n is finite. For any pair of distinct points $x, y \in X_n$ we have $|x - y| \geq \max(h(x), h(y)) \geq 1/n$, so X_n contains $2n^2$ points at most.

3.61 Take four natural numbers a, b, c, d such that $gcd(a, b) = 1$ and gcd $(c, d) = 1$. The product bd is coprime to any element in $N_{a,b} \cap N_{c,d}$: in fact, if $n = a + kb = c + hd$ and p is a prime that divides bd, then either p divides b and then $gcd(p, n) = gcd(p, a) = 1$, or p divides d and so $gcd(p, n) = gcd(p, c) = 1$. Hence

$$\mathbb{N} = N_{1,1}, \qquad N_{a,b} \cap N_{c,d} = \cup\{N_{n,bd} \mid n \in N_{a,b} \cap N_{c,d}\}$$

showing that $\mathcal{B}$ is a basis of a topology $\mathcal{T}$.

Given distinct positive integers n, m, for any prime $p > \max(n, m)$ we have

$$n \in N_{n,p}, \quad m \in N_{m,p}, \quad N_{n,p} \cap N_{m,p} = \emptyset,$$

so that $\mathcal{T}$ is a Hausdorff topology.

Let a, b be coprime, fix a multiple hb of b and let's show that $hb \in N_{c,d}$ implies

$$N_{a,b} \cap N_{c,d} \neq \emptyset.$$

From $hb \in N_{c,d}$ follows that b and d don't have common divisors, so we can find positive integers t, s such that $tb - sd = c - a$. Then

$$a + tb = c + sd \in N_{a,b} \cap N_{c,d}.$$

Take A, B open, non-empty; then there exist four natural numbers a, b, c, d such that $gcd(a, b) = gcd(c, d) = 1$ and

$$N_{a,b} \subset A, \qquad N_{c,d} \subset B.$$

By what we saw above,

$$bd \in \overline{N_{a,b}} \cap \overline{N_{c,d}} \subset \overline{A} \cap \overline{B}.$$

In any metric space (X, d) of cardinality larger than one we may always find two open non-empty sets with disjoint closures, namely

$$\overline{B(x, r)} \subset \{z \in X \mid d(x, z) \leq r\},$$

for $x \in X$ and r a real positive number. Given distinct x, y and a real number $r < d(x, y)/2$, the triangle inequality gives

$$\overline{B(x, r)} \cap \overline{B(y, r)} = \emptyset.$$

3.62 As Y is closed in the Euclidean topology, *a fortiori* it's closed in the lower limit plane. From $Y \cap [t, t + 1[\times [-t, -t + 1[= \{(t, -t)\}$ we deduce that the topology on Y is discrete. For any $t \in \mathbb{R}$ the subset $Y - \{(t, -t)\}$ is closed in X. As function f we can take

$$f(t) = \frac{1}{2} d_{Y - \{(t, -t)\}}(t, -t)$$

and as $h(t)$ any positive number such that

$$[t, t + h(t)[\times [-t, -t + h(t)[\subset B((t, -t), f(t)) .$$

For any integer $n > 0$ the set $\{t \in [-n, n] \mid h(t) > 1/n\}$ contains at most $2n^2 + 1$ elements, so $\{t \in \mathbb{R} \mid h(t) > 0\}$ is countable: contradiction.

16.4 Chapter 4

4.1 B and C are connected, A is disconnected.

4.5 Hint: let $X \subset \mathbb{R}^n$ be the complement of a countable set. For every $p, q \in X$ there exist uncountably many parabolic arcs $\alpha_{p,q,v}$ (as defined in Exercise 4.4) where v is orthogonal to $p - q$.

4.10 Without loss of generality suppose $p = 0$; call B the opposite of A, i.e. $B = \{-x \mid x \in A\}$. Solving the exercise amounts to proving that there is a half-line emanating from 0 that does not meet $A \cup B$; the convexity of A implies that $L \cap A = \emptyset$ or $L \cap B = \emptyset$ for any half-line L from 0. Using polar coordinates, the map

$$f :]0, +\infty[\times \mathbb{R} \to \mathbb{R}^2 - \{0\}, \qquad f(r, \theta) = (r \cos \theta, r \sin \theta),$$

is continuous, so $f^{-1}(A), f^{-1}(B)$ are open. The projection onto the second factor $p :]0, +\infty[\times \mathbb{R} \to \mathbb{R}$ is open, whence $pf^{-1}(A), pf^{-1}(B)$ are open and disjoint in $\mathbb{R}$. By connectedness there exists some $t \in \mathbb{R} - (pf^{-1}(A) \cup pf^{-1}(B))$, so the half-line $r \mapsto (r \cos t, r \sin t)$ does not intersect $A \cup B$.

4.12 Every subset of $\mathbb{Q}$ with at least two points is disconnected. In fact if $X \subset \mathbb{Q}$ and $a, b \in X$, for any irrational number ξ between a and b we can write

$$X = (X \cap] - \infty, \xi[) \cup (X \cap]\xi, +\infty[),$$

a union of open, non-empty, disjoint sets. Therefore every connected component of $\mathbb{Q}$ consists of one point.

4.13 Apply Lemma 4.23 to the family $\{W_i = Z_i \cup Y\}$.

4.14 Let $x_0 \in X - A$ and $y_0 \in Y - B$ be given. The subspaces $\{x_0\} \times Y$ and $X \times \{y_0\}$ are connected and intersect at the point (x_0, y_0). Hence their union $Z = \{x_0\} \times Y \cup X \times \{y_0\}$ is connected. Note that $X \times Y - A \times B$ is the union of all connected spaces $\{x\} \times Y$ and $X \times \{y\}$ for $x \in X - A$ and $y \in Y - B$. These subspaces meet Z, so now it suffices to use Exercise 4.13.

4.18 False: take for instance A_n to be the complement in $\mathbb{R}^2$ of the segment $y = 0$, $|x| \leq n$.

4.20 Denote with $Y \subset X$ the range of $f \colon \,]0, +\infty[\, \to \mathbb{R}^2$, $f(t) = (t^{-1}, \cos(t))$. Clearly Y is path connected, hence connected. To show that X is connected it's enough to show that X is the closure of Y in $\mathbb{R}^2$. The complement of X is a union of three open sets

$$\{x < 0\}, \quad \{|y| > 1\} \quad \text{and} \quad \{(x, y) \mid x > 0, \ y \neq \cos(x^{-1})\},$$

so X is closed in $\mathbb{R}^2$, and now we only need to see that the points in $X - Y$ are adherent to Y. Take $p \in X - Y$ and choose $t > 0$ so that $p = (0, \cos(t))$; then for any neighbourhood U of p and any large enough integer n, the point $f(t + 2\pi n)$ lies in $U \cap Y$.

Now assume, by contradiction, that X is path connected. Choose a path $\alpha \colon [0, 1] \to X$ such that $\alpha(0) = (0, 0)$ and $\alpha(1) \in Y$. Write $\alpha_1, \alpha_2 \colon [0, 1] \to \mathbb{R}$ for the components of α and let c be the maximum of the closed, bounded set $\{t \in [0, 1] \mid \alpha_1(t) = 0\}$. To fix ideas suppose $\alpha_2(c) \geq 0$: the case $\alpha_2(c) \leq 0$ is practically identical. By continuity there is a $\delta > 0$ such that $\alpha_2(t) \geq -1/2$ if $t \in [c, c + \delta]$. But $\alpha_1([c, c + \delta])$ is connected with at least two points, so there exists $\gamma > 0$ such that $[0, \gamma] \subset \alpha_1([c, c + \delta])$. In particular we can find $t \in\,]c, c + \delta]$ such that $\alpha_1(t) > 0$ and $\cos((\alpha_1(t))^{-1}) = -1$. Now observe that $\alpha_1(t) = a > 0$ implies $\alpha(t) = (a, \cos(a^{-1}))$: contradiction.

4.25 Let $\mathcal{A}'$ be the family of complements of elements of $\mathcal{A}$. Then $\mathcal{A}$ has empty intersection precisely when $\mathcal{A}'$ is a cover; moreover, $\mathcal{A}$ has the finite-intersection property iff $\mathcal{A}'$ does not contain finite subcovers.

4.29 The family of closed sets $K_n = f([n, +\infty[)$, $n \in \mathbb{N}$, fulfils Proposition 4.46, so there exist an $x \in \cap_n K_n$. If $f^{-1}(x)$ were finite there would exist an integer n such that $f^{-1}(x) \subset\,]-\infty, n[$, hence $x \notin K_n$.

4.30 For every index $i \in I$ we have $f(x) \leq f_i(x)$, so f_i is bounded; then also f si bounded and has a least upper bound

$$M = \sup_{x \in X} f(x).$$

Consider, for any $i \in I$ and any positive integer n, the open subset

$$U_{i,n} = \{x \in X \mid f_i(x) < M - 1/n\}.$$

Suppose $f(x) < M$ for any $x \in X$, by contradiction. Then for all $x \in X$ there is a positive integer n such that $f(x) < M - 1/n$ and there's also an $i \in I$ such that $f_i(x) < M - 1/n$. Consequently $\{U_{i,n}\}$ is an open cover of X and we can extract a finite subcover

$$X = U_{i_1,n_1} \cup \cdots \cup U_{i_s,n_s}.$$

If $N = \max(n_1, \ldots, n_s)$, then $f(x) < M - 1/N$ for any $x \in X$, violating the definition of M. As counterexample to the existence of a minimum we may consider the maps $g_n : [0, 1] \to [0, 1]$, $n \in \mathbb{N}$,

$$g_n(t) = \begin{cases} 1 - (n-1)t & \text{if } 0 \le t \le 1/n, \\ t & \text{if } 1/n \le t \le 1. \end{cases}$$

4.33 This is a particular case of Exercise 4.29.

4.39 Given $\varepsilon > 0$, the relation $x \sim y \iff x, y$ are ε-*apart* is an equivalence relation and the equivalence classes are open. Hence X is a disjoint union of open cosets, and if X is connected there is one equivalence class only.

If X is compact and disconnected, it is the union of non-empty closed disjoint sets C_1, C_2. The product $C_1 \times C_2$ is compact, so we have

$$r = \min_{x \in C_1, y \in C_2} d(x, y).$$

It is then clear that when $2\varepsilon < r$ the points of C_1 cannot be 'ε-joined' to C_2. For the last item, an example is the subset of $\mathbb{R}^2$ defined by equation $x^2 y = x$.

4.43 X contains matrices other than the identity I, like $2\pi/m$-rotations about a codimension-two subspace. So it suffices to show that the singleton $\{I\}$ is open in X, i.e. there's an open set $U \subset M_{n,n}(\mathbb{R})$ such that $X \cap U = \{I\}$. But $(A - I)(A^{m-1} + \cdots + A + I) = 0$ for any $A \in X$, so take

$$U = \{A \in M_{n,n}(\mathbb{R}) \mid \det(A^{m-1} + \cdots + A + I) \ne 0\}.$$

4.51 Denote by K_n the closed ball of radius n. Argue by contradiction and suppose there is a homeomorphism $f : \mathbb{R}^2 \to \mathbb{R} \times [0, 1]$, then set $D_n = f(K_n)$. The family $\{D_n\}$ is an exhaustion by compact sets and there is an integer N such that $\{0\} \times [0, 1]$ lies in D_N. On the other hand, D_N is compact and its projection onto the first factor is bounded in $\mathbb{R}$. Therefore both open sets $]-\infty, 0[\times [0, 1] - D_N$ and $]0, +\infty[\times [0, 1] - D_N$ are non-empty, and $\mathbb{R} \times [0, 1] - D_N$ is disconnected. But this is inconsistent with the connectedness of $\mathbb{R}^2 - K_N$, so the alleged homeomorphism f cannot exist.

4.54 Hint: consider the exhaustion by compact sets $K_1 \subset K_2 \subset \cdots$ where

$$K_n = \left\{ (x, y, z) \,\middle|\, x^2 + y^2 \leq 1 - \frac{1}{n},\ 0 \leq z \leq n \right\} \cup$$

$$\cup \left\{ (x, y, z) \,\middle|\, x^2 + y^2 \leq n,\ \frac{1}{n} \leq z \leq n \right\}.$$

4.57 Suppose the closed set $S^n - U$ were disconnected. Then $S^n - U$ would read as union of two closed, disjoint and non-empty sets C_1, C_2. As S^n is compact and Hausdorff, also C_1, C_2 are compact, and by Wallace's theorem there are open disjoint sets $A_1, A_2 \subset S^n$ such that $C_1 \subset A_1, C_2 \subset A_2$.

Let $f : \mathbb{R}^n \to U$ be a homeomorphism. As $S^n - (A_1 \cup A_2)$ is a compact set in U, there is a real number r large enough so that $S^n - (A_1 \cup A_2) \subset f(\overline{B(0, r)})$. Write $K = f(\overline{B(0, r)})$; the latter is compact hence closed. Possibly replacing A_i by $A_i - K$ we may assume $A_i \cap K = \emptyset, i = 1, 2$.

The connected set $U - K = f(\mathbb{R}^n - \overline{B(0, r)})$ is contained in $A_1 \cup A_2$; now it suffices to show that

$$(U - K) \cap A_1 = U \cap A_1 \neq \emptyset, \qquad (U - K) \cap A_2 = U \cap A_2 \neq \emptyset$$

to attain an absurd. If $A_1 \cap U = \emptyset$ then $A_1 \subset C_1 \cup C_2$, and since $A_1 \cap C_2 = \emptyset$ it follows $A_1 \subset C_1$. Hence $A_1 = C_1$, in contrast to the connectedness of S^n. Altogether, $U \cap A_1 \neq \emptyset$. In a similar way one proves $U \cap A_2 \neq \emptyset$.

16.5 Chapter 5

5.3 The maps f and g are onto, so also $f \times g$ is surjective. By Lemma 5.4 it's enough to show that $U \subset X \times Z$ open implies $(f \times g)(U)$ open. Let's write U as union of open sets of the canonical basis, say $U = \cup_i A_i \times B_i$. Then $(f \times g)(U) = \cup_i f(A_i) \times g(B_i)$ is an open union.

5.4 Hint: for every connected component $C \subset Y$, prove that $p^{-1}(C)$ is a union of connected components of X.

5.6 By hypothesis f is open, so $f(A^o) \subset f(A)^o$. Then $A^o \subset f^{-1}(f(A)^o) \subset A$, and $A^o = f^{-1}(f(A)^o)$. That $\overline{A}$ is saturated can be proved by passing to complements, keeping in mind that $\overline{A} = X - (X - A)^o$ and the complement of a saturated subset is saturated.

A possible counterexample is the identification of Example 5.5 with saturated sets $A = \{0\} \cup \,]1, 2\pi]$, $B = \,]0, 1]$.

5.8 Call $\pi: \mathbb{R} \to \mathbb{R}/\sim$ the quotient map. As $\pi(1) \neq \pi(-1)$, but 1 and -1 don't have disjoint saturated neighbourhoods, $\mathbb{R}/\sim$ is not Hausdorff. For every $x \in \mathbb{R}$, the restriction

$$\pi:]x - 1, x + 1[\to X$$

is an open immersion, hence any $\pi(x)$ has a neighbourhood homeomorphic to an open interval.

5.12 The only non-trivial part is the third. We have to show that for every open set $U \subset \mathbb{Q}$ such that $0 \in U$, and every open f-saturated set $V \subset \mathbb{R}$ such that $0 \in V$,

$$(f(V) \times U) \cap (f \times Id)(C) = (f \times Id)((V \times U) \cap C) \neq \emptyset.$$

Take $n \in \mathbb{N}$ so that $[0, \sqrt{2}/n] \cap \mathbb{Q} \subset U$, and choose a real $\varepsilon > 0$ sufficiently small for $[n, n + \varepsilon] \subset V$. Since $[n, n + \varepsilon] \times ([0, \sqrt{2}/n] \cap \mathbb{Q})$ intersects the straight line $x + y = n + \sqrt{2}/n$, it follows that $(V \times U) \cap C \neq \emptyset$.

5.15 The inclusion $[0, 1] \hookrightarrow \mathbb{R}$ and the map $e^{2\pi i -}$ factorise through continuous bijections $[0, 1]/\{0, 1\} \to \mathbb{R}/\mathbb{Z} \to S^1$. We know that the restriction of $e^{2\pi i -}$ to $[0, 1]$ is an identification, so the composite $[0, 1]/\{0, 1\} \to \mathbb{R}/\mathbb{Z} \to S^1$ is a homeomorphism. Therefore $[0, 1]/\{0, 1\} \cong \mathbb{R}/\mathbb{Z} \cong S^1$.

The projection $\mathbb{R} \to \mathbb{R}/\mathbb{Z}$ is open and hence $e^{2\pi i -}$ is open, too. The subset $C = \{n + \frac{1}{n} \mid n \in \mathbb{N}, n \geq 2\} \subset \mathbb{R}$ is closed, while the closure of $e^{2\pi i -}(C)$ contains the point $(1, 0)$.

5.31 Hint: use Theorem 5.32 and restrict $A \mapsto \det(A + tI)$ to the subspace of upper-triangular matrices.

16.6 Chapter 6

6.32 Indicate with $p: \mathbb{R} \to \mathbb{R}/\sim$ the quotient map and with $z = p(\mathbb{Z})$ the equivalence class of the integers. Assume z has a countable local basis U_n, $n \in \mathbb{N}$, and choose a number $0 < d_n < 1/2$ such that $[n - d_n, n + d_n] \subset p^{-1}(U_n)$ for each $n \in \mathbb{N}$. The open set

$$\left] -\infty, \frac{1}{2} \right[\bigcup_{n \in \mathbb{N}} \left] n - d_n, n + d_n \right[$$

is saturated, so it corresponds to a neighbourhood of z: the latter, though, doesn't contain any of the neighbourhoods U_n.

6.9 Let X be a first-countable Hausdorff space and C a subset such that $K \cap C$ is closed in K for any compact set K; we'll prove that C is closed in X.

Take $p \in \overline{C}$ and choose a sequence $\{a_n\}$ in C converging to p. We claim $K = \{p\} \cup \{a_n \mid n \in \mathbb{N}\}$ is compact in X. In fact if $K \subset \cup A_i$, A_i open in X, there is an index, say i_0, such that $p \in A_{i_0}$. As $\{a_n\}$ converges to p, $a_n \in A_{i_0}$ for any $n > N$ and some N. For any $n < N$ pick an i_n such that $a_n \in A_{i_n}$, so then

$$K \subset A_{i_0} \cup A_{i_1} \cup \cdots \cup A_{i_N}.$$

By assumption C is closed in K, hence $p \in C$.

6.9 For every neighbourhood U of $(0,0)$ the intersection $U \cap A$ is infinite and then $(0,0)$ is an accumulation point for any onto map $a : \mathbb{N} \to A$. If $b : \mathbb{N} \to A$ converges to $(0,0)$, then for any n we have $b_m \in X - (\{n\} \times \mathbb{N}_0)$ ultimately. Therefore $b(\mathbb{N}) \cap (\{n\} \times \mathbb{N}_0)$ is a finite set. Hence $b(\mathbb{N})$ is closed in X and the sequence b cannot converge to $(0,0)$.

6.11 The metric space $(\mathbb{R}, d)$, $d(x, y) = |x - y|$, is complete and homeomorphic to $(\,]0, 1[\,, d)$, which is not complete.

6.16 If there were two fixed points z_1, z_2, then

$$d(z_1, z_2) = d(f(z_1), f(z_2)) \leq \gamma d(z_1, z_2).$$

As $\gamma < 1$ and $d(z_1, z_2) \geq 0$, we would have $d(z_1, z_2) = 0$, i.e. $z_1 = z_2$.

Let $x \in X$ be a given point. We claim $x_1 = f(x), x_2 = f(x_1), \dots$ is a Cauchy sequence. In fact, for any n

$$d(x_n, x_{n+1}) \leq \gamma^n d(x, x_1),$$

and the triangle inequality, for $m > n$, tells

$$d(x_n, x_m) \leq d(x_n, x_{n+1}) + \cdots + d(x_{m-1}, x_m) \leq d(x, x_1) \sum_{i=n}^{m-1} \gamma^i \leq \gamma^n \frac{d(x, x_1)}{1 - \gamma}.$$

Denote by z the limit of x_n. Since f is continuous,

$$f(z) = \lim f(x_n) = \lim x_{n+1} = z.$$

6.17 Hint: suppose $h : \mathbb{R} \to]0, +\infty[$ has derivative $0 < h' < 1$ (e.g. $h(x) = 1 + \frac{\arctan(x)}{2}$) and consider the map $f(x) = x - h(x)$.

6.20 Suppose, by contradiction, that f was not onto. Choose $x_0 \in X - f(X)$ and write $x_n = f^n(x_0)$ for any $n > 0$. If $h > 0$ is the distance of x_0 from the closed set $f(X)$, then $d(x_0, x_n) \geq h$ for any $n > 0$. By assumption f is an isometry, so $d(x_n, x_m) = d(x_0, x_{m-n}) \geq h$ for all $m > n$; this implies that the sequence x_n does not contain a Cauchy subsequence.

6.27 The answers are:

1. yes;
2. yes (Hint: Exercise 3.61);
3. no (Hint: Baire's theorem).

6.28 Hint: consider the sets

$$F_n = \{x \in \mathbb{R} - A \mid f(x) \geq 1/n\}, \quad n \in \mathbb{N},$$

and use Corollary 6.42 for showing that there's an $a \in A \cap \overline{F_n}$ if n is large enough.

6.33 For every $f \in I$ write $D(f) = \{x \in X \mid f(x) \neq 0\}$. As f is invertible in $C(X, \mathbb{R})$ if and only if $D(f) = X$, the fact that I is a proper ideal forces $D(f) \neq X$ for every $f \in I$.

By contradiction: if $\{D(f) \mid f \in I\}$ were a cover, by compactness we could get a finite subcover $X = D(f_1) \cup \cdots \cup D(f_n)$; if so, then, $g = f_1^2 + \cdots + f_n^2$ would belong in the ideal I and not vanish anywhere.

16.7 Chapter 7

7.2 Consider distinct maps $f, g \colon \mathbb{R} \to \mathbb{R}$ and choose a point $s \in \mathbb{R}$ such that $f(s) \neq g(s)$. Take open disjoint sets $U, V \subset \mathbb{R}$ such that $f(s) \in U$ and $g(s) \in V$. Then f, g respectively belong to the open disjoint sets

$$P(s, U) = \{h \in X \mid h(s) \in U\}, \qquad P(s, V) = \{h \in X \mid h(s) \in V\}.$$

Assume, by contradiction, that the point $0 \in X$, corresponding to the zero map, had a countable local basis of neighbourhoods $\{V_n\}$. The space X being Hausdorff, for any $f \neq 0$ there is an n such that $f \notin V_n$, so $\cap_n V_n = \{0\}$. But for any n there is a finite number of open sets in the subbasis $P(s_1, U_1), \ldots, P(s_k, U_k)$ such that

$$0 \in P(s_1, U_1) \cap \ldots \cap P(s_k, U_k) \subset V_n.$$

Consequently $\{0\}$ would be a countable intersection of open sub-basis sets, which is impossible because $\mathbb{R}$ is not countable.

7.7 Hint: for any strictly increasing function $k \colon \mathbb{N} \to \mathbb{N}$ we can choose a map $a_k \colon \mathbb{N} \to \{-1, 1\}$ such that $a_k(k(n)) = (-1)^n$, for any n. Consider the sequence $\{x_n\}$ in $[-1, 1]^S$ given by

$$x_n \colon S \to [-1, 1], \qquad x_n(a) = a(n).$$

7.23 Let $X = \cup\{U_i \mid i \in I\}$ be a locally finite open cover of a normal space X. Call $\mathcal{A}$ the collection of pairs $(J, \{V_j \mid j \in J\})$ where $J \subset I$ and $\{V_j \mid j \in J\}$ is a family of open sets in X such that $\overline{V_j} \subset U_j$ for any $j \in J$ and $(\cup_{j \in J} V_j) \cup (\cup_{i \notin J} U_i) = X$. The set $\mathcal{A}$ is ordered by extension, in other words $(J, \{V_j \mid j \in J\}) \leq (H, \{W_h \mid h \in H\})$ iff $J \subset H$ and $V_j = W_j$ for all $j \in J$.

The claim is that $\mathcal{A}$ contains maximal elements, so by Zorn's lemma we need to show that every chain $\mathcal{C}$ in $\mathcal{A}$ is bounded from above. So let $\mathcal{C} = \{(J_s, \{V_j \mid j \in J_s\}) \mid s \in S\}$ be a chain and consider

$$(\cup_s J_s, \{V_j \mid j \in \cup_s J_s\})$$

as candidate upper bound. In order to prove that this element belongs in $\mathcal{A}$ the only non-trivial fact to check is that

$$\left(\bigcup_{j \in \cup_s J_s} V_j \right) \cup \left(\bigcup_{j \notin \cup_s J_s} U_j \right) = X.$$

Fix $x \in X$. As the cover $\{U_i\}$ is locally finite, the set $I(x) = \{i \in I \mid x \in U_i\}$ is finite. If $I(x) \not\subset \cup_s J_s$, then $x \in \left(\bigcup_{j \notin \cup_s J_s} U_j \right)$, for otherwise there would be an $s \in S$ such that $I(x) \subset J_s$, and so

$$x \in X - \left(\bigcup_{j \notin J_s} U_j \right) \subset \bigcup_{j \in J_s} V_j \subset \bigcup_{j \in \cup_s J_s} V_j.$$

Take a maximal element $(J, \{V_j \mid j \in J\})$ in $\mathcal{A}$ and let's show $J = I$: this will end the proof. Suppose there is an index $i \notin J$ and consider the closed set

$$A_i = X - \left(\bigcup_{j \in J} V_j \right) \cup \left(\bigcup_{j \notin J \cup \{i\}} U_j \right).$$

Then $A_i \subset U_i$, and the normality hypothesis implies we can find an open set V_i such that $A_i \subset V_i \subset \overline{V_i} \subset U_i$. Hence $(J \cup \{i\}, \{V_j \mid j \in J \cup \{i\}\}) \in \mathcal{A}$, violating the maximality of $(J, \{V_j \mid j \in J\})$.

7.25 We shall prove only the case T3, as T1 and T2 are completely analogous. Let X, Y be T3 spaces, $C \subset X \times Y$ a closed set and $(x, y) \notin C$. There are open sets $U \subset X$, $V \subset Y$ such that $(x, y) \in U \times V \subset X \times Y - C$. Furthermore, there exist open sets $A \subset X$ and $B \subset Y$ such that

$$x \in A \subset \overline{A} \subset U, \qquad y \in B \subset \overline{B} \subset V,$$

whence $(x, y) \in A \times B \subset \overline{A} \times \overline{B} \subset X \times Y - C$.

7.28 Hint: let X be the normal space containing the closed sets Y_n, and A, B two closed sets in X such that $A \cap B \cap Y = \emptyset$. For any n consider the closed sets $A_n = A \cap Y_n$ and $B_n = B \cap Y_n$ and note that $A_n \cap B = A \cap B_n = \emptyset$. Now we can either argue as in the proof of Theorem 7.36, or as in Exercise 8.33.

16.8 Chapter 8

8.2 Hint: let's call $\mathcal{D}$ the family of ideals contained in D. The zero ideal belongs in $\mathcal{D}$, so the family isn't empty. Take $P \in \mathcal{D}$ maximal with respect to the inclusion and suppose $fg \in P$. If, by contradiction, neither ideals $P + (f)$, $P + (g)$ were in D, there would exist four continuous maps $a, b \in A, c, d \in P$ such that $af + c$ and $bg + d$ vanished at finitely many points. Their product would belong to P and yet have only finitely many zeroes: contradiction.

8.7 X has the same cardinality of $X \times X$, so the set of all fibres of $X \times X \to X$ tells that there are families $\mathcal{B}$ of subsets in X satisfying:

1. $|A| = |X|$ for any $A \in \mathcal{B}$;
2. $|A \cap B| < |X|$ for any $A, B \in \mathcal{B}, A \neq B$;
3. $|X| \leq |\mathcal{B}|$.

By Zorn's lemma there is a maximal one $\mathcal{A}$ among them; we want to prove $|X| < |\mathcal{A}|$. Assume, by contradiction, $|X| = |\mathcal{A}|$. Zermelo's theorem gives us a well-ordering $\preceq$ on $\mathcal{A}$ such that

$$|\{B \in \mathcal{A} \mid B \prec A\}| < |A| = |X| = |\mathcal{A}|,$$

for any $A \in \mathcal{A}$. Consequently $|\bigcup \{A \cap B \in \mathcal{A} \mid B \prec A\}| < |A|$, and then $A \not\subset \bigcup \{B \in \mathcal{A} \mid B \prec A\}$. For any $A \in \mathcal{A}$ choose $x_A \in A - \bigcup \{B \in \mathcal{A} \mid B \prec A\}$; clearly $x_A \neq x_B$ if $A \neq B$, so $C = \{x_A \mid A \in \mathcal{A}\}$ has the same cardinality of X. On the other hand $C \cap A \subset \{x_A\} \cup \{x_B \mid B \prec A\}$ for any $A \in \mathcal{A}$, so $|C \cap A| < |X|$, breaching the maximality of $\mathcal{A}$.

8.21 Just for the sake of clarity let's call $B(y, r)$ any open ball in (Y, d) and $\mathcal{B}(f, r)$ any open ball in $(C(X, Y), \delta)$. Let K be compact in X, U an open set in Y and $f \in W(K, U)$. As $f(K)$ is compact, there exists $\epsilon > 0$ such that $d(f(x), y) \geq \epsilon$ for any $x \in K, y \in Y - U$. Therefore

$$f \in \mathcal{B}(f, \epsilon) \subset W(K, U),$$

proving that the metric topology is finer than the compact-open topology. Conversely, take $f \in C(X, Y)$ and $\epsilon > 0$. Then any point $x \in X$ has a compact neighbourhood

$x \in K_x$ such that $f(K_x) \subset B(f(x), \epsilon)$. By compactness X is covered by a finite number of such neighbourhoods, say $X = K_{x_1} \cup \ldots \cup K_{x_n}$, so

$$f \in W(K_{x_1}, B(f(x_1), \epsilon)) \cap \cdots \cap W(K_{x_n}, B(f(x_n), \epsilon)) \subset B(f, 2\epsilon).$$

8.27 Let $U_1, U_2 \subset X \times Y$ be open and non-empty, and take $(x_1, y_1) \in U_1, (x_2, y_2) \in U_2$. By assumption

$$V_1 = \{y \in Y \mid (x_1, y) \in U_1\}, \qquad V_2 = \{y \in Y \mid (x_2, y) \in U_2\}$$

are open and non-empty in Y, so there is a point $y_0 \in Y$ such that $(x_1, y_0) \in U_1$ and $(x_2, y_0) \in U_2$. This implies that $W_1 = \{x \in X \mid (x, y_0) \in U_1\}$ and $W_2 = \{x \in X \mid (x, y_0) \in U_2\}$ are open non-empty sets in X, from which there exists $x_0 \in X$ such that $(x_0, y_0) \in U_1 \cap U_2$.

8.30 Given $x \in X$ denote by $C_x \subset C$ the family of irreducible closed sets containing x; C_x is not empty because it contains the closure of the irreducible space $\{x\}$ (see Lemma 8.26). By Zorn's lemma we just have to prove that any chain $\mathcal{A}$ in C_x is upper bounded. Now, $Y = \cup \{Z \mid Z \in \mathcal{A}\}$ is irreducible since for any open non-empty pair $U, V \subset Y$ there exists $Z \in \mathcal{A}$ such that $U \cap Z \neq \emptyset$ and $V \cap Z \neq \emptyset$. Therefore the closed set $\overline{Y}$ is irreducible and bounds $\mathcal{A}$ from above.

8.32 Hint: for any $a \in \,]0, 1[$ we have

$$g^{-1}([0, a[) = \bigcup_{s<a} X(s), \qquad g^{-1}([0, a]) = \bigcap_{s>a} \overline{X(s)}.$$

8.33 For any positive integer n we can find open sets U_n, V_n with:

$$A \cup \{x \in B \mid f(x) \leq b - 1/n\} \subset U_n, \qquad \overline{U}_n \cap (\{x \in B \mid f(x) \geq b\} \cup C) = \emptyset,$$

$$C \cup \{x \in B \mid f(x) \geq b + 1/n\} \subset V_n, \qquad \overline{V}_n \cap (\{x \in B \mid f(x) \leq b\} \cup A) = \emptyset.$$

The open sets

$$U = \bigcup_n \left(U_n - \bigcup_{i \leq n} \overline{V}_i \right), \qquad V = \bigcup_n \left(V_n - \bigcup_{i \leq n} \overline{U}_i \right),$$

have the required properties.

8.40 If $f: X \to \mathbb{R}$ is a proper map, $K_n = f^{-1}([-n, n])$ defines an exhaustion by compact sets. Suppose conversely that there exists an exhaustion by compact sets $\{K_n\}$; by Urysohn's lemma we have a sequence of continuous maps $f_n: X \to [0, 1]$ such that $f_n = 0$ on K_n and $f_n = 1$ on $X - K^o_{n+1}$. Then $f = \sum_n f_n$ is well defined, continuous and proper.

16.9 Chapter 10

10.2 Consider $x \in X$ and a neighbourhood U of x. We have to show that there's a path connected neighbourhood W of x contained in U. Possibly taking the interior of U we may suppose U open.

Let W be the path component of U containing x: it's sufficient to show that W is open in U and hence a neighbourhood of x. For every $y \in W$ there's a neighbourhood V of y such that any pair of points in V can be joined by a path in U. Hence $V \subset W$.

10.14 Hint: use Example 3.71.

10.15 The map

$$R\colon X \times [0, 1] \to X, \qquad R(x, t) = tx + (1 - t)r(x),$$

where $r(x)$ is the intersection between Y and the straight line through x and $(0, 0, 2)$, is a deformation of X into Y.

10.16 The map $X \times I \to X$, $(x, y, t) \mapsto (tx, ty)$, deforms X into $(0, 0)$.

By contradiction, suppose there is a deformation $R\colon X \times I \to X$ of X into the point $(0, 1)$ and consider the open set

$$U = \{(x, y) \in X \mid y > 0\}.$$

For any rational $a \neq 0$, the points $(0, 1)$, $(a, 1)$ belong in different connected components of U. The open set $R^{-1}(U)$ contains $\{(0, 1)\} \times I$, and by Wallace's theorem there's an open set $V \subset X$ such that $(0, 1) \in V$ and $V \times I \subset R^{-1}(U)$, i.e. $R(x, y, t) \in U$ for any $t \in I$, $(x, y) \in V$. Given an arbitrary rational number $a \neq 0$ such that $(a, 1) \in V$, then, the path $\alpha\colon I \to U$, $\alpha(t) = R(a, 1, t)$, starts at $(0, 1)$ and ends at $(a, 1)$: contradiction.

10.20 Since the inequality $b^2 > 4ac$ can be written $b^2 + (a - c)^2 > (a + c)^2$, by setting $z = b + i(a - c)$, $x = a + c$, we can identify X with the space $\{(z, x) \in \mathbb{C} \times \mathbb{R} \mid |z|^2 > x^2\}$.

The subspace $Y = \{(z, x) \in \mathbb{C} \times \mathbb{R} \mid |z|^2 = 1, x = 0\}$ is homeomorphic to S^1 and a deformation retract of X: a possible deformation is

$$R(z, x, t) = \left(tz + (1 - t)\frac{z}{|z|}, tx\right).$$

10.21 Hint: call $p\colon \mathrm{GL}^+(n, \mathbb{R}) \to X$ the projection on the first $n - 1$ column vectors. Expand the determinant via Laplace's cofactor formula and prove that there exists a continuous $s\colon X \to \mathrm{GL}^+(n, \mathbb{R})$ such that $ps = Id_X$. Then prove that sp is homotopic to the identity.

10.22 Hint: consider the map

$$R: X \times I \to X, \qquad R(f, t)(x) = \max(\|x\|, t) f\left(\frac{x}{\max(\|x\|, t)}\right).$$

16.10 Chapter 11

11.4 We show that α is a constant path; the proof for β and γ is entirely similar. From $\alpha * (\beta * \gamma) = (\alpha * \beta) * \gamma$ follows, in particular, $\alpha(2t) = \alpha(4t)$ for any $t \leq 1/4$. Setting $4t = s$ we obtain

$$\alpha(s) = \alpha(s/2) = \alpha(s/4) = \cdots = \lim_{n \to \infty} \alpha\left(\frac{s}{2^n}\right) = \alpha(0)$$

for any $s \in I$.

11.8 The map $q: [0, 1] \to S^1$, $q(t) = e^{2i\pi t}$, is an identification. By the universal property of identifications the composition with q induces a $1-1$ correspondence between continuous maps $f: S^1 \to X$ and paths $\alpha: [0, 1] \to X$ such that $\alpha(0) = \alpha(1)$.

11.12 Hint: there exists an identification $I^2 \to D^2$ that contracts three sides to one point on the boundary.

11.15 For ease of notation we set $G = i_* \pi_1(Y, y)$ and $K = \ker(r_*)$. By functoriality the retraction r induces a homomorphism $r_*: \pi_1(X, y) \to G$ such that $r_*(a) = a$ for any $a \in G$.

The claim is that if G is normal, then $ab = ba$ for any $a \in G$, $b \in K$. In fact, $r_*(aba^{-1}b^{-1}) = aa^{-1} = 1$, so $aba^{-1}b^{-1} \in K$; on the other hand G is normal, so $ba^{-1}b^{-1} \in G$ and then $a(ba^{-1}b^{-1}) \in G$. Altogether $aba^{-1}b^{-1} \in K \cap G = \{1\}$, whence $ab = ba$. This allows to define a homomorphism

$$\Phi: G \times K \to \pi_1(X, y), \qquad \Phi(a, b) = ab,$$

and we leave the reader to check its invertibility.

11.17 Hint: consider the subspace X, formed by the unit sphere and the coordinate planes in $\mathbb{R}^3$, as the union of open sets

$$A = \{x \in X \mid \|x\| > 0\}, \qquad B = \{x \in X \mid \|x\| < 1\}.$$

11.21 Fix any positive integer N; for any $n \geq N$ we write $f_n: \pi_1(K_N, x) \to \pi_1(K_n, x)$ and $g: \pi_1(K_N, x) \to \pi_1(X, x)$ to mean the homomorphisms induced by the inclusions. Suppose every f_n is an isomorphism, and let us show g is an isomorphism.

g is onto: take $a \in \pi_1(X, x)$ and a loop $\alpha: I \to X$ such that $[\alpha] = a$. Since $\alpha(I)$ is compact, for some sufficiently large integer n we have $\alpha(I) \subset K_n$, so $[\alpha] \in \pi_1(K_n, x)$. But f_n is surjective, so there's a loop $\beta: I \to K_N$ that is path homotopic to α in K_n; then α is path homotopic to β in X, and $a = g([\beta])$.

g is one-to-one: let $\alpha: I \to K_N$ be such that $g([\alpha]) = 0$: this entails there's a path homotopy $F: I^2 \to X$ from α to the constant path. Since $F(I^2)$ is compact there is an $n > N$ such that $F(I^2) \subset K_n$, and α is homotopically trivial in K_n, i.e. $f_n([\alpha]) = 0$. But f_n is injective by hypothesis, so $[\alpha] = 0$.

11.22 Hint: for any closed, bounded interval $J \subset \mathbb{R}$ we set

$$S(J) = \{(x, y, z) \in \mathbb{R}^2 \times J \mid x^2 + y^2 = \sin^2(\pi z)\},$$

$$C(J) = \{(x, y, z) \in \mathbb{R}^2 \times J \mid \sqrt{x^2 + y^2} = \max(0, \sin(\pi z))\}.$$

Using Exercise 11.21 we can prove that if $J \cap \mathbb{Z} \neq \emptyset$ then $S(J)$ and $C(J)$ are simply connected. If J contains just one integer, $S(J)$ is contractible. Now use van Kampen's theorem and induction on the number of integers contained in J. For $C(J)$ the argument is analogous.

16.11 Chapter 12

12.2 (1) From the start E is Hausdorff, so the diagonal Δ is closed in $E \times E$ and hence closed in $E \times_X E$. To prove openness, take $e \in E$ arbitrarily and choose an open set $U \subset E$ such that $e \in U$ and $f: U \to X$ is $1-1$. Then

$$(e, e) \in U \times U \cap E \times_X E = \{(e_1, e_2) \in U \times U \mid p(e_1) = p(e_2)\} \subset \Delta.$$

(2) Let Y be connected and $f, g: Y \to E$ continuous and such that $pf = pg$. Then

$$(f, g): Y \to E \times_X E, \qquad (f, g)(y) = (f(y), g(y)),$$

is continuous, and $(f, g)^{-1}(\Delta)$ is open and closed in Y.

(3) From Example 4.21 it's enough to show U is closed in the open set $V = p^{-1}(p(U))$. Let $y \in V - U$ be any point and $x \in U$ such that $p(y) = p(x)$. As E is Hausdorff there exist open disjoint sets A, B such that $x \in A \subset U$, $y \in B \subset V$; we can suppose $p(B) \subset p(A)$ (otherwise just make B smaller). Then

$$p(U \cap B) = p((U - A) \cap B) \subset (p(U) - p(A)) \cap p(B) = \emptyset,$$

and $U \cap B = \emptyset$ follows.

12.5 Let $s\colon \mathbb{P}^n(\mathbb{R}) \to S^n$ be a hypothetical section, and call $C \subset S^n$ its image. As $\mathbb{P}^n(\mathbb{R})$ is compact and S^n Hausdorff, the subset C and its opposite $D = -C = \{-x \mid x \in C\}$ are closed in S^n. But $C \cup D = S^n$ and S^n is connected for $n > 0$, so $C \cap D \neq \emptyset$. That is to say, there exist $y, z \in \mathbb{P}^n(\mathbb{R})$ such that $s(y) = -s(z)$. Now observe

$$y = p(s(y)) = p(-s(z)) = p(s(z)) = z$$

to reach a contradiction.

12.10 If $b - a \leq 1$ the map is not surjective. If, instead, $b - a > 1$, we can prove that not all fibres have the same cardinality. Let n be the largest integer such that $a + n < b$. If $x \in {]}a, b - n{[}$, the fibre of $e(x)$ consists of the $n + 1$ points $x, x + 1, \dots, x + n$. But as $b - n - 1 \leq a$, the fibre over $e(b - n)$ contains exactly the n points $b - n, b - n + 1, \dots, b - 1$.

12.11 Hint: show that there exists an isomorphism of $\mathbb{Q}$-vector spaces $f\colon \mathbb{R} \to \mathbb{C}$ such that $f(1) = 1$, and then consider the map e.

12.12 Hint: p is open because it's a local homeomorphism, and closed because it maps a compact space to a Hausdorff space. In particular $p(E)$ is open and closed in X, so p is onto. Take $x \in X$; the fibre $p^{-1}(x)$ is discrete and closed in a compact space, hence it is finite. For every $e \in p^{-1}(x)$ choose open sets $e \in U_e \subset E$ and $x \in V_e \subset X$ such that $p\colon U_e \to V_e$ is a homeomorphism. Since E is Hausdorff we can suppose that the U_e are disjoint. Now consider the closed set $C = E - \bigcup_{e \in p^{-1}(x)} U_e$; its image $p(C)$ is closed in X. The open set $V = X - p(C)$ is an admissible open set, and contains x.

12.13 Hint: prove that A can be taken diagonal, up to pre- and post-composing with homeomorphisms. This proof simplifies considerably if we invoke the theory of modules on principal-ideal domains, in particular Smith's normal form for integer matrices.

12.23 Take $k \in \ker(p)$; the map $G \to G, g \mapsto kgk^{-1}$ is a lifting of p.

12.28 Outline of the proof: take a real $R > 0$ sufficiently large and such that the open sets B, H, C are contained in the ball $B(0, R)$. For every pair $(v, t) \in S^2 \times [-R, R]$ define:

1. $b_+(v, t) =$ volume of $B \cap \{x \mid (v \cdot x) \geq t\}$;
2. $b_-(v, t) =$ volume of $B \cap \{x \mid (v \cdot x) \leq t\}$;
3. $h_+(v, t) =$ volume of $H \cap \{x \mid (v \cdot x) \geq t\}$;
4. $h_-(v, t) =$ volume of $H \cap \{x \mid (v \cdot x) \leq t\}$;
5. $c_+(v, t) =$ volume of $C \cap \{x \mid (v \cdot x) \geq t\}$;
6. $c_-(v, t) =$ volume of $C \cap \{x \mid (v \cdot x) \leq t\}$.

The subset $Z \subset S^2 \times [-R, R]$ of points z such that $b_+(z) = b_-(z)$ is closed, and the projection $\pi \colon Z \to S^2$ is bijective hence a homeomorphism; moreover, $(v, t) \in Z$ iff $(-v, -t) \in Z$. Apply Corollary 12.35 to the continuous map

$$f \colon S^2 \to \mathbb{R}^2, \qquad f = g\pi^{-1}, \quad g(z) = (h_+(z) - h_-(z), c_+(z) - c_-(z)).$$

12.31 Consider the continuous mapping

$$g \colon D^2 \to S^1, \qquad g(x) = \frac{x - f(x)}{\|x - f(x)\|}.$$

From the relations $r(x) = x + tg(x), t \ge 0$ and $\|r(x)\|^2 = 1$ descends

$$t^2 + 2t(x \cdot g(x)) + \|x\|^2 = 1, \quad t = -(x \cdot g(x)) + \sqrt{(x \cdot g(x))^2 + 1 - \|x\|^2}.$$

16.12 Chapter 13

13.2 Hint: prove that this space is the base of a 2-sheeted covering with total space $E = S^2 \times (\mathbb{R}^3 - \{0\})$.

13.11 Hint: find continuous maps $S^1 \to GL(n, \mathbb{C})$ and $GL(n, \mathbb{C}) \to S^1$ whose composite is the identity on S^1.

13.12 Take paths $\alpha \colon I \to I^2$ and $\beta \colon I \to I^2$ such that $\alpha(0) = (0, 0), \alpha(1) = (1, 1)$, $\beta(0) = (1, 0)$ and $\beta(1) = (0, 1)$. Consider

$$f \colon I^2 \to \mathbb{R}^2, \qquad f(t, s) = \alpha(t) - \beta(s),$$

and suppose by contradiction that $(0, 0)$ does not lie in the range of f.

Take one side $L \subset I^2$ of the square, and let's prove that the homotopy class of the path $f \colon L \to \mathbb{R}^2 - \{(0, 0)\}$ does not depend neither on α nor on β: to fix ideas, assume $L = \{(t, 0) \mid t \in I\}$ (the other possibilities are similar). Then $f(t, 0) = \alpha(t) - \beta(0) = \alpha(t) - (1, 0)$, so $f(L)$ is contained in the convex subspace

$$\{(x, y) \in \mathbb{R}^2 \mid x \le 0, \ y \ge 0, \ (x, y) \ne (0, 0)\}$$

and $f_{|L}$ is path homotopic to $g(t, 0) = (t - 1, t)$. Repeating this argument for the other three sides will end up saying that the loop $f \colon \partial I^2 \to \mathbb{R}^2 - \{(0, 0)\}$ is path homotopic to $g \colon \partial I^2 \to \mathbb{R}^2 - \{(0, 0)\}$, where

$$g(t, 0) = (t - 1, t), \qquad g(1, s) = (s, 1 - s),$$

$$g(t, 1) = (t, t - 1), \qquad g(0, s) = (s - 1, -s).$$

But g generates the fundamental group of $\mathbb{R}^2 - \{(0,0)\}$, whereas $f_{|\partial I^2}$ is homotopically trivial because it extends to the whole square.

13.15 Pick $a \in A$ and set $e = \tilde{f}(a)$, $y = q(a)$, $x = F(y) = p(e)$. Functoriality gives a commutative diagram of group homomorphisms

$$
\begin{array}{ccc}
\pi_1(A,a) & \xrightarrow{\tilde{f}_*} & \pi_1(E,e) \\
\downarrow{\scriptstyle q_*} & & \downarrow{\scriptstyle p_*} \\
\pi_1(Y,y) & \xrightarrow{F_*} & \pi_1(X,x).
\end{array}
$$

By assumption q_* is onto, so

$$F_* \pi_1(Y,y) = F_* q_* \pi_1(A,a) = p_* \tilde{f}_* \pi_1(A,a) \subset p_* \pi_1(E,e).$$

Theorem 13.18 says there's unique lift $\tilde{F}: Y \to E$ of p such that $\tilde{F}(y) = e$. Since $\tilde{f}$ and $\tilde{F}q$ are lifts of $p\tilde{f} = Fq$ and $\tilde{f}(a) = \tilde{F}q(a) = e$, by connectedness $\tilde{f} = \tilde{F}q$.

13.16 Observe that $\alpha = \beta * \beta$, where $\beta(t) = \alpha(t/2)$.

13.30 Every discrete subspace of the compact space $(S^1)^3$ is finite, so any covering map $(S^1)^3 \to G/\Gamma$ must have finite degree. Now, G is simply connected and the fundamental group of G/Γ is isomorphic to Γ, so the question has a negative answer if we show that Γ does not contain Abelian subgroups of finite index. But for that we can consider the matrices

$$A = \begin{pmatrix} 1 & 1 & 0 \\ 0 & 1 & 0 \\ 0 & 0 & 1 \end{pmatrix}, \qquad B = \begin{pmatrix} 1 & 0 & 0 \\ 0 & 1 & 1 \\ 0 & 0 & 1 \end{pmatrix},$$

and note that $A^n B^n \neq B^n A^n$ for any $n > 0$.

16.13 Chapter 14

14.1 Denote by A the complement of

$$\{(x,y,z) \mid y = 0, \ x^2 + z^2 = 1\} \cup \{(x,y,z) \mid y = z = 0, \ x \geq 1\}$$

in $\mathbb{R}^3$. To prove $\pi_1(A) = \mathbb{Z}$ we consider the simply connected open set

$$B = \{(x,y,z) \mid x^2 + y^2 + z^2 > 1\}.$$

As $A \cap B$ is simply connected, the inclusion $A \subset A \cup B$ induces an isomorphism of the fundamental groups: now observe that $A \cup B$ is the complement of a circle, so $\pi_1(A) = \mathbb{Z}$.

14.5 Let G be an arbitrary non-Abelian group and $\tau, \sigma \in G$ such that $\sigma\tau \neq \tau\sigma$ (for example G could be the group of permutations of three elements and σ, τ distinct transpositions). The universal property guarantees the existence of a group homomorphism $\phi : F_S \to G$ such that $\phi(a) = \sigma$ and $\phi(c) = \tau$ for any $c \in S - \{a\}$. Therefore $\phi(ab) \neq \phi(ba)$, and then $ab \neq ba$.

14.18 The complement of the semi-axes retracts by deformation to

$$S^2 - \{(1, 0, 0), (0, 1, 0), (0, 0, 1)\}.$$

The latter is, under stereographic projection, homeomorphic to the complement of two points in $\mathbb{R}^2$. Therefore the fundamental group is free on two generators.

14.22 Hint: take $p : \mathbb{C} \to \mathbb{C}$, $p(z) = z^n$, and consider the covering

$$\mathbb{C} - p^{-1}(\{0, 1\}) \xrightarrow{\;p\;} \mathbb{C} - \{0, 1\}.$$

14.30 Every tree is contractible, hence simply connected. At the same time any graph arises by attaching one-cells to a tree. Every tree has Euler-Poincaré characteristic 1, and attaching one-cells lowers it by 1 each time.

14.34 We define $\mu_n = \{z \in \mathbb{C} \mid z^n = 1\}$ and recall $S^3 = \{w \in \mathbb{C}^2 \mid \|w\| = 1\}$. The group μ_n acts by multiplication on S^3 in a properly discontinuous way, and the quotient is a lens space L_n with fundamental group μ_n. The free product $\mu_3 * \mu_2$ is therefore the fundamental group of $L_3 \cup L_2 / \sim$, where $\sim$ identifies a point in L_2 to a point in L_3 (see Exercise 14.16).

The group $\mu_3 \times \mu_2$ acts properly discontinuously on

$$X = S^3 \times \mu_2, \qquad Y = \mu_3 \times S^3,$$

according to

$$(\xi, \eta)(u, \rho) = (\xi u, \eta\rho), \qquad (\xi, \eta)(\gamma, v) = (\xi\gamma, \eta v),$$

where $(\xi, \eta) \in \mu_3 \times \mu_2$, $(u, \rho) \in X$, $(\gamma, v) \in Y$. Fix a point $p \in S^3$ and consider $E = X \cup Y / \sim$, where the relation identifies $\alpha(p, 1) \in X$ with $\alpha(1, p) \in Y$, for any $\alpha \in \mu_3 \times \mu_2$. The fundamental group of E is free on two generators, while $\mu_3 \times \mu_2$ acts properly discontinuously and the quotient is homeomorphic to $L_3 \cup L_2 / \sim$.

This same reasoning may be employed to prove that the kernel of the natural homomorphism $\mu_n * \mu_m \to \mu_n \times \mu_m$ is free on $(n - 1)(m - 1)$ generators.

16.14 Chapter 15

15.16 Hint: use the Cayley transform, the principle of identity of polynomials and Example 14.16 to show that for any reduced word $a^{s_1} b^{r_1} \ldots$ written in the letters a, b, a^{-1}, b^{-1}, the set of pairs (A, B) of special orthogonal matrices with $A^{s_1} B^{r_1} \cdots = I$ is nowhere dense and closed. Baire's theorem guarantees D is dense.

15.17 Every continuous map $f \colon \mathbb{P}^2(\mathbb{R}) \to \mathbb{P}^2(\mathbb{R})$ lifts to a continuous map of the universal coverings: that is to say, there's a commutative diagram of continuous maps

$$
\begin{array}{ccc}
S^2 & \xrightarrow{\;g\;} & S^2 \\
\downarrow & & \downarrow \\
\mathbb{P}^2(\mathbb{R}) & \xrightarrow{\;f\;} & \mathbb{P}^2(\mathbb{R})
\end{array}
$$

Corollary 15.23 implies there's an $x \in S^2$ such that $g(x) = \pm x$.

Index

© Springer International Publishing Switzerland 2015
M. Manetti, *Topology*, UNITEXT - La Matematica per il 3+2 91,
DOI 10.1007/978-3-319-16958-3

Printed in the United States
By Bookmasters